Advances in Industrial Control

For further volumes:
www.springer.com/series/1412

Francesco Vasca • Luigi Iannelli
Editors

Dynamics and Control of Switched Electronic Systems

Advanced Perspectives for Modeling, Simulation and Control of Power Converters

Springer

Editors
Francesco Vasca
Dipartimento di Ingegneria, Grace: Group for Research
Universita del Sannio in Benevento
Benevento
Italy

Luigi Iannelli
Dipartimento di Ingegneria, Grace: Group for Research
Universita del Sannio in Benevento
Benevento
Italy

ISSN 1430-9491 e-ISSN 2193-1577
Advances in Industrial Control
ISBN 978-1-4471-2884-7 e-ISBN 978-1-4471-2885-4
DOI 10.1007/978-1-4471-2885-4
Springer London Dordrecht Heidelberg New York

British Library Cataloguing in Publication Data
A catalogue record for this book is available from the British Library

Library of Congress Control Number: 2012935755

Printed on acid-free paper

Springer is part of Springer Science+Business Media (www.springer.com)

To our friend Ulf
To our families

Series Editors' Foreword

The series *Advances in Industrial Control* aims to report and encourage technology transfer in control engineering. The rapid development of control technology has an impact on all areas of the control discipline. New theory, new controllers, actuators, sensors, new industrial processes, computer methods, new applications, new philosophies, ..., new challenges. Much of this development work resides in industrial reports, feasibility study papers, and the reports of advanced collaborative projects. The series offers an opportunity for researchers to present an extended exposition of such new work in all aspects of industrial control for wider and rapid dissemination.

On reading the recent IEEE Control Systems Society report *The Impact of Control Technology: Overview, Success Stories and Research Challenges* (www.ieeecss.org/main/IoCT-report) it was interesting to be reassured that the three tenets of industrial control engineering we have long espoused still appear to hold. We have always thought that industrial applications of control involve detailed domain knowledge, the application of tools from the corpus of generic techniques in modelling, control systems design, and validation held together by a significant input of engineering skill and ingenuity to modify and adapt generic techniques to the specific domain of application and make the transition to a successful implementation.

In many cases, it is the demands and constraints of applications that motivates new theoretical developments and extensions of existing generic control concepts and methods within a given domain. This *Advances in Industrial Control* volume *Dynamics and Control of Switched Electronic Systems: Advanced Perspectives for Modeling, Simulation and Control of Power Converters* edited by Francesco Vasca and Luigi Iannelli illustrates this convincingly by showing the very significant recent contributions that system theorists and control engineers have made to the field of power electronics. The monograph also exemplifies the way in which our three tenets of industrial control engineering, domain knowledge, control systems techniques, and engineering skill and ingenuity come together to create a valuable contribution to the industrial control field.

The problem of modeling and control of power electronics systems is that they have circuit topologies that include continuous-time elements like resistors, inductors, capacitors, and voltage and current sources that are interfaced with electronic devices like diodes, and electronic switches such as thyristors, transistors, and MOSFETs. This leads to hybrid system types that involve continuous-time system behaviour mixed with discrete-time/logic switching behaviour. As with most branches of technology, the desire to use the power of computer-based simulation technology leads to the need for mathematical models. Since these devices are used to control other devices, it is a short step to the use of classical and advanced control methods in the designs. However, a key challenge is to introduce more analytical and computer-based approaches without losing sight of the real-world applications and the practical limitations and constraints that arise.

To achieve a nicely balanced and well-organised overview of developments in this power electronics field, the book's editors have structured the presentations according to four topic themes:

- Part I—*Inspiring Applications*—introduces the reader to the essential "domain knowledge" and looks at the intricacies of pulse-width modulation and commercial circuit topologies amongst other topics.
- Part II—*Mathematical Modelling Perspectives*—includes four chapters of modelling approaches for power electronic systems.
- Part III—*Advanced Control Strategies*—includes control approaches based on hybrid systems/hybrid feedback, sliding-mode control, model predictive control and adaptive PI control; together they demonstrate how different control approaches can make an impact in this field.
- Part IV—*Simulating Controlled Converters: A Challenging Issue*—is the closing group of chapters and covers offline simulation, discussion of numerical methods, test examples, and real-time simulation tools.

Each chapter is well structured with a clear statement of what is to be covered, illustrative examples, insightful concluding remarks, and useful reference sections. The comprehensive presentation makes for a self-contained volume on recent contributions to power electronic systems from the control community that should interest control theorists and control engineers and bring a new readership of researchers and engineers from the electronics community to the *Advances in Industrial Control* series.

Industrial Control Centre
Glasgow
Scotland, UK

M.J. Grimble
M.A. Johnson

Preface

Switched electronic systems are used in a huge number of everyday domestic and industrial utilities: liquid crystal displays, home appliances, lighting, personal computers, power plants, transportation vehicles and so on. Efficient operations of all such applications depend on the essential "hidden work" done by switched electronic systems, whose behaviour is determined by a suitable interconnection and control of analogue and digital devices.

From the engineering point of view, most switched electronic systems can be classified as power converters. These systems can be viewed as networks composed of semiconductor switches, e.g., thyristors, transistors and diodes, along with passive elements, e.g., inductors, capacitors and resistors, and current–voltage sources. Such systems play a fundamental role in all those environments where regulation, control and conversion of the electrical energy is a key issue. In recent years, continuously more demanding efficiency and quality constraints are imposed on the electrical energy in applications such as electrical drives, power systems and transportation vehicles. This has determined a renewed research interest in the study of formal approaches for the analysis and control of power converters. In this context the analysis, design and control of switched electronic systems is a key issue, and the use of advanced mathematical tools can be an effective approach for tackling increasing requirements in performance and efficiency.

From the mathematical point of view, switched systems represent a useful framework for modelling power converters. A typical approach for the analysis of power electronics converters consists of assuming idealised characteristics of the electronic switches. This approach yields to an important advantage: neglecting some details of the device behaviour does not eclipse the basic operation of the circuit, making easier the analysis and, sometimes, reducing the computational load for numerical simulations. Ideal characteristics of the switching devices allow one to model power electronic converters as switched systems. The network evolves through multiple topologies (or modes) depending on the discrete states of the switching devices (conducting or blocking). Unfortunately, the commutation conditions of the electronic switches, also in open-loop operating conditions, can depend on the state variables, and the switched model eventually becomes rather complex also for simple converter topologies. Then, idealisation of power converters introduces problems not commonly encountered when analysing generic switched models or non-switched electrical networks. In this sense the analysis of switched electronic systems can represent a source for new ideas and benchmarks also for the more general class of switched and hybrid systems.

This book presents an overview of the recent research advances on modelling, simulation and control of switched electronic systems. Contents organisation, illustration of realistic examples and numerical simulations provide the reader with specific quick references and a mathematically based state-of-the-art on the analysis and control design techniques for switched power converters. The book is organised in four parts. Part I provides some inspiring preliminaries coming from the power electronics community. Non-conventional converters topologies, advanced modulation techniques, commercial solutions for power converter controllers and applications to electrical energy grid guide the reader throughout the comprehension and exploration of most promising techniques used in power electronics which could inspire some practically-based directions for theoretical investigations on switched systems. Part II is dedicated to the most promising mathematical modelling frameworks suitable for switched electronic systems: continuous-time averaging, differential algebraic equations, linear complementarity models, energy- and power-based approaches. In Part III some modern control strategies for power converters are presented dealing with averaging, sliding modes, model predictive control, Hamiltonian systems. Challenges in numerical simulations of switched electronic systems are proposed in Part IV by presenting an overview of most common commercial simulation tools, numerical integrations of complementarity models and real-time techniques.

The book will be readily appreciated by engineers, mathematicians and physicists working on system and circuit theory, control systems development, electronic and energy conversion systems design:

- Several practical examples originating from power electronics stimulate control engineers and mathematicians to look at open issues in modelling, simulation and control of switched systems.
- A good balance between rigorous analysis and intuitive illustrations easily allows electronic engineers to become familiar with advanced mathematical techniques suitable for electronic systems.
- Examples on the modelling, simulation and control of power converters can inspire industrial engineers to the solution of related analysis and design problems in complex energy conversion systems.

We wish the reader an interesting and insightful journey throughout the book!

Benevento, Italy

Francesco Vasca
Luigi Iannelli

Contents

List of Contributors

Vincent Acary INRIA Rhône-Alpes, Centre de recherche Grenoble, St Ismier Cedex, France

Stefan Almér Automatic Control Laboratory, ETH Zürich, Zürich, Switzerland

Gianluca Angelone Department of Engineering, University of Sannio, Benevento, Italy

Jean Bélanger Opal-RT Technologies, Montréal, Québec, Canada

Domingo Biel Institute of Industrial and Control Engineering, Universitat Politècnica de Catalunya, Barcelona, Spain

Kanat Camlibel Institute of Mathematics and Computer Science, University of Groningen, Groningen, The Netherlands

Chin Chang Power Management, SEMTECH Corporation, Camarillo, CA, USA

Christian Dufour Opal-RT Technologies, Montréal, Québec, Canada

Gerardo Escobar ABB Switzerland Ltd, Baden-Dattwil, Switzerland

Enric Fossas Institute of Industrial and Control Engineering, Universitat Politècnica de Catalunya, Barcelona, Spain

Luc-André Grégoire Opal-RT Technologies, Montréal, Québec, Canada

Michael Hernandez-Gomez Laboratory of Signals and Systems, CNRS, Gif-sur-Yvette, France

Ngai-Man Ho ABB Switzerland Ltd, Baden-Dattwil, Switzerland

Luigi Iannelli Department of Engineering, University of Sannio, Benevento, Italy

Dimitri Jeltsema Delft Institute of Applied Mathematics, Delft University of Technology, Delft, The Netherlands

Ulf Jönsson Department of Mathematics, Royal Institute of Technology (KTH), Stockholm, Sweden

Françoise Lamnabhi-Lagarrigue Laboratory of Signals and Systems, CNRS, Gif-sur-Yvette, France

Sébastien Mariéthoz Automatic Control Laboratory, ETH Zürich, Zürich, Switzerland

Dragan Nešić Department of Electrical and Electronic Engineering, The University of Melbourne, Parkville, Victoria, Australia

Romeo Ortega Laboratory of Signals and Systems, CNRS, Gif-sur-Yvette, France

Tarek Ould Bachir École Polytechnique de Montréal, Montreal, Québec, Canada

Carmen Pedicini Department of Engineering, University of Sannio, Benevento, Italy

Sami Pettersson ABB Switzerland Ltd, Baden-Dattwil, Switzerland

Jacquelien M.A. Scherpen Faculty of Mathematics and Natural Sciences, University of Groningen, Groningen, The Netherlands

John Schönberger Plexim GmbH, Zürich, Switzerland

Jian Sun Department of Electrical, Computer, and Systems Engineering, Rensselaer Polytechnic Institute, Troy, NY, USA

Siew-Chong Tan Department of Electronic and Information Engineering, Hong Kong Polytechnic University, Kowloon, Hong Kong; Experimental Power Grid Center, Institute of Chemical and Engineering Sciences, Agency for Science, Technology and Research, Singapore, Singapore

Andrew R. Teel Department of Electrical Engineering, University of California Santa Barbara, Santa Barbara, CA, USA

Stephan Trenn Department for Mathematics, University of Kaiserslautern, Kaiserslautern, Germany

Chi Kong Tse Department of Electronic and Information Engineering, Hong Kong Polytechnic University, Kowloon, Hong Kong

Francesco Vasca Department of Engineering, University of Sannio, Benevento, Italy

Wei Wang Department of Electrical and Electronic Engineering, The University of Melbourne, Parkville, Victoria, Australia

Part I
Inspiring Applications

Chapter 1
Converter Topologies

Siew-Chong Tan and Chi Kong Tse

1.1 Introduction

Switching power converters are in existence for almost five decades. While their research and development have been extensively pursued, and their findings exhaustively applied to the modern world of electronics and industrial applications, the teaching and dissemination of their understandings in books and lecture notes are still confined to the most primitive types of converters known to engineers. These include the simplest and most fundamental types of topologies, namely the voltage buck, boost, and buck–boost converters and their transformer-isolated versions, namely, the forward and flyback converters. Discussions covering interesting aspects of the more complicated high-order converters, namely the Ćuk, SEPIC, and zeta converters have also been commonly furnished in the literature [1–3, 5].

In this chapter on "Converter Topologies", however, we will be a little more adventurous in our exploration by offering an exclusive coverage on a few less commonly known converters that can be derived using fundamental circuit theory and common engineering insights. We believe that these converters possess unique features that may give them a competitive edge in applications related to the emerging types of energy sources and loads. For example, the family of fundamental current converters that will be discussed may be useful for applications where the energy source and load behave like current sources and sinks. The family of cascaded

S.-C. Tan (✉) · C.K. Tse
Department of Electronic and Information Engineering, Hong Kong Polytechnic University, Kowloon, Hong Kong
e-mail: ensctan@polyu.edu.hk

C.K. Tse
e-mail: encktse@polyu.edu.hk

S.-C. Tan
Experimental Power Grid Center, Institute of Chemical and Engineering Sciences, Agency for Science, Technology and Research, Singapore, Singapore
e-mail: tan_siew_chong@ices.a-star.edu.sg

F. Vasca, L. Iannelli (eds.), *Dynamics and Control of Switched Electronic Systems*, Advances in Industrial Control,
DOI 10.1007/978-1-4471-2885-4_1, © Springer-Verlag London Limited 2012

quadratic converters that will be discussed would be useful for applications that require a wide conversion range, e.g. in ultra-low-voltage devices.

Concurrent to the objective of introducing some less-known types of converters, this chapter also provides from a circuit-theory perspective, how fundamental converters are derived, how they are related, and how they can be integrated to result in more complicated types of converters. In particular, we will start the chapter with how voltage buck, boost, and buck–boost converters can be derived from the reconfiguration of a switching-inductor cell with its input voltage source and output voltage sink. We will then move on to discuss how the less-known family of current buck, boost, and buck–boost converters can be derived from the reconfiguration of a switching-capacitor cell with its input current source and output current sink. The dual relationship between the family of voltage converters and current converters is then reviewed.

Following that, we move on to discuss how the family of fourth-order Ćuk, SEPIC, and zeta converters can be derived using a two-inductor–two-switch circuit cutset approach. In connection to this, we will also provide a glimpse at a missing circuit from this set of converters that has been absent from the literature. Next, on the topic of fourth-order converters, we will illustrate how the same family of converters can be derived from cascading two fundamental converters. From that, we demonstrate that using the same technique of cascading two fundamental converters, different types of quadratic converters with wide conversion ratio can be obtained [4]. Lastly, the family of fourth-order converters that are derived by adding filters to the second-order converters will be discussed. Before continuing our exploration, an overview of how the different converter topologies are related using our method of classification is shown in Fig. 1.1.

1.2 Minimum Configurable Switching Storage Structures

Switching power conversion encompasses two types of conversions, namely voltage conversion and current conversion. Power converters which electronically convert an energy source of one voltage level to another voltage level are known as voltage converters. In this respect, current converters are power electronics that convert an energy source of one current level to another current level. The basic mechanism of performing an ideally lossless voltage or current conversion involves the rapid repeated connection and disconnection of the voltage or current source to an appropriate storage device, namely an inductor or a capacitor, in a controlled manner via electronic switches, such that the energy is transferred at the appropriate voltage or current level before being released to the output load in the desired form. In configuring such power converters, it is important that voltage sources are never connected directly to a capacitor and that current sources are never connected directly to an inductor. From a theoretical viewpoint, there are two simplest configurable switching storage structures possible, namely a star-connected two-switch inductor cell and a delta-connected two-switch capacitor cell. We will next give a discussion on these storage cells and how they are configured to give the various possible types of power converters, some of which are already well known to us.

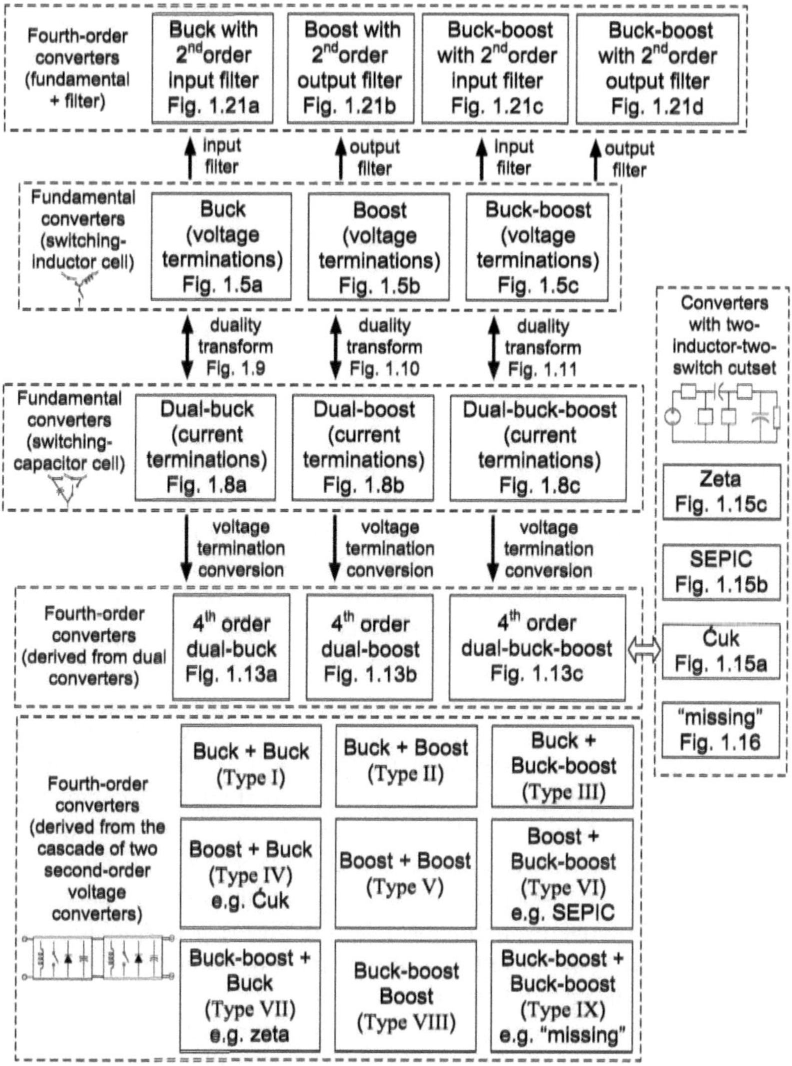

Fig. 1.1 Overview of the relationships of converter topologies

1.2.1 Switching-Inductor Cell

The *switching-inductor cell* comprises a storage inductor and two switches arranged in a star connection as depicted in Fig. 1.2. Since current flowing through an inductor must be continuous and cannot become zero instantaneously, the two switches must be designed to operate complementarily

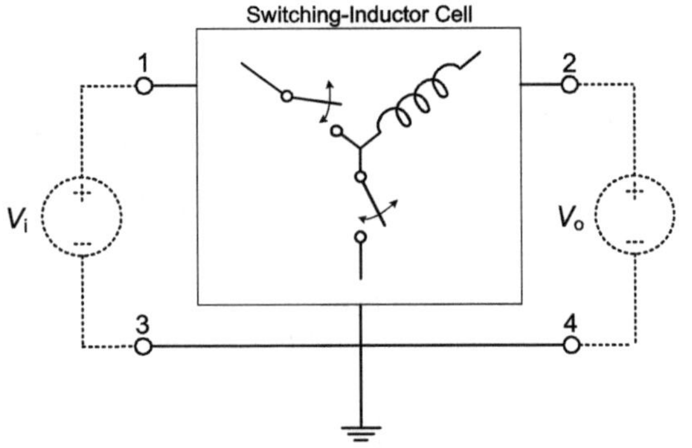

Fig. 1.2 DC–DC voltage converter using the switching-inductor cell

such that a closed-circuit continuous path will always exist for the flow of the inductor current. For this reason, this switching-inductor cell represents the minimum configurable switching-inductor structure possible. To qualify for a lossless conversion process, the switching-inductor cell must be implemented with an input voltage source together with an output voltage sink (see Fig. 1.2), which forms the basis for a voltage-to-voltage converter.

As illustrated in Fig. 1.3, there are three possible ways of configuring the switching-inductor cell while still maintaining the star-connection structure. Each can be realised by swapping the positions of any two of the three circuit elements. For proper operation of such a configuration with its input source and output sink, there should be no instance when the two switches are concurrently turned on. This will cause a direct short circuit of either the source, the sink, or the source with the sink for the respective configurations given in Fig. 1.3a–c. Additionally, when there is current flowing in the inductor, there should be no instance when both the switches are simultaneously turned off.

By replacing the voltage sink in Fig. 1.2 with a more realistic and practical circuit representation in the form of a resistor–capacitor sink as illustrated in Fig. 1.4, and by considering the three possible configurations of the switching-inductor cell given in Fig. 1.3, three fundamental types of voltage converters can be obtained. They are the voltage *buck*, *boost*, and *buck–boost converters*, which represent the fundamental topologies of the voltage converters. The schematics of these converters are given in Fig. 1.5a–c. For economical reasons, one of the two switches is replaced by a diode. The primary difference between the topologies is that each converter has a different voltage conversion ratio M such that $M(D) = V_o/V_i$, where D is the duty cycle and is a function of the turn-on time T_{ON} and the period T which is the sum of turn-on and turn-off time periods, $T_{\mathrm{ON}} + T_{\mathrm{OFF}}$, of the active switch S, i.e. $D = T_{\mathrm{ON}}/(T_{\mathrm{ON}} + T_{\mathrm{OFF}})$. For example, the voltage conversion ratios $M(D)$ for the respective converters when they are operating in the continuous inductor conduction

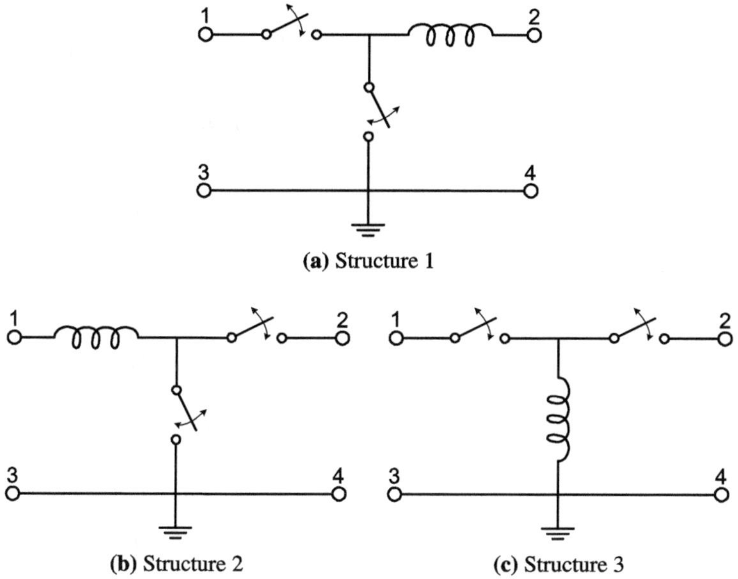

Fig. 1.3 The three possible configurations of the switching inductor cell

Fig. 1.4 Voltage sink replaced by resistor–capacitor sink

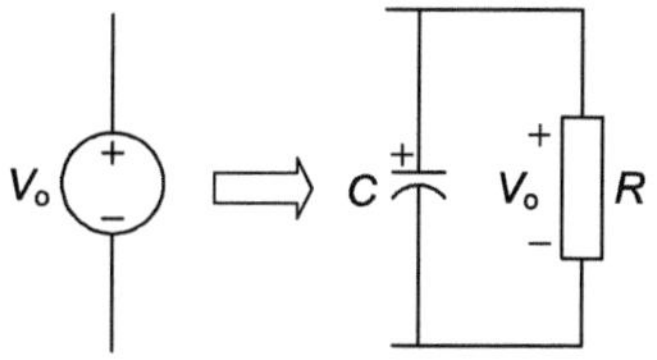

mode (current flow of the inductor never falls to zero) are

$$\frac{V_o}{V_i} = D \qquad \text{(voltage buck converter)}, \tag{1.1}$$

$$\frac{V_o}{V_i} = \frac{1}{1-D} \qquad \text{(voltage boost converter)}, \tag{1.2}$$

$$\frac{V_o}{V_i} = \frac{D}{1-D} \qquad \text{(voltage buck–boost converter)}. \tag{1.3}$$

From these equations, one can expect that since the duty ratio is confined as $0 < D < 1$, the voltage buck converter can only produce an output voltage smaller than the input voltage, i.e. $V_o < V_i$; a voltage boost converter can only produce an output voltage bigger than the input voltage, i.e. $V_o > V_i$; whereas a voltage buck–boost converter will have $V_o < V_i$ when $0 < D < 0.5$ and $V_o > V_i$ when $0.5 < D < 1$. In the case of discontinuous inductor conduction mode, the converters inherit a different set of conversion ratios. Nevertheless, their properties of stepping up or stepping down voltage levels remain the same regardless of the conduction

Fig. 1.5 The three fundamental DC–DC voltage converters

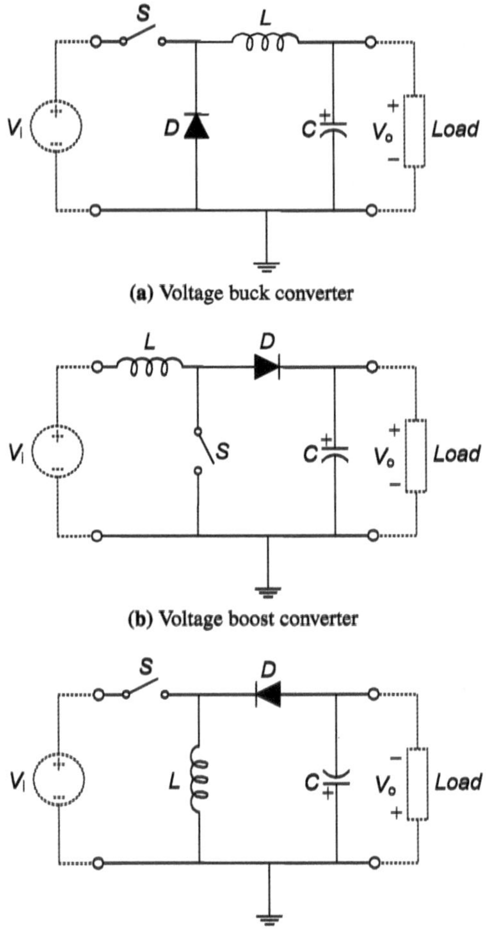

(a) Voltage buck converter

(b) Voltage boost converter

(c) Voltage buck–boost converter

mode. Finally, since the family of fundamental converters are well known, the details of their operations will not be further discussed. We move on to the switching-capacitor cell and its family of current converters.

1.2.2 Switching-Capacitor Cell

The *switching-capacitor cell* comprises a storage capacitor and two switches arranged in a delta connection as depicted in Fig. 1.6. The voltage of a capacitor has to be maintained continuous and cannot become zero instantaneously. The presence of the two switches is important in the sense that they must be

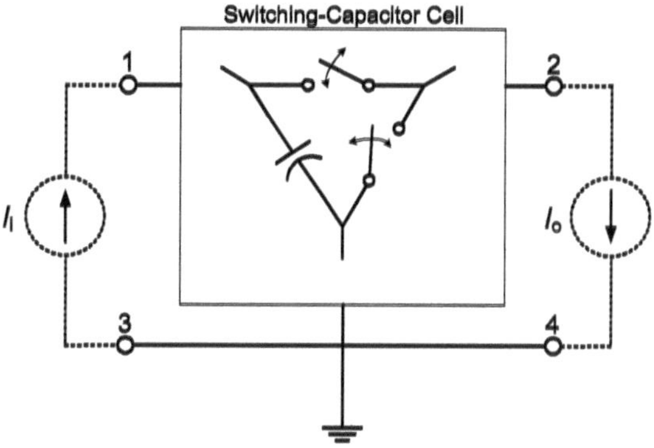

Fig. 1.6 DC–DC current converter using the switching-capacitor cell

synchronised to operate complementarily to ensure that there is no circumstance that the capacitor has a direct short circuit. This gives the minimum configurable switching-capacitor structure. For lossless energy transfer, the switching-capacitor cell must be implemented with an input current source and an output current sink. This forms the basis for the current-to-current converters.

As illustrated in Fig. 1.7, there are three possible ways of configuring the switching-capacitor cell while still maintaining the delta-connection structure. Each can be realised by swapping the positions of any two of the three circuit elements. For proper operation of such a configuration with its input source and output sink, there should be no instance when the two switches are concurrently turned off. This will cause a direct open circuit of either the source (in case (a)) or the sink (in case (b)), or a short circuit between the current source and the current sink (in case (c)) for the respective configurations given in Fig. 1.7a–c. Additionally, when there is stored charge in the capacitor, there should be no instance when both the switches are simultaneously turned on, thus avoiding a capacitor short circuit.

By considering the three possible configurations of the switching-capacitor cell given in Fig. 1.7, three fundamental types of current converters can be obtained. They are the current buck, boost, and buck–boost converters, which are in reality the dual counterparts of the voltage converters given in Fig. 1.5a–c. These converters represent the fundamental topologies in the family of current converters. The schematics of these converters are given in Fig. 1.8a–c. For economical reasons, one of the two switches can be replaced by a diode. Here, each of these topologies possesses a different current conversion ratio M, such that $M(D') = I_o/I_i$, where $D' = 1 - D$ and $D = T_{\mathrm{ON}}/(T_{\mathrm{ON}} + T_{\mathrm{OFF}})$ represents the duty cycle of S. The current conversion ratios $M(D')$ for the respective converters when they are operating in the continuous voltage conduction mode (voltage of the capacitor never falls to zero) are

$$\frac{I_o}{I_i} = D' \qquad \text{(current buck converter)}, \tag{1.4}$$

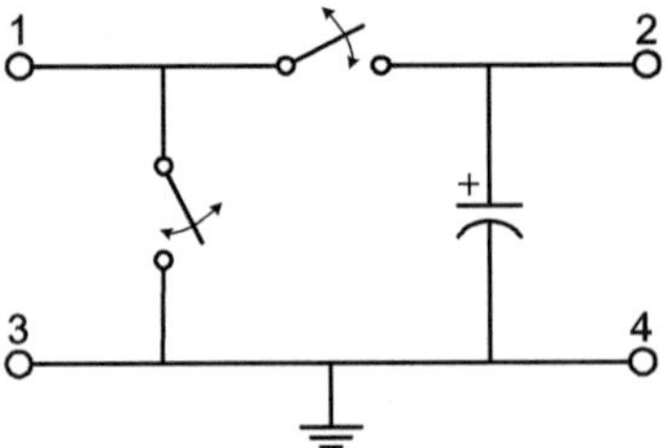

(a) Structure 1

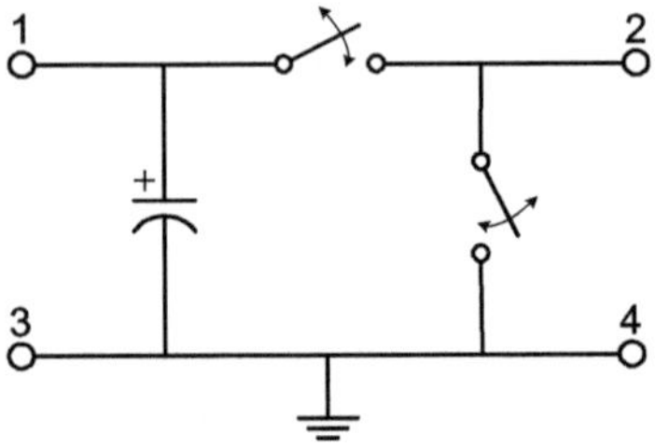

(b) Structure 2

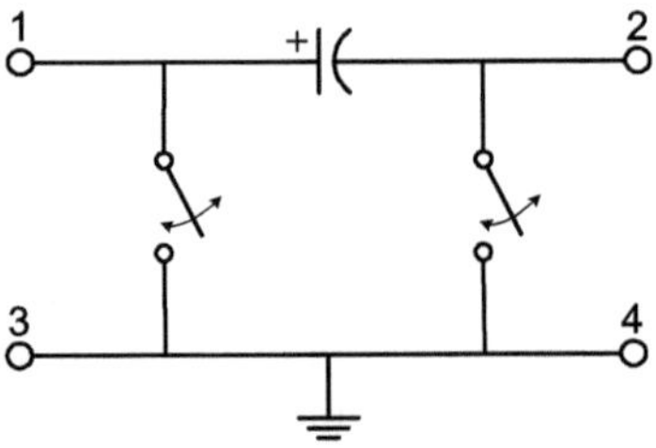

(c) Structure 3

Fig. 1.7 The three possible configurations of the switching-capacitor cell

$$\frac{I_o}{I_i} = \frac{1}{1 - D'} \quad \text{(current boost converter)}, \tag{1.5}$$

$$\frac{I_o}{I_i} = \frac{D'}{1 - D'} \quad \text{(current buck–boost converter)}. \tag{1.6}$$

It is clear from the equations that for a buck current converter, the output current is always smaller than its input current, i.e. $I_o < I_i$; for the current boost converter, output current is always bigger than the input current, i.e. $I_o > I_i$; and for the current buck–boost converter, $I_o < I_i$ when $0 < D' < 0.5$ and $I_o > I_i$ when $0.5 < D' < 1$.

1.2.3 Duality of Voltage and Current Converters

The family of current converters are basically the dual counterparts of the voltage converters. So far, we have demonstrated the formulation of these converters from basic circuit rules and framework of a switching-inductor cell and a switching-capacitor cell. Interestingly, using the principle of circuit duality, it is possible to extract the same family of current converters from the voltage converters, and conversely, the family of voltage converters from the current converters.

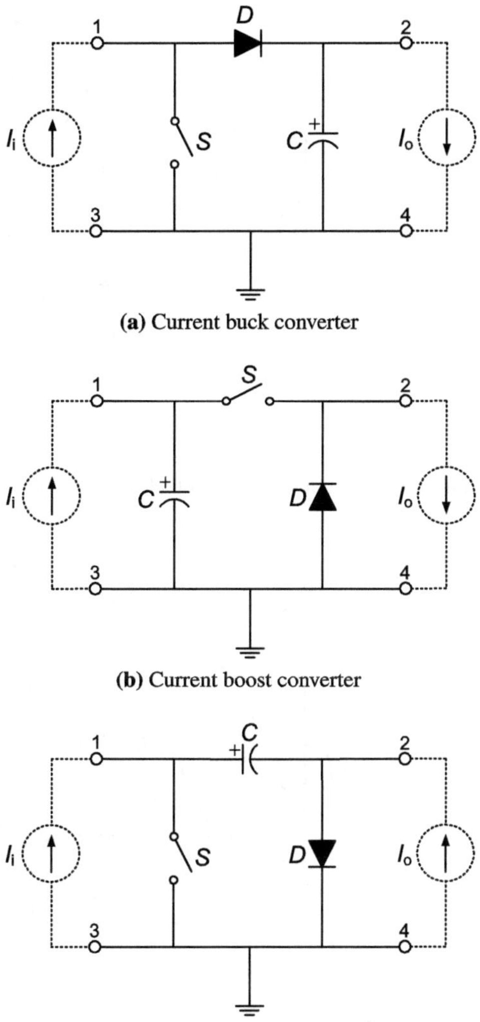

(a) Current buck converter

(b) Current boost converter

(c) Current buck–boost converter

Fig. 1.8 The three fundamental DC–DC current converters

A pictorial illustration of the transformation of the family of voltage DC–DC converters into the family of current DC–DC converters are given in the following figures, Figs. 1.9, 1.10, 1.11. Here, the converters drawn in dotted lines are the family of voltage converters and those drawn in solid lines are the family of current converters. To understand the process of the transformation, we recall that in circuit duality, a voltage source would be replaced by a current source and vice versa, an inductor would be replaced by a capacitor and vice versa, a resistor by a conductance and vice versa, a closed switch by an open switch and vice versa, a turn-on signal by a turn-off signal and vice versa. The process involves the applications of these rules.

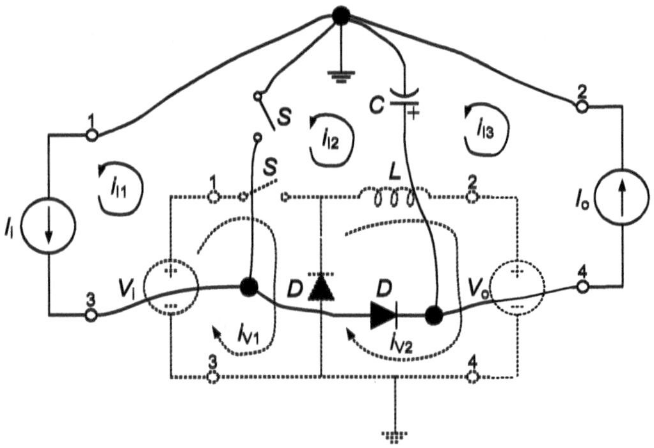

Fig. 1.9 Dual transformation of a buck converter

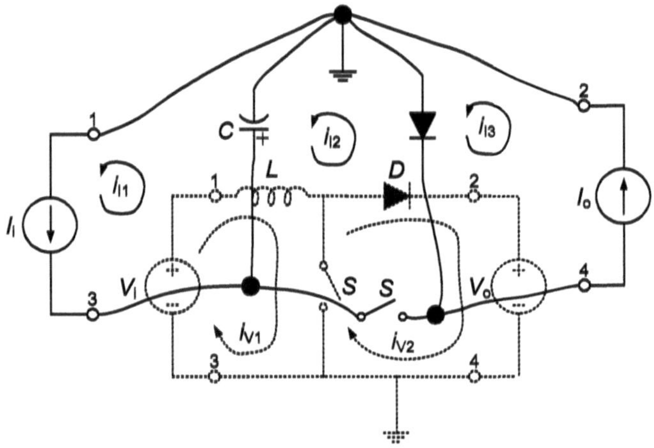

Fig. 1.10 Dual transformation of a boost converter

1.2.4 Practical Current DC–DC Converters

While the reported family of current DC–DC converters in Fig. 1.8a–c are theoretically feasible, in reality, it is uncommon that practical current source and current sink are readily available. Hence, a more practical scenario would be to replace the current sources and current sinks with their circuit equivalent of voltage sources and sinks, which are respectively shown in Fig. 1.12a–b. This will result in an interesting family of current buck, boost, and buck–boost converters that are based on voltage sources and sinks, as depicted in Fig. 1.13a–c. Note that the derived converters possess two inductors and two capacitors and they must be operated only in the continuous inductor conduction mode such that the currents of both the induc-

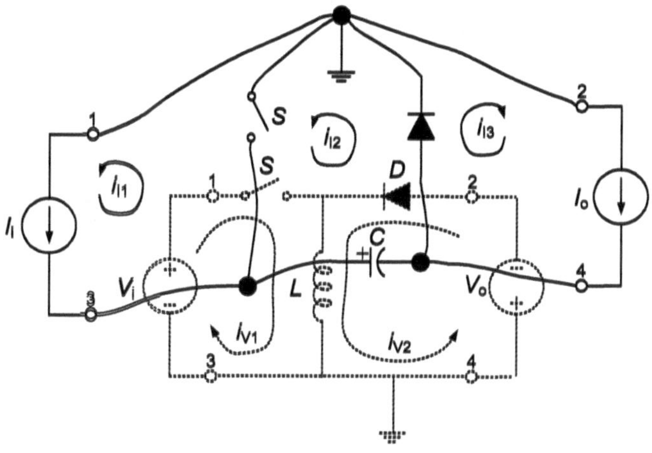

Fig. 1.11 Dual transformation of a buck–boost converter

tors are always continuous (to preserve the original property of a current source and current sink). The presented current buck and boost converters and their modified versions, as illustrated in Figs. 1.8a–c, 1.13a–b, are relatively unknown topologies which may be interesting for new emerging applications. Coincidentally, the modified buck–boost converter given in Fig. 1.13c happens to be the well-known Ćuk converter, which will be discussed in the following.

1.3 Fourth-Order Converters

The Ćuk, SEPIC, and zeta converters are a family of fairly interesting fourth-order converters that were introduced in the 1980s. They have been classified as fourth-order converters for the reason that each of these converters comprises four independent storage elements involving two inductors and two capacitors that are interactively configured with the voltage source and sink via a switch and a diode.

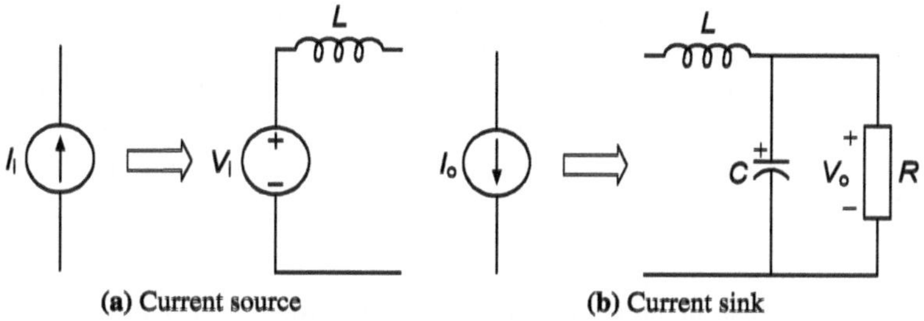

Fig. 1.12 (a) Current source replaced by an equivalent of voltage source and series inductor and (b) current sink replaced by a resistor–capacitor sink and series inductor

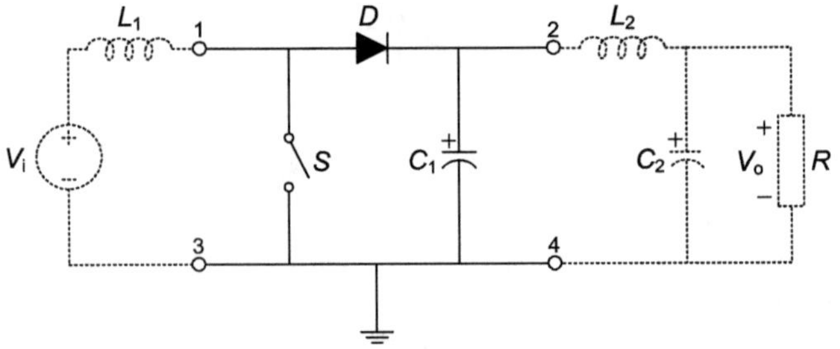

(a) Modified current buck converter

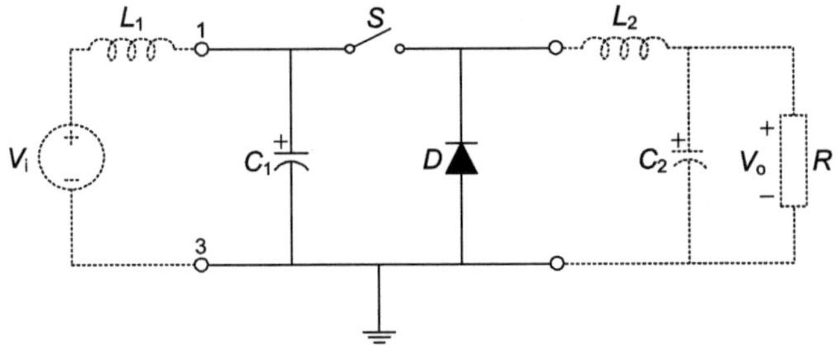

(b) Modified current boost converter

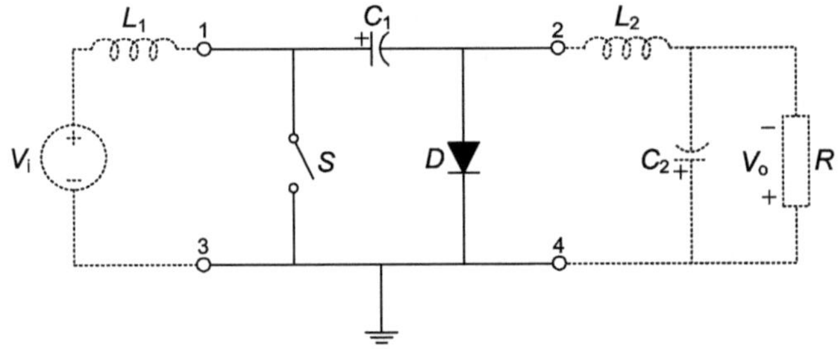

(c) Modified current buck–boost converter

Fig. 1.13 The three fundamental DC–DC current converters with their input current sources and output current sinks replaced by equivalent voltage source and sink circuits

From a topological viewpoint, fourth-order converters have twice the number of storage elements and are more complicated than the fundamental voltage buck, boost, and buck–boost converters, which are primarily second-order converters. Nevertheless, the fourth-order converters possess unique features which renders them preferential in certain applications over the second-order converters.

Next, we will illustrate how this family of fourth-order converters can be obtained from a two-inductor–two-switch circuit cutset. We will also reveal the identity of a "missing" fourth-order converter that has been absent from the literature.

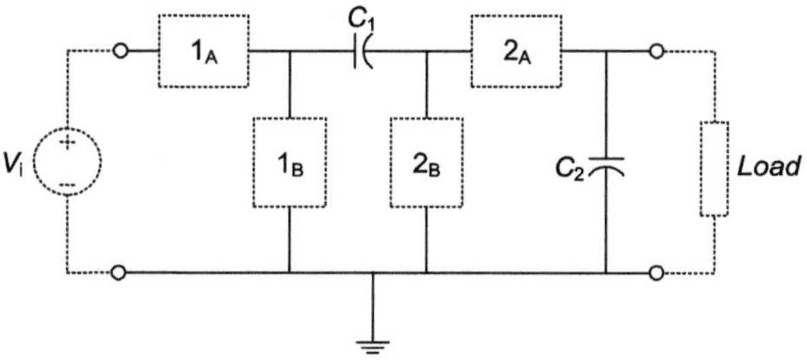

Fig. 1.14 Generic fourth-order converter with two-inductor–two-switch cutset

1.3.1 Two-Inductor–Two-Switch Circuit Cutset

Based on the two-inductor–two-switch cutset, the family of fourth-order converters can be portrayed in the generic structure as depicted in Fig. 1.14. In this diagram, the four boxes in dotted lines which are labelled as 1_A, 1_B, 2_A, and 2_B represent either an inductor or a switch. The structure itself is known as the two-inductor–two-switch circuit cutset. To ensure that there will be no short-circuit operation of the voltage source and the voltage sink, the cutset must be configured such that if component 1_A is a switch, then component 1_B must be an inductor, and vice versa. Likewise, if component 2_A is a switch, then component 2_B must be an inductor, and vice versa. With these constraints, there exist only four possible configurations in this two-inductor–two-switch circuit cutset. One of these possible configurations is: 1_A is an inductor, 1_B is a switch, 2_A is an inductor, and 2_B is a diode. This makes the *Ćuk converter* given in Fig. 1.15a. A second configuration is to assign 1_A as an inductor, 1_B as a switch, 2_A as an inductor, and 2_B as a diode. This gives the *SEPIC converter*, which is shown in Fig. 1.15b. The third configuration is to assign 1_A as a switch, 1_B as an inductor, 2_A as an inductor, and 2_B as a diode, to give what is known as the *zeta converter*, as depicted in Fig. 1.15c. For all the three topologies, the voltage conversion ratios $M(D)$ for the continuous inductor conduction mode are exactly that of a buck–boost converter, i.e.

$$\frac{V_o}{V_i} = \frac{D}{1-D}. \tag{1.7}$$

The difference between these converters and the voltage buck–boost converter is that for the latter, both the input current and the output current are pulsating in nature. However, for the Ćuk converter, both the input current and the output current are non-pulsating. For the SEPIC converter, the input current is non-pulsating but the output current is pulsating. For the zeta converter, the input current is pulsating but the output current is non-pulsating.

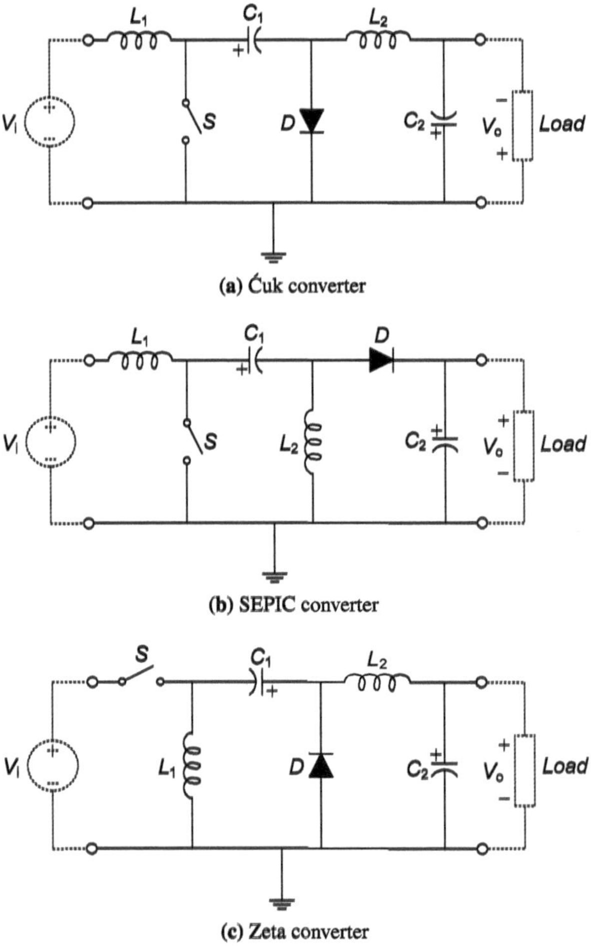

Fig. 1.15 Known types of fourth-order converters with the two-inductor–two-switch cutset

The three converters described above are the only known types of fourth-order converters with this cutset. A "missing" fourth-order converter that can be derived from the same cutset has been absent from the literature. This "missing" converter would be the remaining configuration of the cutset which is to set 1_A as a switch, 1_B as an inductor, 2_A as a diode, and 2_B as an inductor. Figure 1.16 shows the "missing" converter. Since this converter is rarely mentioned, a simple discussion of its operation will be provided.

There are three possible operating states for this "missing" converter. The equivalent circuit diagrams of the converter for the various operating states are given in Fig. 1.17a–c. During ON state (see Fig. 1.17a) when switch S is turned on, L_1 will be energised by voltage source V_1 and the energy stored in C_1 will be transferred to L_2. Diode D is turned off and C_2 will supply the load. There are two possible oper-

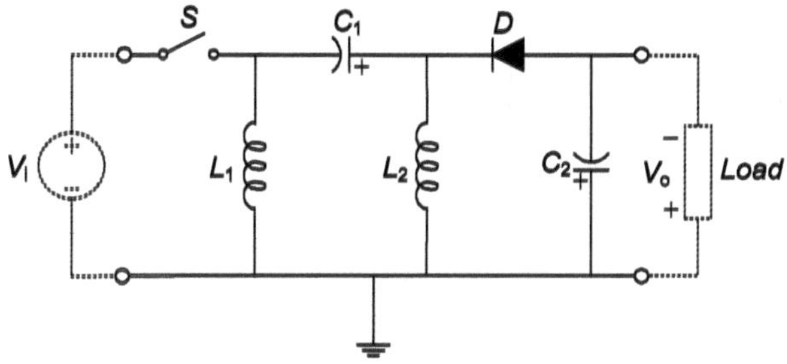

Fig. 1.16 The "missing" fourth-order converter

ating states when switch S is turned off. At the instance S is turned off, the converter enters the OFF state (see Fig. 1.17b). Here, the energy stored in L_1 is transferred to C_1 and the energy stored in L_2 will be transferred to C_2 and the load via diode D. This process will continue until the next ON state or when there is no more energy stored in L_2, i.e. the current of L_2 reaches zero. When this happens, a third operating state with L_2 being reversely energised by L_1 takes place. From Fig. 1.17c, in this state, L_1 will concurrently charge C_1 while reversely energising L_2. The diode D is turned off and the load is supplied by C_2. Similar to the other three topologies in the cutset, the voltage conversion ratio $M(D)$ of the "missing" converter in the continuous inductor conduction mode is the same as that of a buck–boost converter,

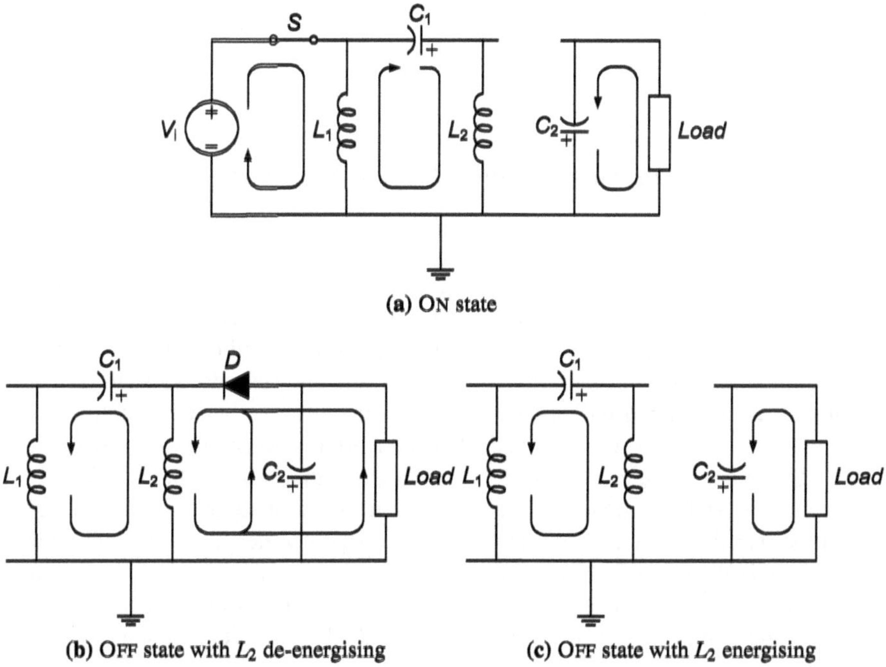

Fig. 1.17 The three possible operating states of the "missing" fourth-order converter

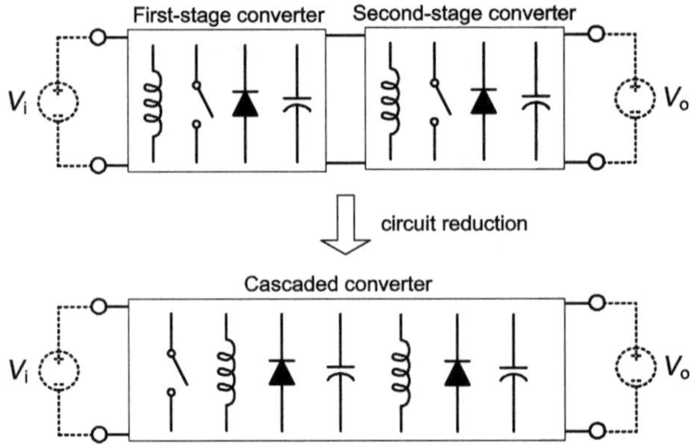

Fig. 1.18 Fourth-order voltage converter derived from the cascade of two second-order voltage converters

i.e.

$$\frac{V_o}{V_i} = \frac{D}{1-D}. \tag{1.8}$$

In fact, this converter behaves exactly like a fundamental voltage buck–boost converter with pulsating input and pulsating output currents. This may explain its redundancy in the family of buck–boost converters since unlike the three other topologies, the "missing" converter does not actually solve the issue of pulsating current at the input or output.

1.3.2 Cascading of Two Second-Order Converters

In the previous part, it is illustrated that by using the two-inductor–two-switch cutset, a family of four fourth-order converters can be derived. Interestingly, the same family of fourth-order converters can be obtained by combining two fundamental second-order converters in cascade and then doing a circuit reduction to result in a cascaded converter with only one active switch S, as shown in Fig. 1.18. For example, the Ćuk converter is essentially an equivalent of a first-stage boost converter that is cascaded with a second-stage buck converter. The SEPIC converter is an equivalent of a first-stage boost converter that is cascaded with a second-stage buck–boost converter. The zeta converter is an equivalent of a first-stage buck–boost converter that is cascaded with a second-stage buck converter. The "missing" converter is an equivalent of a first-stage buck–boost converter that is cascaded with a second-stage buck–boost converter.

Since there are three fundamental voltage converters, namely the buck, boost, and buck–boost converters, there will be in all nine possible types of cascaded fourth-order converters, which are hereby labelled as Types 1–9, that can be obtained from the cascading of two converters (refer to Fig. 1.1):

1 First-stage buck plus second-stage buck;
2 First-stage buck plus second-stage boost;
3 First-stage buck plus second-stage buck–boost;
4 First-stage boost plus second-stage buck (e.g. Ćuk);
5 First-stage boost plus second-stage boost;
6 First-stage boost plus second-stage buck–boost (e.g. SEPIC);
7 First-stage buck–boost plus second-stage buck (e.g. zeta);
8 First-stage buck–boost plus second-stage boost;
9 First-stage buck–boost plus second-stage buck–boost (e.g. "missing").

Figure 1.19 illustrates how the Ćuk converter can be derived from the Type 4 configuration of a first-stage boost and a second-stage buck converter. As shown in the figure, the diode and capacitor of the boost converter are first interchanged and the buck converter is vertically flipped before both the converters are integrated. The points where the two converters are cascaded remain the same with A connecting A' and B connecting B'. Since the active switches S_{Buck} and S_{Boost} share a common driving signal and their positions are overlapped when the two converters are integrated, only one switch S will be required. For the same reason, one of the diodes D_{Buck} or D_{Boost} can be removed. Using the same approach of swapping component's positions and eliminating a redundant switch and diode, the other three converters, namely SEPIC, zeta, and the "missing" converter can also be derived. In fact, with such an approach, all other types of fourth-order converters, including the quadratic converters, can be derived.

1.3.3 Quadratic Converters

The *quadratic converters* are a class of single-switched cascaded converters that have a wide output-to-input conversion range and their conversion ratios $M(D)$ have a quadratic dependence on the duty cycle D.

For example, a Type 1 topology in quadratic form, which is a buck converter in cascade with a second buck converter, will have a conversion ratio $M(D)$ comprising the multiplication of the conversion ratio of the first-stage buck converter of $M_1(D_1) = D_1$ and the second-stage buck converter of $M_2(D_2) = D_2$, i.e. $M(D) = M_1(D_1) \times M_2(D_2) = D_1 D_2$. Figure 1.20a shows an example of a Type 1 passive-buck active-buck quadratic converter. With only one active switch, this Type 1 buck–buck quadratic converter has a conversion ratio $M(D) = D^2$. Functionally, with a driving signal operating with duty cycle of 0.5, a conversion

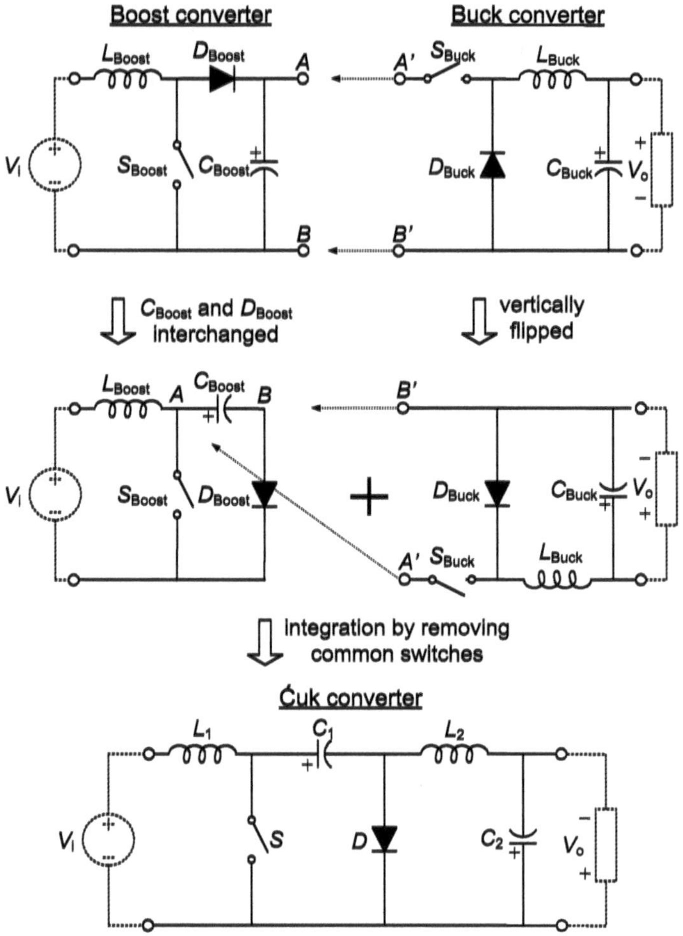

Fig. 1.19 Ćuk converter derived from the cascade of a boost and a buck converter

ratio of 0.25 can be achieved with this converter. Such kind of quadratic effect on the duty cycle significantly extends the conversion range of a converter without the use of a transformer.

Similarly, by combining a buck and a buck–boost converter, a Type 3 passive-buck active-buck–boost quadratic converter can be obtained as depicted in Fig. 1.20b. The conversion ratio of a Type 3 quadratic converter is $M(D) = D^2/(1-D)$. By combining two buck–boost converters, a Type 9 passive-buck–boost active-buck–boost quadratic converter can be obtained as shown in Fig. 1.20c. The conversion ratio of a Type 9 quadratic converter is $M(D) = D^2/(1-D)^2$. Out of the nine available types of cascaded converters, there are seven types which can possibly become quadratic converters, if properly configured, as given in Table 1.1. The remaining two, namely Types 2 and 4, are non-quadratic converters with a conversion ratio that is similar to a buck–boost converter.

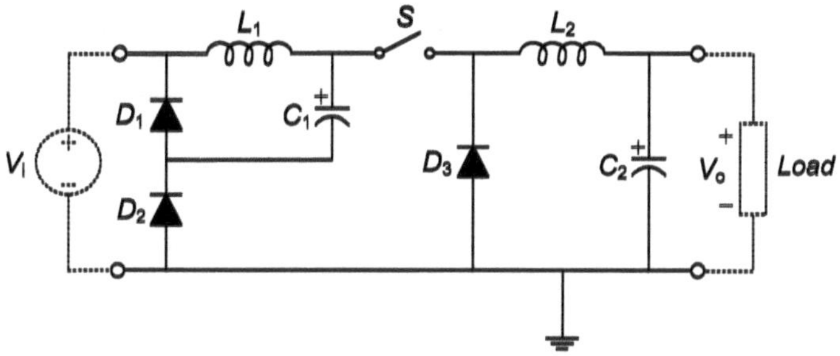

(a) Passive-buck-active-buck-quadratic converter

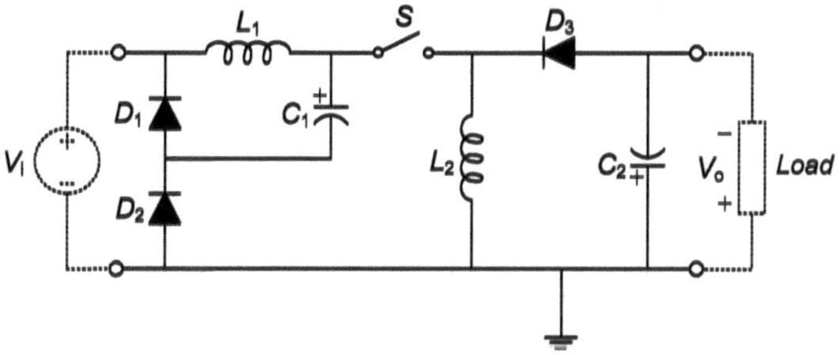

(b) Passive-buck active-buck–boost quadratic converter

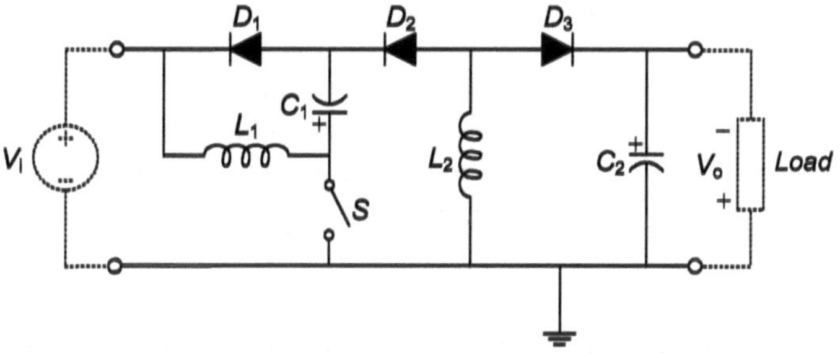

(c) Passive-buck–boost active-buck–boost quadratic converter

Fig. 1.20 Three quadratic converters derived from the cascade of two second-order converters

Importantly, the aforementioned quadratic converters resulted from the cascade of two fundamental converters are possible only if the integration of the converters is done so that the individual converters in cascade still retain their respective power conversion properties after being cascaded. In other words, the overall conversion ratio of the cascaded converter must be a full cascade (multiplication) of the original conversion ratio of the two individual converters, i.e. $M(D) = M_1(D)M_2(D)$. Otherwise, a quadratic power conversion cannot be achieved with the cascaded converter. For example, in the case of the SEPIC, zeta, and the "missing" converter, which are of Types 6, 7, and 9, respectively, the conversion ratios are of non-quadratic form, i.e. $M(D) = D/(1 - D)$, which makes them effectively a buck–boost/boost–buck converter. The reason for this is that the integration of these con-

Table 1.1 Conversion ratio and description of the nine cascaded converters

Classification	Converters in Cascade	Conversion Ratio $M(D)$	Description
Type 1	buck + buck	D^2	quadratic
Type 2	buck + boost	$D/(1-D)$	non-quadratic
Type 3	buck + buck–boost	$D^2/(1-D)$	quadratic
Type 4	boost + buck	$D/(1-D)$	non-quadratic
Type 5	boost + boost	$1/(1-D)^2$	quadratic
Type 6	boost + buck–boost	$D/(1-D)^2$	quadratic
Type 7	buck–boost + buck	$D^2/(1-D)$	quadratic
Type 8	buck–boost + boost	$D/(1-D)^2$	quadratic
Type 9	buck–boost + buck–boost	$D^2/(1-D)^2$	quadratic

verters has been performed to the extent that the circuit reduction has resulted in the distortion of the original conversion property of the individual converters in cascade. For the SEPIC converter, which is derived from the cascade of boost and buck–boost converters, the supposed conversion ratio would be in the quadratic form of $M(D) = D^2/(1-D)$. However, circuit integration and simplification lead to an overall cascaded (SEPIC) converter that has lost one component of its step-down property D, making its conversion ratio of $M(D) = D/(1-D)$. The same explanation applies to the zeta and the "missing" converters.

Finally, it is worth emphasising that the quadratic converters are useful in applications where the difference between the input source and the output sink is too significant for the required power conversion to be achieved by a conventional converter, and when the use of transformer-based converter is not preferred. For details of the quadratic converters, readers are referred to the work in [4].

1.3.4 Adding of Second-Order Filter to Fundamental Converters

Lastly, we will introduce another kind of fourth-order converters that can be obtained from fundamental converters through the addition of second-order inductive-capacitive $L_{\text{filt}}C_{\text{filt}}$ filters. The function of $L_{\text{filt}}C_{\text{filt}}$ in these converters is for filtering and not energy storage. Hence, these components are typically small in capacity and size relative to the main inductor L and capacitor C of the converter. Consequently, $L_{\text{filt}}C_{\text{filt}}$ tackles only the filtration of high-frequency signals. They do not have any real effect on the actual process of power conversion. However, their addition to second-order converters would still mean that the converters possess four independent storage elements. Therefore, they are still technically classified as fourth-order converters.

In any case, the placement of second-order filters is strategically conducted to minimise the pulsating flow of currents. For the voltage buck converter, which has a pulsating input current, a second-order filter is placed at the input side of the

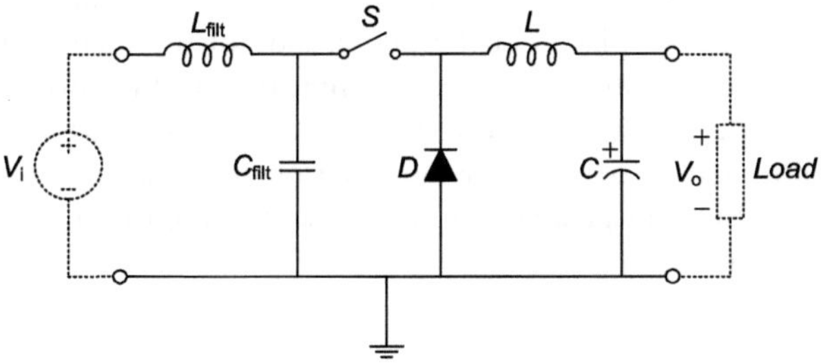

(a) Buck voltage converter with input filter

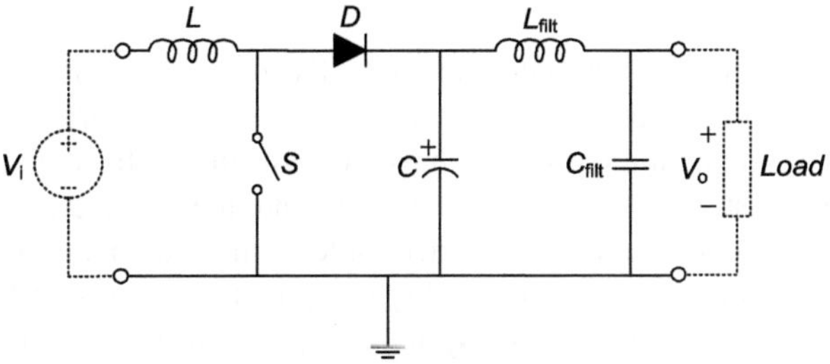

(b) Boost voltage converter with output filter

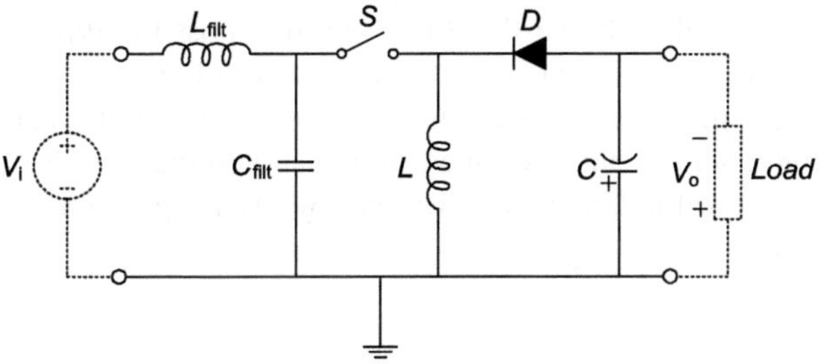

(c) Buck-boost voltage converter with input filter

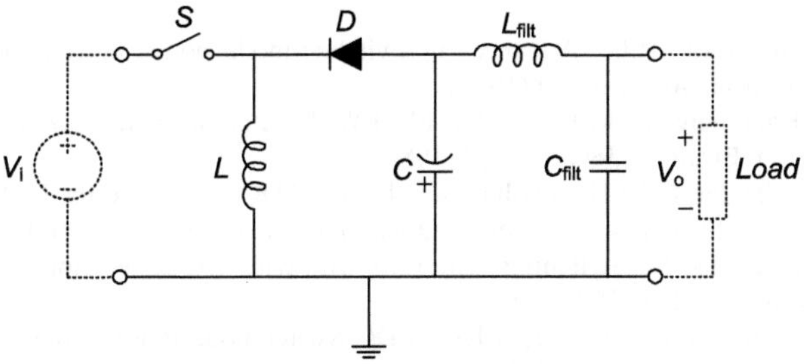

(d) Buck–boost voltage converter with output filter

Fig. 1.21 Fourth-order converters derived from the adding of second-order filter to fundamental converters

converter, as shown in Fig. 1.21a, to reduce the high-frequency current flow from the source. For the voltage boost converter, which has a pulsating output current, the second-order filter is placed at the output side of the converter, as shown in Fig. 1.21b, to reduce the high-frequency current flow to the sink. For the voltage buck–boost converter, which has both its input and output currents pulsating, the second-order filter can exist at either the input or output side of the converter, as shown in Fig. 1.21c–d, respectively, to reduce either high-frequency current flow from the source or to the sink.

1.4 Summary

Some relatively unknown, but interesting, DC–DC power converter topologies have been discussed in this chapter. The derivation of such converters and their relationships with fundamental converters have been discussed from a circuit-theoretic viewpoint. It is illustrated that current converters that may be useful for applications requiring current sources and sinks are derivable from fundamental voltage converters by using the principle of circuit duality or by using a proposed framework of a switching-capacitor cell. Based on a systematic procedure of deriving fourth-order Ćuk, SEPIC, and zeta converters from a two-inductor–two-switch-circuit cutset, we show that there is actually a missing topology from this family of fourth-order converters. We have also shown that there are nine possible types of fourth-order converters that can be obtained from the cascading of two fundamental converters. Out of these nine types of cascaded converters, seven types can be configured as quadratic converters. Finally, we have demonstrated that fourth-order converters can be obtained from fundamental converters by inserting second-order inductive-capacitive filters.

References

1. Ćuk, S., Middlebrook, R.D.: Advances in switched-mode power conversion—Part 1. IEEE Trans. Ind. Electron. **30**(1), 10–19 (1993)
2. Jozwik, J.J., Kazimierczuk, M.K.: Dual SEPIC PWM switching-mode DC/DC power converter. IEEE Trans. Ind. Electron. **36**(1), 64–70 (1989)
3. Maasey, R.P., Snyder, E.C.: High voltage single-ended DC–DC converter. In: Proc. of the IEEE Power Electronics Specialist Conference, Orlando, Florida, USA, pp. 156–159 (1977)
4. Maksimovic, D., Ćuk, S.: Switching converters with wide DC conversion range. IEEE Trans. Power Electron. **6**(1), 151–157 (1991)
5. Severns, R.P., Bloom, G.E.: Modern DC-to-DC Switchmode Power Converter Circuits. Van Nostrand Reinhold, New York (1985)

Chapter 2
Pulse-Width Modulation

Jian Sun

2.1 Form and Function

Pulse-width modulation (PWM) is the basis for control in power electronics. The theoretically zero rise and fall time of an ideal PWM waveform represents a preferred way of driving modern semiconductor power devices. With the exception of some resonant converters, the vast majority of power electronic circuits are controlled by PWM signals of various forms. The rapid rising and falling edges ensure that the semiconductor power devices are turned on or turned off as fast as practically possible to minimise the switching transition time and the associated switching losses. Although other considerations, such as parasitic ringing and electromagnetic interference (EMI) emission, may impose an upper limit on the turn-on and turn-off speed in practical situations, the resulting finite rise and fall time can be ignored in the analysis of PWM signals and processes in most cases. Hence only ideal PWM signals with zero rise and fall time will be considered in this chapter.

Pulse-width modulation can take different forms [3]. The pulse frequency is one of the most important parameters when defining a PWM method and can be either constant or variable. A constant-frequency (CF) PWM signal can be produced simply by comparing a reference signal, $r(t)$, with a carrier signal, $c(t)$, as depicted in Fig. 2.1a. The binary PWM output can be mathematically written as

$$b_{\mathrm{pwm}}(t) = \mathrm{sgn}\big[r(t) - c(t)\big], \tag{2.1}$$

where 'sgn' is the sign function.

Three types of carrier signals are commonly used in constant-frequency PWM:

1. Sawtooth Carrier, reported in Fig. 2.1b: The leading (rising) edge of PWM output occurs at fixed instants in time while the position of the trailing (falling) edge is

J. Sun (✉)
Department of Electrical, Computer, and Systems Engineering, Rensselaer Polytechnic Institute, Troy, NY 12180-3590, USA
e-mail: jsun@rpi.edu

F. Vasca, L. Iannelli (eds.), *Dynamics and Control of Switched Electronic Systems*, Advances in Industrial Control,
DOI 10.1007/978-1-4471-2885-4_2,

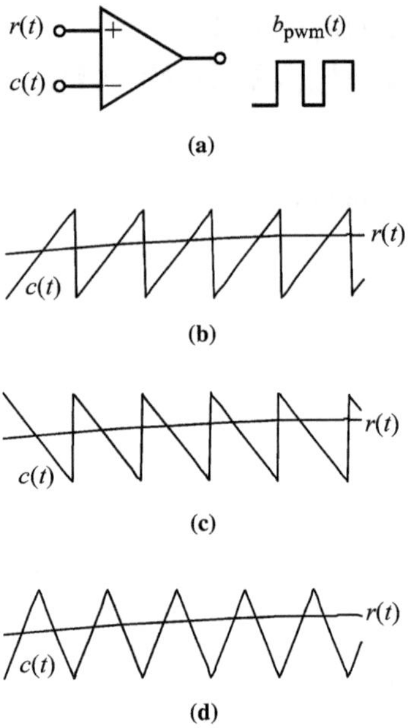

Fig. 2.1 Constant-frequency PWM implemented by a comparator with different carrier signals

modulated as the reference signal level varies. Hence the method is also called constant-frequency *trailing-edge modulation*.

2. Inverted Sawtooth Carrier, reported in Fig. 2.1c: The trailing (falling) edge of PWM output occurs at fixed instants in time while the position of the leading (rising) edge is modulated as the reference signal level varies. The method is usually referred to as constant-frequency *leading-edge modulation*.
3. Triangle Carrier, reported in Fig. 2.1d: Both the leading edge and the trailing edge of the PWM output is modulated. The rising and falling edge of the triangle are usually symmetric so that the pulse is centred within a carrier cycle when the reference is a constant. The method is called constant-frequency *double-edge modulation*.

Trailing-edge modulation is most common in DC–DC converters. As will be discussed in the next section, double-edge modulation eliminates certain harmonics when the reference is a sine wave, and is a preferred method for AC–DC and DC–AC converters where the PWM reference contains a sinusoidal component. A combination of synchronised leading-edge and trailing-edge modulation has also been used to control a boost single-phase power factor correction (PFC) converter and a buck DC–DC converter to reduce ripple in the intermediate DC bus capacitor [25].

The illustrations in Fig. 2.1 assumed analog implementation. When digital implementation is used, the reference is usually sampled at a regular frequency and the carrier can be replaced by a counter/timer. To avoid multiple switching transition within a carrier cycle, the reference should be sampled at the point where the carrier reaches its peak or valley [9]. Pulse-width modulation using such sampled references is called *regular-sampling* PWM. To distinguish from such sampled PWM,

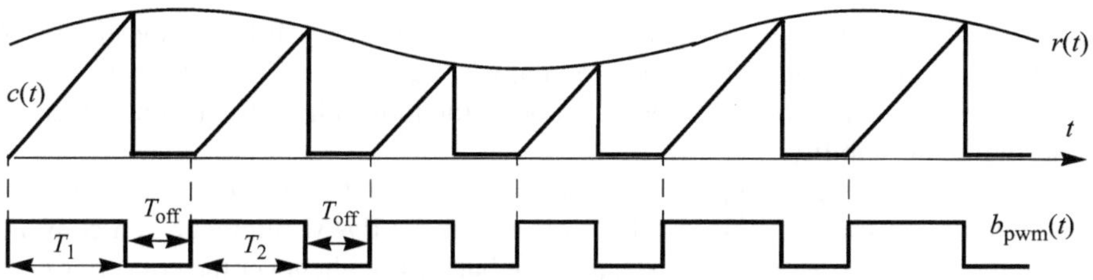

Fig. 2.2 Variable-frequency PWM with constant OFF-time

the analog version discussed before is also called *natural-sampling* PWM in the literature. With a sawtooth or inverted sawtooth carrier, samples are usually taken at the beginning of a carrier cycle. With a triangle carrier, on the other hand, the reference can be sampled either once at the peak of the triangle or twice at both the peak and the valley of the triangle; the former is referred to as *symmetrical sampling*, while the latter is called *asymmetrical sampling* due to the fact that the rising and falling edge of the triangle are compared with different samples of the reference. Regular-sampling PWM is usually used in high power inverters and rectifiers and will not be further discussed in this chapter. On the other hand, the effects of sampling can be incorporated into the PWM spectral models by modifying the double Fourier integral to be presented in the following sections [9].

Variable-frequency (VF) PWM, although not as popular as CF PWM, has also been used in practice. Three common variations of VF PWM are: (a) constant OFF-time, variable ON-time; (b) constant ON-time, variable OFF-time; and (c) hysteretic control. Figure 2.2 depicts constant OFF-time, variable ON-time PWM using a sawtooth-like carrier signal. The switch is turned on after a fixed OFF-time, T_{off}, at which point the sawtooth signal also starts to rise at a constant rate. The switch is turned off again when the sawtooth signal intersects with the reference, at which point the carrier signal is reset to zero. The switch is then kept OFF for a fixed time (T_{off}) again before the next switching cycle starts. As can be seen, the ON-time changes with the reference and the switching frequency increases with the decrease in the reference level, resulting in a variable frequency operation when the reference varies. Constant ON-time, variable OFF-time can be implemented in a similar manner. A popular application of the VF PWM is in boundary-mode control of boost power factor correction (PFC) converters, where the switch operates with a constant ON-time and is turned on as soon as the diode current reduces to zero, resulting in an average input current that is proportional to the input voltage [22]. Hysteretic control is usually applied in conjunction with a current or voltage regulator, and doesn't involve an explicit PWM process. It will not be further discussed in this chapter.

One common concern about VF PWM is the difficulty associated with the design of input and output filters. The filter corner frequency would have to be selected based on the lowest possible switching frequency in order to provide the required attenuation for ripple and EMI under all operation conditions. This usually leads to a conservative design with significant volume and cost penalties. On the other hand, both constant ON-time and constant OFF-time VF PWM exhibit a leading phase angle in their small-signal dynamic transfer functions, as will be demonstration in Sect. 2.3 of this chapter. Since the modulator is part of the feedback control loop, such a leading phase boosts the phase angle of the loop at high frequencies, thereby improving the stability.

The reference signal depends on the application and is usually independent of the modulation method. For DC–DC converters, the PWM reference is a constant when the converter operates in a steady state but varies whenever the converter goes through a transient. The spectral characteristics of such a PWM waveform with constant frequency and constant duty ratio can be readily determined by Fourier analysis. To characterise the dynamic behaviour of a modulator in DC–DC converters, the reference can be assumed to consist of a DC value corresponding to the steady-state duty ratio and a sinusoidal component representing a small-signal perturbation. A small-signal model can be obtained for the modulator by computing the component in the PWM output at the frequency of the sinusoidal perturbation. Such small-signal models will be developed in Sect. 2.3 for both constant-frequency and variable-frequency PWM.

For AC–DC and DC–AC converters, the reference signal typically contains at least one sinusoidal component at the fundamental frequency of the AC input or output of the converter. For poly-phase (e.g. three-phase) converters, each phase will have a separate reference and their sinusoidal components are shifted from each other by the same phase angle that separates the input or output phase voltages. Often, the PWM references also contain harmonics of the fundamental component. This is the case, for example, in three-phase converters where triple harmonics can be purposely injected into the PWM references to increase the utilisation of the DC voltage, that is, to maximise the AC voltages that can be produced from a given DC voltage source before the modulator saturates [20].

For the reasons stated above as well as for the purpose of developing general PWM models, the reference signal, $r(t)$, is this chapter is assumed to consist of a DC and a single-frequency sinusoidal component in general:

$$r(t) = R_0 + R_1 \cos(2\pi f_1 t + \theta_1). \tag{2.2}$$

This will be used in the spectral analysis of different PWM methods in the next two sections. Each of the amplitudes R_0 and R_1 can be set to zero depending on the specific applications under study. Additional harmonics can also be included in (2.2) if needed.

Mathematical analysis will be presented for different PWM processes in the following sections to develop a more in-depth understanding of their characteristics and to provide models that can be used for different design purposes. Section 2.2 introduces double Fourier series as a general method to characterise constant-frequency PWM processes. Spectral characteristics of different PWM methods are also compared using the double Fourier series models. Section 2.3 discusses small-signal modelling of both constant-frequency and variable-frequency pulse-width modulator for dynamic modelling and control design of DC–DC converters. The section also studies the effects of interleaved PWM of multiple converter modules and uses the analytical spectral models to characterise the ripple cancellation effects under different interleaving strategies. Section 2.4 discusses peak current control, a control method widely used in DC–DC converters where the inductor current is used in place of the carrier signal for PWM control.

2.2 PWM Spectral Modelling and Characteristics

This section presents the development of mathematical models for constant-frequency PWM signals where the reference is defined by (2.2). All three types of carrier signals—sawtooth, inverted sawtooth and triangle—will be considered. The objective is to obtain analytical models that describe the PWM output spectrum when each type of carrier signal is used. Such spectral models are especially important for AC–DC and DC–AC converter applications where the PWM spectrum is directly related to the harmonics in the input or output voltage and current of the converter.

A periodic function can be represented by a Fourier series. A PWM waveform is periodic if the carrier frequency, denoted as f_c hereafter, and the fundamental frequency f_1 of the reference defined in (2.2) are commensurable, that is, if the frequency ratio f_c/f_1 is a rational number. In the special case when f_c is a multiple of f_1, Fourier analysis can be performed on the PWM waveform over a period corresponding to one cycle of the reference, that is, a time interval equal to $1/f_1$. However, even in such a special case, carrying out the Fourier analysis is not an easy undertaking because each switching point in the PWM waveform is defined by a trigonometric equation involving both the sinusoidal reference defined by (2.2) and a linear function of time representing the carrier signal. Since such equations can only be solved numerically, the Fourier analysis results will be specific to the case analysed and have to be redeveloped if any parameter involved changes. In the more likely case where the carrier frequency is not a multiple of (but still commensurable with) the reference fundamental frequency, Fourier analysis would have to be performed over an interval corresponding to multiple fundamental cycles, making the numerical problem quickly intractable.

The double Fourier series method introduced below solves this problem by reformulating the mathematics that define the PWM waveform. Instead of the original function of time involving two different frequencies, a new two-dimensional function is introduced where the carrier and the reference fundamental frequency each corresponds to an independent variable. The two-dimensional function is decomposed using a double Fourier series which is then converted back into a function of time involving both frequencies to give the spectrum of the original PWM waveform.

2.2.1 Double Fourier Series Method

Double Fourier analysis is a generalisation of the conventional Fourier analysis to two-dimensional functions. According to mathematical theory [24], a two-dimensional function $f(x, y)$ that is periodical in both x and y, with a period equal to 2π in both axes, can be represented by a double Fourier series in the form of

$$f(x,y) = \frac{A_{00}}{2} + \sum_{n=1}^{+\infty}(A_{0n}\cos ny + B_{0n}\sin ny) + \sum_{m=1}^{+\infty}(A_{m0}\cos mx + B_{m0}\sin mx)$$
$$+ \sum_{m=1}^{+\infty}\sum_{n=\pm 1}^{\pm\infty}\big(A_{mn}\cos(mx+ny) + B_{mn}\sin(mx+ny)\big), \tag{2.3}$$

where the coefficients are defined as follows:

$$A_{mn} = \frac{1}{2\pi^2}\int_0^{2\pi}\int_0^{2\pi} f(x,y)\cos(mx+ny)\,\mathrm{d}x\,\mathrm{d}y, \tag{2.4}$$

$$B_{mn} = \frac{1}{2\pi^2}\int_0^{2\pi}\int_0^{2\pi} f(x,y)\sin(mx+ny)\,\mathrm{d}x\,\mathrm{d}y. \tag{2.5}$$

Double Fourier series method was first used to model PWM signals in for communication applications [2]. Reference [13] presented spectral models of different PWM signals using this method. Application of the method in power electronics first appeared in [5], and more systematically in [9]. The book [3] by Black provides a detailed explanation of the double Fourier analysis method as applied to constant-frequency trailing-edge PWM. The review below follows the presentation of [3] in order to develop an understanding of the original and elegant idea introduced in [2].

The first step is to reformulate the PWM process such that it can be described by a two-dimensional function. This is illustrated in Fig. 2.3 for constant-frequency trailing-edge modulation. Figure 2.3a shows the sawtooth carrier and the reference signal, and Fig. 2.3b shows the resulting PWM output. Figure 2.3c is generated by stacking multiple carrier and reference sets $\{(c(t), r(t)\}$, $\{C_m + c(t), C_m + r(t)\}$, $\{2C_m + c(t), 2C_m + r(t)\}$, ..., where C_m is the peak-peak amplitude of the carrier signal. Line AB is an extension of the first slope of the original sawtooth signal. The horizontal lines l_1, l_2, ..., separate the stacked carriers and are spaced from each other by C_m. Note that the intersection points of line AB with these lines and the stacked references are the same as those between the original carrier and reference shown in Fig. 2.3a. Hence they define the same PWM waveform. This provides an alternative way to define, mathematically, the trailing-edge PWM process.

A similar representation can be developed for other PWM methods. Figure 2.4 shows how this is done for double-edge modulation using a triangle carrier. Unlike in the case with a sawtooth carrier, an alternate sequence of the original carrier and reference and their inverse are stacked in this case. In other words, the stacked carriers are $c(t), 2C_m - c(t), 2C_m + c(t), 4C_m - c(t), \ldots$, and the stacked references are $r(t)$, $2C_m - r(t)$, $2C_m + r(t)$, $4C_m - r(t)$, ... Line AB is an extension of the rising edge of the first triangle of the original carrier, $c(t)$. Note again that the intersection points of line AB with the stacked references are the same as those between the original carrier and reference.

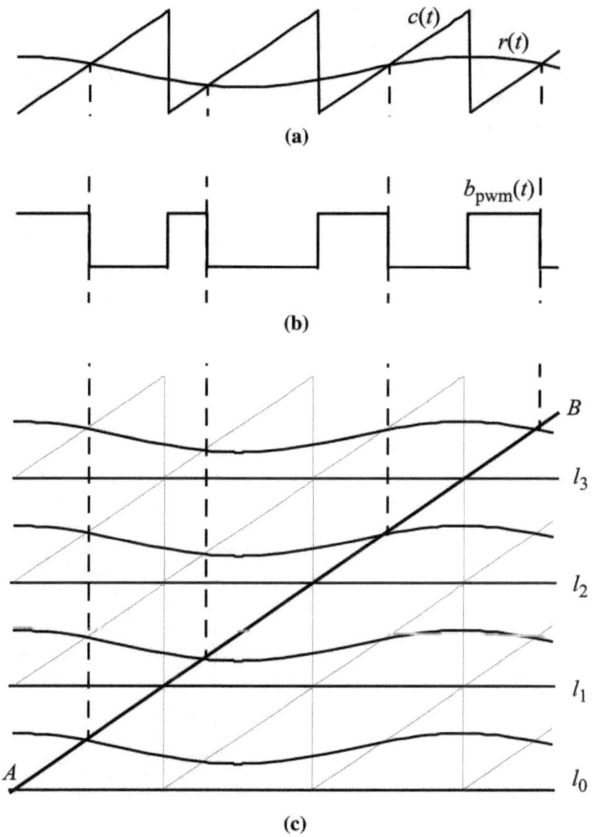

Fig. 2.3 CF trailing-edge modulation and its alternative representation for the development of a double Fourier series

The second step in the development of a double Fourier series representation is to define a two-dimensional function based on the alternative arrangement introduced above. Take double-edge modulation as example. Redefine the horizontal axis in Fig. 2.4c as x axis and let $x = 2\pi f_1 t + \theta_1$. Additionally, redefine the vertical axis as y axis and let $y = 2\pi f_c t + \theta_c$ so that the height of the triangle in the original coordinate system corresponds to π in y. θ_1 and θ_c are the initial phase angle of the reference and the carrier at $t = 0$, respectively. This new x–y coordinate system is illustrated in Fig. 2.5a where the stacked reference waveforms are represented by the edge of the shadowed areas. Imagine now a three-dimensional graph defined by a function $z = f(x, y)$ which is equal to $+1$ when the point (x, y) lies inside any of the shadowed areas, and -1 otherwise, as illustrated in Fig. 2.5b. It is obvious that is periodic in both x and y, with a common period equal to 2π. Therefore, $f(x, y)$ can be represented by a double Fourier series. Note that the double Fourier coefficients are defined analytically and no numerical values are involved up to this point.

Line AB in the x–y–z coordinate system is defined implicitly by $x = 2\pi f_1 t + \theta_1$ and $y = 2\pi f_c t + \theta_c$. Imagine now a vertical plane is erected along the line, indicated by the dashed line in Fig. 2.5b. This plane cuts across the three-dimensional graph

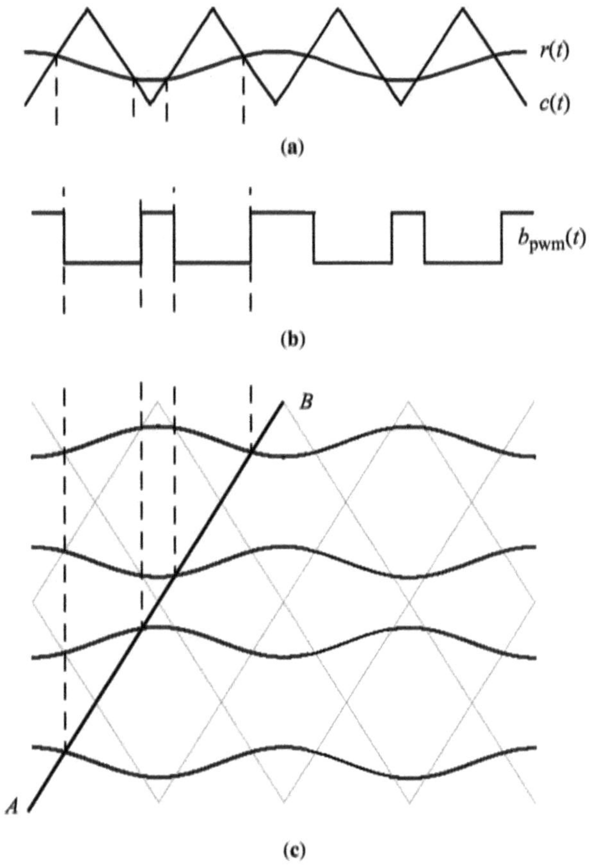

Fig. 2.4 CF double-edge modulation and its alternative representation for the development of a double Fourier series

defined above, and the intersection area, defined mathematically by

$$f(2\pi f_1 t + \theta_1, 2\pi f_c t + \theta_c), \tag{2.6}$$

has the same shape as the PWM waveform. Therefore, a spectral model is obtained for the PWM waveform when x and y in the double Fourier series representation of $f(x, y)$ are replaced by $x = 2\pi f_1 t + \theta_1$ and $y = 2\pi f_c t + \theta_c$, respectively. Note again that there is no need to explicitly define the PWM waveform and no limitations are imposed on the relationship between the carrier and the reference fundamental frequency.

With some minor modification of the two-dimensional function, the same procedure can be applied to other PWM methods. Although natural-sampling PWM has been assumed in the development so far, the method can be applied to regular-sampled PWM by using the sampled form instead of the original references.

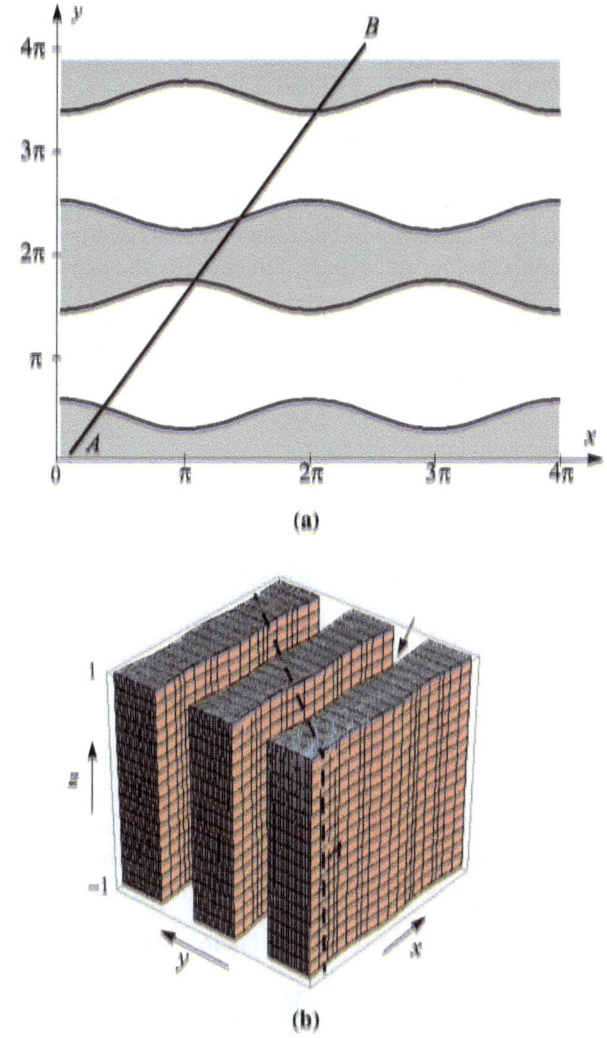

Fig. 2.5 Definition of a two-dimensional function $f_k(x, y)$ for double Fourier analysis

2.2.2 PWM Spectral Models

Double Fourier series models are given below for commonly used constant-frequency PWM methods. The reference is assumed to be defined by (2.2) in general. For the convenience of application in different power electronic circuits, we will consider

- *Unipolar modulation*: The carrier changes between 0 and a positive peak (C_m); the reference is always positive; and the PWM output switches between 0 and 1; and

- *Bipolar modulation*: The carrier is symmetric about zero, with a amplitude equal to C_m; the reference is a sine wave without DC offset; and the PWM output switches between $-1/2$ and $+1/2$.

The angular frequency of the carrier and the reference will be denoted $\omega_c (= 2\pi f_c)$ and $\omega_1 (= 2\pi f_1)$, respectively, and will be used in place of f_c and f_1 when it is convenient. The initial phase angles of the carrier and the reference (denoted by θ_1 and θ_c) will be assumed nonzero in general. Two additional parameters as defined below are also used to simplify the expressions:

- The average duty ratio:

$$D = \frac{R_0}{C_m}; \tag{2.7}$$

- The modulation index:

$$M = \frac{2R_1}{C_m}. \tag{2.8}$$

Each spectral model also involves Bessel functions of the first kind, $J_n(z)$, defined as follows:

$$J_n(z) = \frac{\mathrm{j}^{-n}}{\pi} \int_0^{2\pi} \mathrm{e}^{\mathrm{j}z\cos\theta} \mathrm{e}^{\mathrm{j}n\theta} \, d\theta. \tag{2.9}$$

This function arises from the double integral involved in the definition of the double Fourier series coefficients (2.4) and (2.5). The argument z can be a real or complex number, and the function is real-valued if z is real. The other argument, n, is real and will always be an integer in the applications discussed here. Characteristics of the Bessel functions will be reviewed in the next subsection.

1. Unipolar Trailing-Edge Modulation, see Fig. 2.6:

$$\begin{aligned} b_{\mathrm{pwm}}(t) &= D + \frac{M}{2}\cos(\omega_1 t + \theta_1) \\ &\quad + \sum_{m=1}^{+\infty} \frac{1}{m\pi} \big\{ \sin\big[m(\omega_c t + \theta_c)\big] \\ &\quad - J_0(m\pi M) \sin\big[m(\omega_c t + \theta_c) - 2mD\pi\big] \big\} \\ &\quad + \sum_{m=1}^{+\infty} \sum_{n=\pm 1}^{\pm\infty} \frac{J_n(m\pi M)}{m\pi} \\ &\quad \times \sin\left[\frac{n\pi}{2} - m(\omega_c t + \theta_c) - n(\omega_1 t + \theta_1) + 2mD\pi\right]. \end{aligned} \tag{2.10}$$

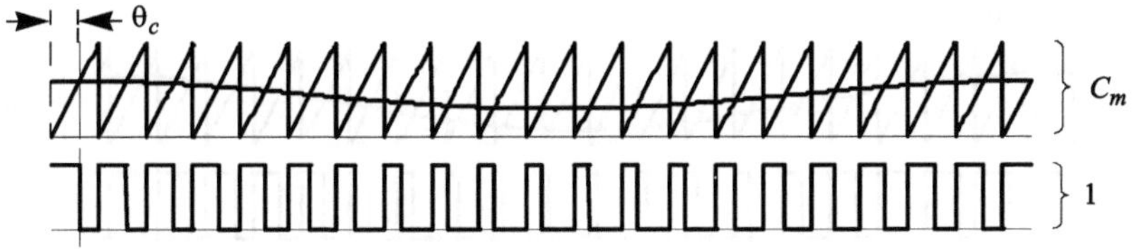

Fig. 2.6 Unipolar trailing-edge modulation using a sawtooth carrier

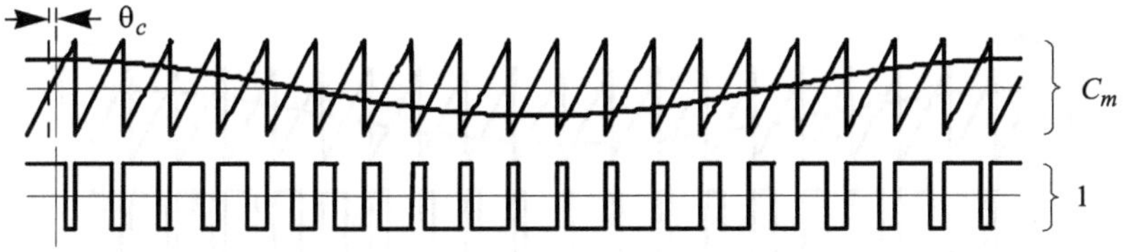

Fig. 2.7 Bipolar trailing-edge modulation using a sawtooth carrier

2. Bipolar Trailing-Edge Modulation, see Fig. 2.7 with $R_0 = 0$:

$$\begin{aligned} b_{\text{pwm}}(t) = {} & \frac{M}{2}\cos(\omega_1 t + \theta_1) \\ & + \sum_{m=1}^{+\infty} \frac{1}{m\pi}\big[\cos(m\pi) - J_0(m\pi M)\big]\sin\big[m(\omega_c t + \theta_c)\big] \\ & + \sum_{m=1}^{+\infty}\sum_{n=\pm 1}^{\pm\infty} \frac{J_n(m\pi M)}{m\pi}\sin\left[\frac{n\pi}{2} - m(\omega_c t + \theta_c) - n(\omega_1 t + \theta_1)\right]. \end{aligned} \tag{2.11}$$

3. Unipolar Leading-Edge Modulation, see Fig. 2.8:

$$\begin{aligned} b_{\text{pwm}}(t) = {} & D + \frac{M}{2}\cos(\omega_1 t + \theta_1) \\ & - \sum_{m=1}^{+\infty} \frac{1}{m\pi}\big\{\sin\big[m(\omega_c t + \theta_c)\big] \\ & - J_0(m\pi M)\sin\big[m(\omega_c t + \theta_c) + 2mD\pi\big]\big\} \\ & - \sum_{m=1}^{+\infty}\sum_{n=\pm 1}^{\pm\infty} \frac{J_n(m\pi M)}{m\pi} \\ & \times \sin\left[\frac{n\pi}{2} - m(\omega_c t + \theta_c) - n(\omega_1 t + \theta_1 + \pi) - 2mD\pi\right]. \end{aligned} \tag{2.12}$$

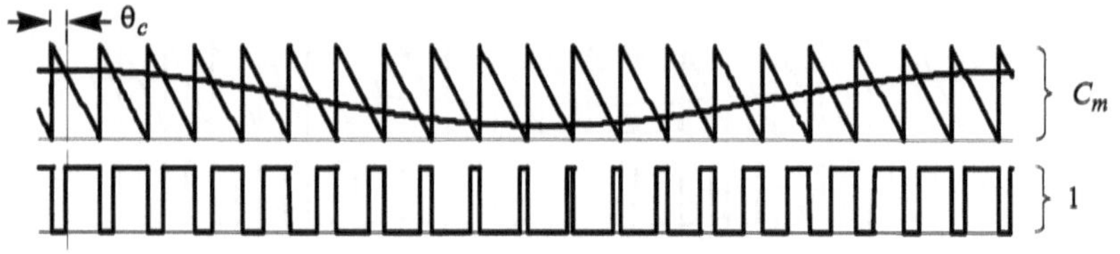

Fig. 2.8 Unipolar leading-edge modulation using a sawtooth carrier

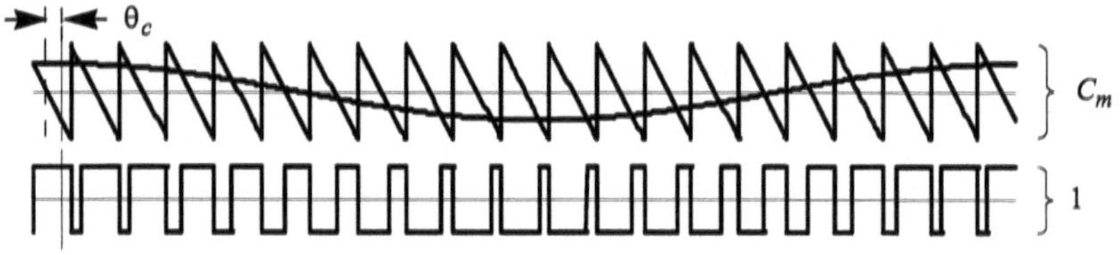

Fig. 2.9 Bipolar leading-edge modulation using a sawtooth carrier

4. Bipolar Leading-Edge Modulation, see Fig. 2.9 with $R_0 = 0$:

$$\begin{aligned}
b_{\mathrm{pwm}}(t) = {} & \frac{M}{2}\cos(\omega_1 t + \theta_1) \\
& - \sum_{m=1}^{+\infty} \frac{1}{m\pi}\big[\cos(m\pi) - J_0(m\pi M)\big]\sin\big[m(\omega_c t + \theta_c)\big] \\
& - \sum_{m=1}^{+\infty} \sum_{n=\pm 1}^{\pm\infty} \frac{J_n(m\pi M)}{m\pi} \\
& \times \sin\left[\frac{n\pi}{2} - m(\omega_c t + \theta_c) - n(\omega_1 t + \theta_1 + \pi)\right]. \qquad (2.13)
\end{aligned}$$

5. Unipolar Double-Edge Modulation, see Fig. 2.10:

$$\begin{aligned}
b_{\mathrm{pwm}}(t) = {} & D + \frac{M}{2}\cos(\omega_1 t + \theta_1) \\
& + \sum_{m=1}^{+\infty} \frac{2}{m\pi} J_0\left(\frac{m\pi M}{2}\right)\sin(Dm\pi)\cos\big[m(\omega_c t + \theta_c)\big] \\
& + \sum_{m=1}^{+\infty} \sum_{n=\pm 1}^{\pm\infty} \frac{2}{m\pi} J_n\left(\frac{m\pi M}{2}\right)\sin\frac{(2Dm + n)\pi}{2} \\
& \times \cos\big[m(\omega_c t + \theta_c) + n(\omega_1 t + \theta_1)\big]. \qquad (2.14)
\end{aligned}$$

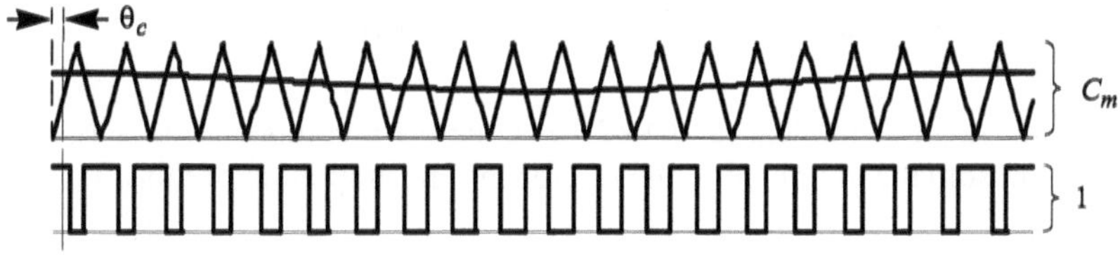

Fig. 2.10 Unipolar leading-edge modulation using a sawtooth carrier

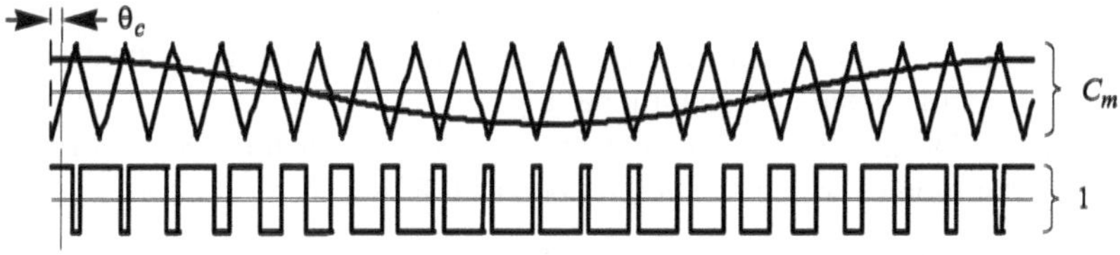

Fig. 2.11 Bipolar double-edge modulation using a triangle carrier

6. Bipolar Double-Edge Modulation, see Fig. 2.11 with $R_0 = 0$:

$$
\begin{aligned}
b_{\mathrm{pwm}}(t) = {} & \frac{M}{2}\cos(\omega_1 t + \theta_1) \\
& + \sum_{m=1}^{+\infty} \frac{2}{m\pi} J_0\left(\frac{m\pi M}{2}\right) \sin\frac{m\pi}{2} \cos\left[m(\omega_c t + \theta_c)\right] \\
& + \sum_{m=1}^{+\infty} \sum_{n=\pm 1}^{\pm\infty} \frac{2}{m\pi} J_n\left(\frac{m\pi M}{2}\right) \sin\frac{(m+n)\pi}{2} \\
& \times \cos\left[m(\omega_c t + \theta_c) + n(\omega_1 t + \theta_1)\right].
\end{aligned}
\tag{2.15}
$$

The double Fourier analysis method can also be used to develop spectral models for regular-sampled PWM by replacing the reference functions by their sampled version [9]. The method is also applicable when the reference contains more than one sine component (as long as it is periodic). However, the double integrals cannot be simplified by using the Bessel functions when the reference contains more than one sine component. Nevertheless, the double integral formulation in that case still provides a simple way to define the spectral model and can be evaluated numerically when quantitative analysis is needed.

2.2.3 Spectral Characteristics of PWM Converters

Each of the spectral models (2.10)–(2.13) contains a component that is in-phase with the sinusoidal component in the reference. The amplitude of this component,

$$\frac{M}{2} = \frac{R_1}{C_m},$$

is equal to that of the reference scaled down by the amplitude of the carrier. A DC component also exists in the PWM output under unipolar modulation, and its ratio to the DC component in the reference is equal to the reciprocal of the amplitude of the carrier, the same as that for the sine component. This linear relationship is an important characteristic of constant-frequency PWM and will be further discussed in the next section on small-signal modelling.

Besides the DC and the fundamental sine component, each PWM waveform contains many harmonics which can be grouped as follows:

1. Carrier frequency and its harmonics, at frequencies $f = mf_c, m = 1, 2, \ldots, +\infty$. These components exist regardless whether the reference is constant or variable.
2. Sideband harmonics of the carrier and its harmonics, at frequencies $f = mf_c + nf_1, n = \pm 1, \pm 2, \ldots, \pm\infty$.

The sideband harmonics only appear when the reference is modulated. The amplitude of the sideband harmonic at frequency $f = mf_c + nf_1$ is equal to

$$\frac{J_n(m\pi M)}{m\pi} \tag{2.16}$$

for leading- or trailing-edge modulation, and to

$$\frac{2}{m\pi} J_n\left(\frac{m\pi M}{2}\right) \tag{2.17}$$

for double-edge modulation. Note that (2.17) can be converted to (2.16) by redefining the parameter M for double-edge modulation. Additionally, since

$$\left|J_n(x)\right| = \left|J_{-n}(x)\right| \tag{2.18}$$

the sideband harmonic at frequency $f = mf_c + nf_1$ has the same magnitude as that at frequency $f = mf_c - nf_1$ for any integer m and n; in other words, sideband harmonics appear in pairs and are symmetric about each carrier harmonic. Figure 2.12 depicts Bessel functions $J_0(x)$ through $J_{10}(x)$ to show their basic characteristics.

Unipolar modulation is usually used in DC–DC converters where the reference is constant except during transient. The sideband harmonics in this case are instrumental in explaining the time-varying behaviour in small-signal analysis when the perturbation frequency approaches half the switching frequency, which will be discussed in the next section. Bipolar modulation, on the other hand, is commonly used in PWM inverters and rectifiers. A basic building block for such converters is the

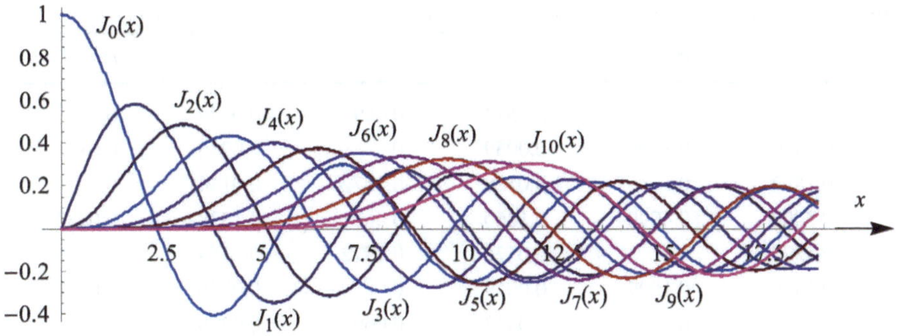

Fig. 2.12 A plot of Bessel functions $J_0(x)$ through $J_{10}(x)$

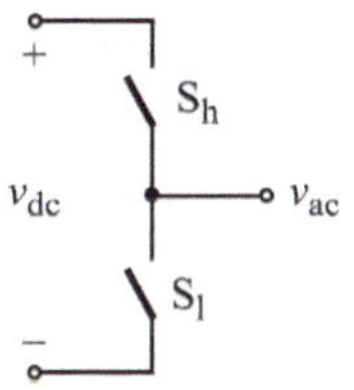

Fig. 2.13 A half-bridge circuit serving as a basic building block for AC–DC interface

half-bridge circuit depicted in Fig. 2.13. The AC terminal is alternately connected to the positive and the negative terminal of the DC link depending on whether the upper-side switch, S_h, or the lower-side switch, S_l, is turned on. Usually the two switches are controlled by complementary gate signals such that at any time one and only one switch conducts. If the middle of the DC link is used as the reference point and a bipolar PWM signal $v_{\rm pwm}(t)$ defined in the previous subsection is applied to the upper switch, the DC and the AC voltages are related to each other by

$$v_{\rm ac}(t) = v_{\rm dc}(t)v_{\rm pwm}(t). \tag{2.19}$$

With a constant DC-link voltage, $v_{\rm ac}$ is proportional to the PWM signal such that a spectral model can be readily obtained for $v_{\rm ac}$ from the bipolar PWM spectral models developed in the previous subsection. Therefore, the performance of the three bipolar PWM methods can be compared on the basis of their spectral models (2.11), (2.13) and (2.15). To that end, note first that the carrier harmonics and their sideband harmonics produced by trailing-edge bipolar PWM are equal in amplitude to the corresponding harmonics in leading-edge modulation. Hence these two methods have the same performance as far as harmonics are concerned. To compare them with double-edge bipolar modulation, the amplitudes of the first three carrier harmonics ($m = 1, 2, 3$) and their sideband components are tabulated in Table 2.1 for $n = \pm 1, \pm 2, \pm 3$. The comparison shows that double-edge modulation produces far fewer harmonics. In particular, the harmonic at frequency $f = mf_c + nf_1$ is zero if $m + n$ is even because the amplitude of the harmonic

Table 2.1 Comparison of carrier and sideband harmonics predicted by (2.11) and (2.15)

n	$m=1$, (2.11)	$m=1$, (2.15)	$m=2$, (2.11)	$m=2$, (2.15)	$m=3$, (2.11)	$m=3$, (2.15)
−3	0.0697	0	0.0573	0.0697	0.0279	0
−2	0.1426	0.1099	0.0059	0	0.0236	0.0881
−1	0.1572	0	0.0526	0.1572	0.0154	0
0	0.3008	0.4090	0.1860	0	0.1338	0.0853
1	0.1572	0	0.0526	0.1572	0.0154	0
2	0.1426	0.1099	0.0059	0	0.0236	0.0881
3	0.0697	0	0.0573	0.0697	0.0279	0

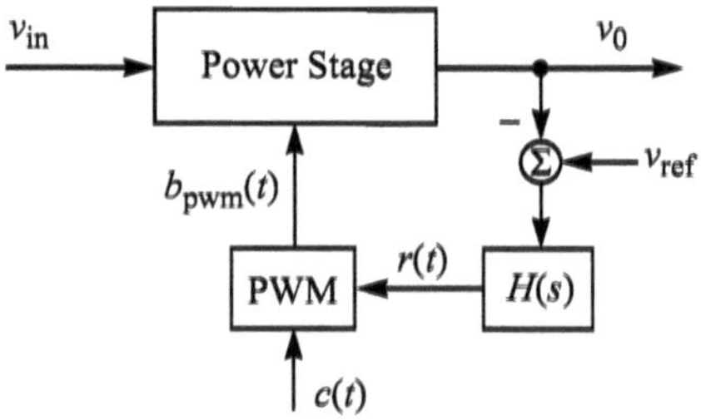

Fig. 2.14 A pulse-width modulator as part of a feedback control system

includes the term

$$\sin\frac{(m+n)\pi}{2} \tag{2.20}$$

which is zero when $m+n$ is even. In other words, double-edge modulation eliminates all even-order sideband harmonics of even-order carrier harmonics, as well as odd-order sideband harmonics of odd-order carrier harmonics. The reduced harmonic content is a significant advantage of double-edge modulation and makes the method more attractive for AC–DC and DC–AC converter application.

2.3 Small-Signal Modelling

Pulse-width modulator is an integral part of feedback control in power electronics. The reference signal, $r(t)$, used in the spectral analysis in the previous section is usually the output of a feedback controller (compensator), as depicted in Fig. 2.14 where $H(s)$ denotes the feedback controller transfer function. The modulation process itself, as demonstrated by the spectral models presented in the previous section, is highly nonlinear. Since most practical designs use linear control, a linear, time-invariant model that captures the small-signal behaviour of a modulator is needed for converter-level analysis and control design. Such small-signal models are developed in this section for both constant-frequency and variable-frequency PWM. The second topic treated here is interleaving—the operation of multiple pulse-width

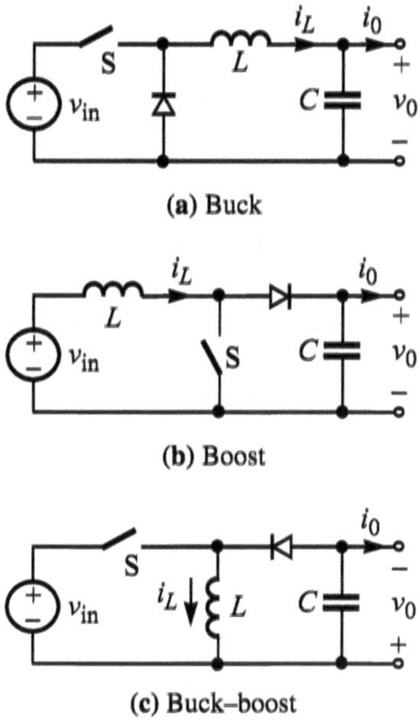

Fig. 2.15 Basic DC–DC converter topologies

modulators with phase-shifted carrier signals. We will briefly discuss the application of interleaving in parallel- and series-connected converters, and use the spectral models developed in the previous section to characterise the harmonica cancellation effects resulting from different interleaving arrangements.

2.3.1 Small-Signal Modelling of Constant-Frequency PWM

A PWM-controlled converter can be described by a set of linear differential equations in each conduction state of the switches if all components are linear and the switches (including diodes) are assumed ideal. The number of possible conduction states depends on the number of switches and the operation pattern of the converter. Figure 2.15 depicts three basic DC–DC converter circuits each of which uses one switch and one diode. In the continuous conduction mode (CCM) of operation [23], that is, when the inductor current flows continuously, the diode conducts whenever the switch is OFF, such that there are only two possible conduction states over a switching cycle: (a) the switch is ON and the diode is OFF; (b) the switch is OFF and the diode is ON. With the inductor current (i_L) and the output capacitor voltage (v_O) modelled as state variables, each of these converters can be described by two state-space models as follows where $x = (i_L, v_O)^\top$, $u = (v_{in}, i_O)^\top$:

$$\dot{x} = A_0 x + B_0 u \tag{2.21}$$

when the switch is OFF, and

$$\dot{x} = A_1 x + B_1 u \tag{2.22}$$

when the switch is ON. Denoting the unipolar PWM signal that drives the switch as $s(t)$, that is, $s(t) = 0$ when the switch is OFF and $s(t) = 1$ when the switch is ON, one can combine the two equations as follows:

$$\dot{x} = \left[A_0 + (A_1 - A_0)s(t)\right]x + \left[B_0 + (B_1 - B_0)s(t)\right]u. \tag{2.23}$$

This model is not amenable to linear control design due to the time-varying control function $s(t)$ being multiplied with the state and input variables. A standard technique to deal with such time-varying problem is averaging [6]. In this method, the right-hand side of (2.23) is replaced by its "moving" average over a switching cycle, resulting in the so-called state-space averaged model where the binary switching function $s(t)$ is replaced by its algebraic average over a switching cycle, that is, the duty ratio d of the switch [6]:

$$\frac{\mathrm{d}}{\mathrm{d}t}\langle x\rangle = \left[A_0 + (A_1 - A_0)d\right]\langle x\rangle + \left[B_0 + (B_1 - B_0)d\right]u. \tag{2.24}$$

The duty ratio is the control variable in the new averaged model, which is nonlinear in general due to the presence of the product between the duty ratio and the state variables.[1] A linear model suitable for linear control design can be obtained by conventional small-signal linearisation about a given operation point. Since it is the modulator reference, $r(t)$, that is actually controlled, a transfer function from the modulator input to the duty ratio of its PWM output is needed to complete such a small-signal model. This transfer function is usually referred to as the *modulator gain* and can be developed in either the time or the frequency domain.

To develop the modulator gain in the time domain, assume the reference consists of a constant, R_0, corresponding to the DC operation point of the converter about which the small-signal model is being developed. The corresponding steady-state duty ratio is denoted as D. Consider now that a small-signal perturbation, $\hat{R}$, is added to the reference. It's a matter of simple algebra to show that the corresponding perturbation in the duty ratio is

$$\hat{d} = \frac{\hat{R}}{C_m} \tag{2.25}$$

for any of the constant-frequency PWM methods reviewed in Sect. 2.1. Therefore, the gain of a CF pulse-width modulator is the reciprocal of the amplitude, C_m, of the carrier signal regardless what form of carrier is used.

To develop a small-signal modulator model in the frequency domain, consider a reference defined by (2.2) where the DC component, R_0, corresponds to the steady-state operation point and the second term defines a small-signal perturbation at frequency f_1. Each of the spectral models (2.10)–(2.15) developed in the previous

[1] An exception is the buck converter and its derivatives where switching only changes the connection of the input voltage to the converter. Therefore, the only nonlinear term in the averaged model is the product between the duty ratio and the input voltage, and the model is linear if the input voltage is assumed constant.

section contains a component at the same frequency and is in-phase with the perturbation in $r(t)$. The amplitude of this perturbation term is equal to

$$\frac{M}{2} = \frac{R_1}{C_m} \tag{2.26}$$

which shows again that the modulator gain is $1/C_m$ regardless the form of the carrier signal.

The discussion above did not consider possible sideband components that could appear at the perturbation frequency and affect the small-signal response of the modulator. Based on (2.10)–(2.15), the frequency of the sideband components can be written as

$$mf_c + nf_1, \quad m = 1, 2, \ldots, +\infty, \; n = \pm 1, \pm 2, \ldots, \pm\infty. \tag{2.27}$$

When the perturbation frequency, f_1, is at half the carrier frequency, for example, the frequency of the sideband component corresponding to $m = 1$ and $n = -1$ is

$$f = f_c - f_1 = 2f_1 - f_1 = f_1, \tag{2.28}$$

which is the same as the perturbation frequency. For unipolar trailing-edge modulation, this sideband component is

$$-\frac{J_{-1}(\pi M)}{\pi} \cos[\omega_1 t - \theta_1 + \theta_c - 2D\pi].$$

The coincidence of such a sideband component with the response of the modulator at the perturbation frequency presents several problems for linear analysis:

1. The amplitude of the sideband component is a nonlinear function of the perturbation;
2. The phase difference to the reference, $2\theta_1 - \theta_c + 2D\pi$, is not constant and depends on the initial phase of the perturbation as well as the carrier.

The nonlinear amplitude can be replaced by a linear approximation of the Bessel function under small-signal assumption, that is, by assuming $M \ll 1$. However, the variable phase angle indicates that the response of the modulator at the perturbation frequency is dependent of the initial phase angle of the perturbation. Such time-varying behaviour violates the underlying assumption for linear analysis and represents a fundamental limitation of the small-signal linear model and the associated linear control design techniques for PWM converters.

In general, a sideband component at frequency $mf_c + nf_1$ will appear at the perturbation frequency, f_c, if the latter is such that

$$f_1 = \frac{mf_c}{1-n}, \quad m = 1, 2, \ldots, +\infty, \; n = 0, -1, -2, \ldots, -\infty. \tag{2.29}$$

However, since the amplitude of the sideband component at frequency is proportional to Bessel function $J_n(x)$ and the lowest-order term in the Taylor expansion of

$J_n(x)$ is proportional to

$$\frac{x^{|n|}}{2^{|n|}n!}, \tag{2.30}$$

the contribution of such a sideband component to the modulator response at the perturbation frequency can be ignored for the purpose of small-signal analysis except for the cases of $n = 0$ and $n = -1$, in other words, when the perturbation frequency is a multiple of half the carrier frequency:

$$f_1 = \frac{mf_c}{2}, \quad m = 1, 2, \ldots, +\infty. \tag{2.31}$$

The above equation indicates that the lowest frequency at which small-signal response of a pulse-width modulator is affected by sideband components and is no longer time-invariant is half the carrier frequency. This places an upper frequency limit on the validity of averaged models of PWM converters. Nonlinear control methods that overcome this limit of linear models are discussed in other chapters of this book.

Averaging effectively removes the switching ripple in the responses of a converter. If necessary, the ripple can be recovered from the averaged model. A method [16] to reconstruct switching ripple from the averaged model of a PWM converter will be discussed in Sect. 2.4 in conjunction with modelling of peak-current control.

The straightforward state-space averaging method as used above is only applicable to PWM converters operating in the continuous conduction mode. In the discontinuous conduction mode (DCM), the inductor current returns to zero in each switching cycle such that the diode only conducts for a portion of the interval in which the switch is OFF. This leads to a third interval over a switching cycle in which both the switch and the diode are OFF. The state-space model of the converter in this third interval needs to be included in the averaged model. Additionally, since the length of the conduction interval of the diode depends on several parameters and is not constant, a so-called duty ratio constraint is needed to relate the duty ratio of the diode to other variables in order to completely define the averaged model. These are discussed in reference [23], which also includes analytical small-signal models for the three basic converters shown in Fig. 2.15.

A major focus of power electronics research in the 1980s and 1990s was the development of various resonant-type converters. The basic idea of resonant converters is to create a resonance in the voltage or current of a switch by using additional passive and/or active devices. The resonant response causes the switch voltage or current to naturally return to zero, thereby allowing the switching operation to occur under zero-voltage or zero-current conditions to significantly reduce the switching transition loss. Averaged modelling of resonant converters is complicated by the presence of additional time constants of the resonant circuit. Singular perturbation theory and time-scale separation techniques can be applied to deal with such multiple-time-scale systems and to develop averaged models that can be used for linear control design [21]. The modulator model presented above can still be used. Some resonant converters, such as quasi-resonant and load-resonant converters, require variable-frequency PWM control, which will be discussed in the next subsection.

The control method depicted in Fig. 2.14 is essentially the voltage-mode control, in which the switch duty ratio is derived directly from the error signal of the output voltage. The duty ratio to output voltage transfer function, which can be obtained from the small-signal linearised averaged model, is of second-order for second-order converters such as those depicted in Fig. 2.15, with highly under-damped resonant poles. It is difficult to achieve fast control for such second-order systems because the crossover frequency of the voltage control loop usually has to be kept below the power stage resonant frequency in order to ensure sufficient stability margin. An additional difficulty of voltage-mode control for boost and buck–boost converters is the existence of a right half-plane (RHP) zero in the duty ratio to output voltage transfer function, which is usually at a frequency even lower than the power stage resonant frequency. Current-mode control, in which an additional loop is closed for the inductor current, can be used to solve these problems. Section 2.4 will present one of such current control methods, the peak-current control.

2.3.2 Small-Signal Modelling of Variable-Frequency PWM

The modulator gain developed in the previous subsection is valid for constant-frequency PWM. Variable-frequency modulators exhibit unique features in their small-signal behaviour and are modelled here.

Consider first the constant-OFF-time VF PWM, which is illustrated in Fig. 2.16 for two cases:

1. The reference $r(t)$ is constant;
2. The reference $r(t)$ consists of a constant and a small-signal perturbation, as defined by (2.2).

The resulting PWM outputs are shown as v_1 and v_2, respectively. By definition, the small-signal gain of the modulator is the ratio of the component in v_2 that is at the perturbation frequency, f_1, to the amplitude of the perturbation, R_1. For the purpose of small-signal modelling, it is assumed that the perturbation is much smaller than the constant, that is, $R_1 \ll R_0$. Additionally, the following variables are defined:

- The up-slope of the ramp signal, $c(t)$, is constant and is denoted as m_0.
- The steady-state ON-time (when $R_1 = 0$) is denoted as T_{ON}. It can be determined from Fig. 2.16 that

$$T_{ON} = \frac{R_0}{m_0}.$$

- The constant OFF-time is equal to T_{OFF}. Denote $1/(T_{ON} + T_{OFF})$ by f_c, which is the steady-state carrier frequency of the modulator when the perturbation is not present ($R_1 = 0$).

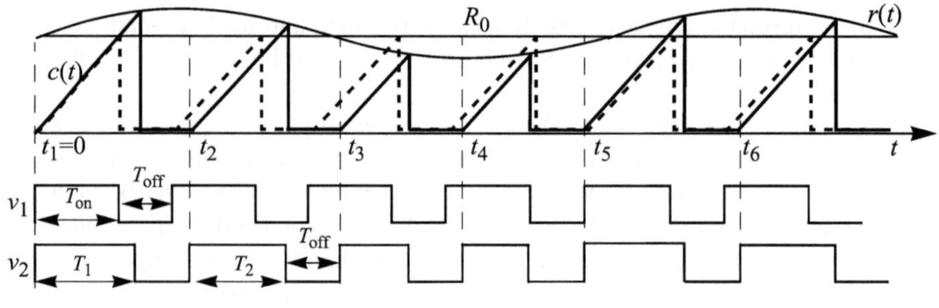

Fig. 2.16 Constant-OFF-time modulation. v_1: PWM output when the reference is a constant; v_2: PWM output when the reference $r(t)$ contains a sinusoidal perturbation

Using these notations, reference [14] developed the small-signal gain of the VF constant-OFF-time modulator as follows

$$\frac{1}{m_0(T_{ON}+T_{OFF})} \sin\frac{sT_{OFF}}{j2}\left(\sin\frac{s(T_{ON}+T_{OFF})}{j2}\right)^{-1} e^{\frac{sT_{ON}}{2}}, \tag{2.32}$$

where $j=\sqrt{-1}$ is the imaginary unit. Compared to the constant-frequency modulator gain developed in the previous subsection, this gain is much more complicated and exhibits nonlinear characteristics that are frequency dependent.

Defining a function $K_1(s)$ as

$$K_1(s) = \frac{T_{ON}+T_{OFF}}{T_{OFF}} \sin\frac{sT_{OFF}}{j2}\left(\sin\frac{s(T_{ON}+T_{OFF})}{j2}\right)^{-1}, \tag{2.33}$$

the modulator gain (2.33) can be written as

$$\Gamma_1(s) = \frac{T_{OFF}}{m_0(T_{ON}+T_{OFF})^2} K_1(s) e^{\frac{sT_{ON}}{2}}. \tag{2.34}$$

Note that $\Gamma_1(s)$ consists of three parts:

1. $T_{OFF}/[m_0(T_{ON}+T_{OFF})^2]$, which is independent of the frequency and represents the DC gain of the modulator. The DC gain can also be derived by taking the derivative of the steady-state duty ratio with respect to the modulator input, R_0 (note that $T_{ON}=R_0/m_0$):

$$\frac{d}{dR_0}\left(\frac{T_{ON}}{T_{ON}+T_{OFF}}\right) = \frac{d}{dR_0}\left(\frac{R_0/m_0}{R_0/m_0+T_{OFF}}\right) = \frac{T_{OFF}}{m_0(T_{ON}+T_{OFF})^2}.$$

   This can be compared with the gain of constant-frequency pulse-width modulators, which, using the notations introduced here, can be written as $[m_0(T_{ON}+T_{OFF})]^{-1}$.

2. The second part, represented by $K_1(s)$, is frequency-dependent but is always real-valued, hence does not contribute to the phase response of the modulator gain. Its magnitude is close to unity except at frequencies approaching half the switching frequency.

3. The third part in (2.34) does not contribute to the amplitude of the gain. It corresponds to a phase lead that increases linearly with the frequency as well as the ON-time, T_{ON}.

The phase lead is the most interesting part of the VF PWM gain and represents an advantage of the modulation method because it helps to increase the phase margin of the control loop. The phase lead is most significant at high frequencies and/or when T_{ON} is large. For example, at one tenth of the switching frequency, the phase lead will be 14.4° for $T_{\mathrm{ON}}/(T_{\mathrm{ON}}+T_{\mathrm{OFF}})=0.8$. This may have significant impact on loop design, especially when wide bandwidth is required.

The modulator gain for constant ON-time control can be obtained by exploiting the duality between constant ON-time and constant OFF-time control. To this end, assume that the same signal $r(t)$ as shown in Fig. 2.16 is now used to modulate the OFF-time while the ON-time is fixed at T_{ON}. Denoting the resulting constant ON-time PWM waveform as $\overline{p}(t, T_{\mathrm{ON}})$ and the original constant OFF-time PWM waveform as $p(t, T_{\mathrm{OFF}})$, one can see that $\bar{p}(t, T_{\mathrm{ON}}) = 1 - p(t, T_{\mathrm{OFF}})$ if T_{ON} under constant ON-time PWM is equal to T_{OFF} under constant OFF-time PWM. Therefore, the small-signal gain of a constant ON-time VF modulator can be obtained by exchanging T_{ON} and T_{OFF} in (2.32) and multiplying the resulting expression by -1, as given below:

$$\frac{-1}{m_0(T_{\mathrm{ON}}+T_{\mathrm{OFF}})}\sin\frac{sT_{\mathrm{ON}}}{\mathrm{j}2}\left(\sin\frac{s(T_{\mathrm{ON}}+T_{\mathrm{OFF}})}{\mathrm{j}2}\right)^{-1}\mathrm{e}^{\frac{sT_{\mathrm{OFF}}}{2}}. \tag{2.35}$$

Like in the case of constant OFF-time PWM, we can define a function

$$K_2(s) = \frac{T_{\mathrm{ON}}+T_{\mathrm{OFF}}}{T_{\mathrm{ON}}}\sin\frac{sT_{\mathrm{ON}}}{\mathrm{j}2}\left(\sin\frac{s(T_{\mathrm{ON}}+T_{\mathrm{OFF}})}{\mathrm{j}2}\right)^{-1}, \tag{2.36}$$

and rewrite (2.35) as follows:

$$\Gamma_2(s) = \frac{-T_{\mathrm{ON}}}{m_s(T_{\mathrm{ON}}+T_{\mathrm{OFF}})^2}K_2(s)\mathrm{e}^{\frac{sT_{\mathrm{OFF}}}{2}}. \tag{2.37}$$

Similar to $K_1(s)$, $K_2(s)$ is approximately equal to unity except at high frequencies. The phase of consists of two parts: a 180° phase reversal (due to the minus sign in front of the expression) and a phase lead that is proportional to the OFF-time. Hence it can be concluded that both constant OFF-time and constant ON-time VF PWM creates a phase lead which is beneficial for control stability.

Note that the term

$$\sin\frac{s(T_{\mathrm{ON}}+T_{\mathrm{OFF}})}{\mathrm{j}2} \tag{2.38}$$

appears as a denominator in both $K_1(s)$ and $K_2(s)$. This term becomes zero at frequencies

$$s = \frac{\mathrm{j}2k\pi}{T_{\mathrm{ON}}+T_{\mathrm{OFF}}}, \tag{2.39}$$

that is, at multiples of the steady-state PWM frequency $f_c = 1/(T_{ON} + T_{OFF})$. This is, however, not a practical problem since the use of the small-signal model, as discussed in the previous section, should be limited to half the PWM frequency.

2.3.3 Interleaved PWM

When multiple PWM converters are connected in series or in parallel, it is advantageous to operate all converters at the same frequency but with certain phase shift among their PWM signals. The technique is referred to as interleaving and has been widely used in power electronics. The benefit of interleaving can be understood by an example shown in Fig. 2.17, where two identical buck converters operate in parallel with the same input and output voltage. Both converters operate with the same carrier frequency and reference (hence also duty ratio), but the two carrier signals are shifted from each other by half a carrier cycle. The resulting ripples (without the DC component) of the two inductor currents are shown at the bottom of Fig. 2.17. The output voltage is treated as a constant, and the two inductors are assumed identical. As can be seen, the combined current, i_L, that charges the output capacitor has much smaller ripple, and the ripple is cancelled completely when the duty ratio is 0.5.

Similar effects can be observed in the input current, i_{in}, of the two parallel buck converters. The same technique can be applied to other DC–DC converters. In general, with N identical converters operating in the parallel and their carrier signals being phase-shifted from each other by one Nth of a carrier cycle, the fundamental frequency of the combined input or output current is Nf_c, where f_c is the carrier frequency individual converters are switched at. This ripple cancellation effect can be exploited to reduce input and output filtering requirements.

The effects of interleaving on current ripple can also be analysed in the frequency domain by using the PWM spectral models developed in the previous section.

Consider again the parallel buck converters shown in Fig. 2.17. Response of each of the inductor currents can be written as

$$L_1 \frac{\mathrm{d}}{\mathrm{d}t} i_{L1} = v_{\text{in}} s_1(t) - v_{\text{O}},$$

$$L_2 \frac{\mathrm{d}}{\mathrm{d}t} i_{L2} = v_{\text{in}} s_2(t) - v_{\text{O}},$$

where $s_1(t)$ and $s_2(t)$ are the switching functions of switch S_1 and S_2 as defined in Fig. 2.17. Assuming $L_1 = L_2 = L$, we obtain the response of the combined current i_L as follows:

$$L \frac{\mathrm{d}}{\mathrm{d}t} i_L = v_{\text{in}} \left[s_1(t) + s_2(t) \right] - 2v_{\text{O}}. \tag{2.40}$$

With constant input and output voltages, (2.40) indicates that the cancellation of harmonics in current i_L is due to the cancellation of harmonics in the PWM functions

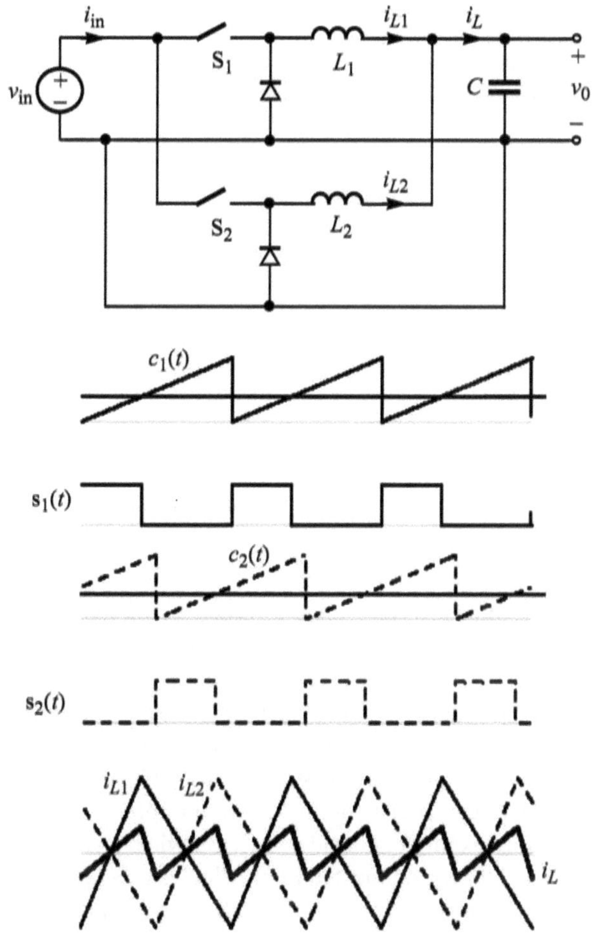

Fig. 2.17 Two buck converters operate in parallel and the resulting ripple cancellation effect through interleaving

$s_1(t)$ and $s_2(t)$. Consider the unipolar trailing-edge PWM spectral model (2.10), which simplifies to

$$b_{pwm}(t) = D + \sum_{m=1}^{+\infty} \frac{2\sin(mD\pi)}{m\pi} \cos\left[m(\omega_c t + \theta_c) - mD\pi\right]$$

when the reference is constant ($M = 0$). Without losing generality, assume the initial phase angle of the first carrier signal, $c_1(t)$, is zero, that is, $\theta_{c1} = 0$. The second carrier signal, $c_2(t)$, is delayed from $c_1(t)$ by half a carrier cycle, hence $\theta_{c2} = -\pi$. Therefore,

$$s_1(t) = D + \sum_{m=1}^{+\infty} \frac{2\sin(mD\pi)}{m\pi} \cos[m\omega_c t - mD\pi], \tag{2.41}$$

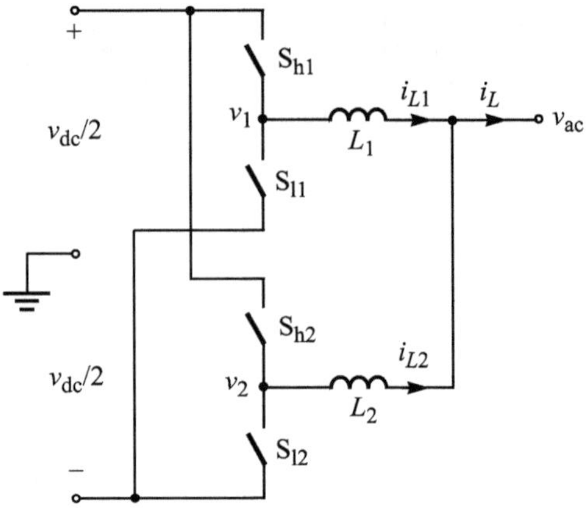

Fig. 2.18 Two half-bridge converters connected in parallel through inductors on the AC side

$$s_2(t) = D + \sum_{m=1}^{+\infty} \frac{2\sin(mD\pi)}{m\pi} \cos\big[m\omega_c t - m(D+1)\pi\big], \tag{2.42}$$

which indicates that all odd-order harmonics of the carrier are eliminated in $s_1(t) + s_2(t)$; in other words, the lowest-order harmonic in the combined output current is at twice the carrier frequency.

The frequency-domain analysis presented above can be generalised to any number of parallel converters as well as to other converter topologies. It is especially effective for interleaved AC–DC and DC–AC converters where the varying ripple pattern makes it difficult to study the ripple cancellation effects in the time domain.

As an example, consider the case when two of the half-bridge converter shown in Fig. 2.13 are connected in parallel, as depicted in Fig. 2.18. The DC input terminals are connected together directly. To prevent short circuit, an inductor has to be inserted at the output of each converter before the two can be connected in parallel. Assume each pair of switches are controlled by a bipolar double-edge modulator. Further, assume the reference for both modulators is $r(t) = R_1\cos(2\pi f_1 t + \theta_1)$, and the triangle carriers have the same frequency and amplitude, but are phase shifted from each other by half a carrier cycle. Without losing generality, assume the initial phase angle of both the reference and the first carrier is zero. Based on the analyses presented in Sect. 2.2, the voltage produced by each half bridge relative to the middle of the DC link is

$$v_1(t) = \frac{v_{dc}}{2}\left[\frac{M}{2}\cos(\omega_1 t) + \sum_{m=1}^{+\infty} \frac{2}{m\pi} J_0\left(\frac{m\pi M}{2}\right)\sin\frac{m\pi}{2}\cos(m\omega_c t) + \sum_{m=1}^{+\infty}\sum_{n=\pm1}^{\pm\infty} \frac{2}{m\pi} J_n\left(\frac{m\pi M}{2}\right)\sin\frac{(m+n)\pi}{2}\cos(m\omega_c t + n\omega_1 t)\right], \tag{2.43}$$

$$v_2(t) = \frac{v_{dc}}{2}\left[\frac{M}{2}\cos(\omega_1 t) + \sum_{m=1}^{+\infty}\frac{2}{m\pi}J_0\left(\frac{m\pi M}{2}\right)\sin\frac{m\pi}{2}\cos\bigl[m(\omega_c t - \pi)\bigr]\right.$$
$$+\sum_{m=1}^{+\infty}\sum_{n=\pm 1}^{\pm\infty}\frac{2}{m\pi}J_n\left(\frac{m\pi M}{2}\right)\sin\frac{(m+n)\pi}{2}$$
$$\left.\times\cos\bigl[m(\omega_c t - \pi) + n\omega_1 t\bigr]\right]. \tag{2.44}$$

Similar to the case of two parallel buck DC–DC converters, harmonic cancellation between i_{L1} and i_{L2} can be studied based on harmonic cancellation between v_1 and v_2. By comparing the two expressions given above, it can be easily seen that interleaving cancels all odd-order harmonics of the carrier and all their sideband components. Further, since the amplitude of any even-order carrier harmonic is zero, the only harmonics remained in $v_1(t) + v_2(t)$ are the following odd-order sideband components of even-order carrier harmonics:

$$\frac{4v_{dc}}{2}\sum_{m=2,4,\dots}^{+\infty}\sum_{n=\pm 1,\pm 3,\dots}^{\pm\infty}\frac{1}{m}J_n\left(\frac{m\pi M}{2}\right)\sin\frac{(m+n)\pi}{2}\cos(m\omega_c t + n\omega_1 t). \tag{2.45}$$

The above analysis can be extended to any number of parallel converters. The same analysis and ripple cancellation principle also apply to converters connected in series. In general, with N identical converters operating in either parallel or series, and their carrier shifted from each other by one Nth of a carrier cycle, all harmonics except those at N-multiples of the carrier frequency and their sideband components will be eliminated.

The interleaving method discussed so far confines the carrier phase shift to one Nth of a carrier cycle when N modules are connected. It has been recognised recently [15] that this may not be optimal. For example, the frequency range in which conducted electromagnetic interference (EMI) is measured by many EMI standards is from 150 kHz to 30 MHz. EMI filter size is typically driven by the lowest carrier harmonic within this frequency range. With three converters in parallel and each switching at 50 kHz, interleaving as discussed before would eliminate the first and the second carrier harmonics as well as their sideband components, but leaves the third carrier harmonic and its sideband components unaffected. Since the third carrier harmonic in this case (150 kHz) is right at the starting frequency of EMI measurement, interleaving doesn't help to reduce the EMI filter size.

For the above example, it would be more advantageous to reduce the third carrier harmonic and its sideband components. There are other applications where it is desirable to selectively reduce certain carrier harmonics and their sideband components. This can be achieved by the newly developed *asymmetric interleaving* method [1], in which the carrier phase shift is not confined to one Nth of a carrier cycle and can vary from one pair of converters to another. To distinguish from this new method, the interleaving method discussed before is called *symmetric inter-*

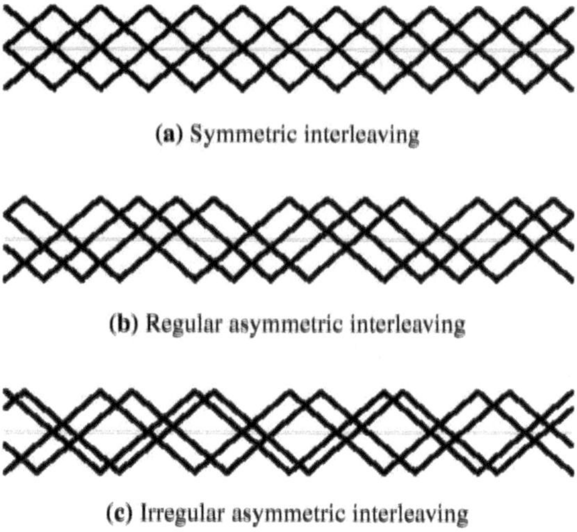

Fig. 2.19 Interleaving of four triangle carrier signals

leaving. Figure 2.19 illustrates the basic concept of asymmetric interleaving, which is further divided into two subcategories:

1. *Regular Asymmetric Interleaving*: Like symmetric interleaving, the phase shifts between adjacent converters are still equal in this case but they don't add to one carrier period. The carrier initial phase angle of the kth converter can be written as

$$\theta_{ck} = (k-1)\Delta\theta_c + \theta_{c1}, \quad k = 1, 2, \ldots, N, \tag{2.46}$$

   where $\Delta\theta_c$ is a constant,
2. *Irregular Asymmetric Interleaving*: The carrier phase shifts are not equal and don't necessarily add to one carrier period.

The harmonic cancellation effects of asymmetric interleaving among N parallel or series converters can be characterised by using the harmonic scaling factor defined below where is the initial phase angle of the kth converter [15]:

$$\kappa_m = \frac{1}{N}\sum_{k=1}^{N} e^{jm\theta_{ck}}. \tag{2.47}$$

The scaling factor allows the combined harmonic spectrum of N parallel or series converters to be determined from that of one converter by simply multiplying each harmonic by $N\kappa_m$. Note that κ_m applies to the mth carrier harmonic as well as all of its sideband components. Under symmetric interleaving, it can be verified that

$$\kappa_m = \begin{cases} 1 & \text{if } m = iN, \\ 0 & \text{if } m \neq iN, \end{cases} \tag{2.48}$$

Table 2.2 Harmonic scaling factors under regular asymmetric interleaving for the case of $N = 4$

m	$\Delta\theta_c$						
	18°	22.5°	30°	45°	75°	90°	110°
1	0.939	0.906	0.837	0.653	0.205	0	0.196
2	0.769	0.653	0.433	0.000	0.224	0	0.262
3	0.524	0.318	0.000	0.271	0.271	0	0.837
4	0.250	0.000	0.250	0.000	0.433	1	0.133
5	0.000	0.213	0.224	0.271	0.958	0	0.086
6	0.182	0.271	0.000	0.000	0.000	0	0.433
7	0.267	0.180	0.224	0.653	0.126	0	0.583
8	0.250	0.000	0.250	1.000	0.250	1	0.163
9	0.149	0.180	0.000	0.653	0.653	0	0.000
10	0.000	0.271	0.433	0.000	0.837	0	0.925

where i is any integer. In general, κ_m varies between 0 and 1, with 0 indicating complete cancellation of the mth carrier harmonic and all of its sideband components. Table 2.2 tabulates the value of κ_m under regular asymmetric interleaving with different values of $\Delta\theta_c$ for the case of $N = 4$.

2.4 Peak Current Control

The principle of peak-current control is illustrated in Fig. 2.20. Compared to the voltage-mode control depicted in Fig. 2.14, the inductor current is used here in place of the carrier signal to generate the PWM signal. The method works as follows: The switch is turned on by a clock signal at the beginning of each switching cycle. The sensed inductor current is compared to the output of the voltage compensator, denoted as i_p in Fig. 2.20, and the switch is turned off when the two become equal. For stability reasons, a compensation ramp with a slope equal to m_c is usually added to the sensed inductor current such that the inductor current is effectively compared to the reference minus the compensation ramp.

Peak-current control offers several practical advantages over voltage-mode control for DC–DC converters and other applications. Since the peak of the inductor current is directly controlled, it provides a built-in over-current protection for the switch and other components. The use of the inductor current in the pulse-width modulation process effectively introduces an inner feedback control loop that changes the control characteristics and simplifies the design of the output voltage feedback compensator. Additionally, the method is easy to implement since there is no need for an external ramp (carrier) signal for PWM. For these reasons, most power supplies used in practice are controlled by this method.

Peak-current control belongs to the general category of current-mode or current-programmed control [8]. Other current-mode control methods include average current control [7], nonlinear carrier control [11] and nonlinear average current con-

Fig. 2.20 Principle of peak-current control

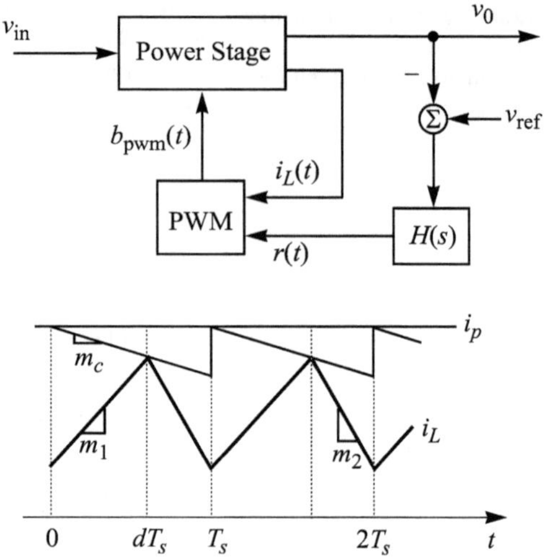

trol [18]. These methods are particularly useful for rectifiers with active power factor correction (PFC). Modelling of peak-current control is included in this chapter because of its direct involvement in the pulse-width modulation process.

2.4.1 Averaged Modelling and Duty Ratio Constraint

The state-space averaged model developed in the previous section is still valid under peak-current control. However, to complete the model, a duty ratio constraint that relates the duty ratio, d, to other variables used in the averaged model is needed. Such a duty ratio constraint can be developed from the principle of the control method, and there are several different models in the literature. Early work on this subject developed the duty ratio constraint by assuming that the inductor current waveform is in a steady state [10]. With reference to Fig. 2.20,[2] the average of the inductor current, $\langle i_L \rangle$, can be related to other variables by the following equation if one assumes that the inductor current returns to its initial value at the end of each switching cycle, that is, when the current waveform has reached its steady state:

$$\langle i_L \rangle = i_p - \left(m_c + \frac{m_1}{2} \right) d T_s. \tag{2.49}$$

It was pointed out in [26] that using the steady-state current waveform is inconsistent with the objective of developing a dynamic model. Indeed, the resulting averaged model only predicts the DC and low-frequency behaviour of peak-current controlled DC–DC converters [17]. The duty ratio constraint proposed in [26] doesn't involve

[2]Note that all slopes defined in Fig. 2.20 (m_c, m_1 and m_2) are positive variables.

steady-state assumption and is given below:

$$\langle i_L \rangle = i_p - m_c d T_s + \frac{T_s}{2}\left[m_1 d^2 + m_2(1-d)^2\right]. \tag{2.50}$$

This results in improved accuracy of the averaged model in the high-frequency region. However, neither model can predict the subharmonic (ripple, switching) instability that is characteristic of peak-current control [8]. As a remedy for this limitation, an additional second-order transfer function representing the so-called sample-and-hold effects was proposed in [12] as part of the modulator gain, which complicates the model and has been a source of confusion.

The subharmonic instability problem can be explained with reference to the current waveform depicted in Fig. 2.20. Denote the value of the inductor current at the beginning of the kth switching cycle as $i_L[k]$. Without a compensation ramp, that is, when $m_c = 0$, the inductor current reaches the reference, i_p, when the switch turns off. Hence,

$$i_L[k] = d[k] m_1 T_s + i_p, \tag{2.51}$$

from which the duty ratio, $d[k]$, in the kth switching cycle can be found:

$$d[k] = \frac{i_p - i_L[k]}{m_1 T_s}. \tag{2.52}$$

To determine the value of the inductor current at the end of the kth switch cycle, denoted as $i_L[k+1]$, note that

$$i_L[k+1] = i_p - \big(1 - d[k]\big) m_2 T_s. \tag{2.53}$$

Substituting (2.52) into (2.53), we have

$$i_L[k+1] = -\frac{m_2}{m_1} i_L[k] + \left(1 + \frac{m_2}{m_1}\right) i_p - m_2 T_s. \tag{2.54}$$

This first-order sampled-data model indicates that the inductor current response is unstable if

$$-\frac{m_2}{m_1} \tag{2.55}$$

lies outside the unit circle, that is, if

$$\frac{m_2}{m_1} > 1. \tag{2.56}$$

To relate this instability condition to usual operation parameters of a converter, note that $m_1 d T_s = m_2(1-d)T_s$ holds when the inductor current is in steady-state such that

$$\frac{m_2}{m_1} = \frac{d}{1-d}, \tag{2.57}$$

which indicates that a peak-current controlled converter is inherently unstable when the duty ratio is larger than 50%.

2.4.2 Ripple Reconstruction from Averaged Models

Duty ratio constraints that enable an averaged model to predict subharmonic instability under peak-current control are introduced here. The method is based on reconstructing the ripple current of the inductor from the state-space averaged model and using the estimated ripple to define the duty ratio. The ripple estimation method is developed from a mathematical procedure called KBM (Krylov–Bogoliubov–Mitropolsky) algorithm [4]. The KBM algorithm is reviewed in the following while characteristics of the resulting averaged models will be presented in the next subsection.

Denote, for any of the second-order DC–DC converters shown in Fig. 2.15, the instantaneous inductor current as $i_L(t)$ and its local average as $\langle i_L(t)\rangle$. Under CCM operation, the converter can be described by the following piece-wise linear model

$$\frac{\mathrm{d}}{\mathrm{d}t}x = f(t,x,u) = \begin{cases} A_0 x + B_0 u, & \text{if } t \in [0, dT_s], \\ A_1 x + B_1 u, & \text{if } t \in [dT_s, T_s], \end{cases} \tag{2.58}$$

where, as in the previous section, x is the vector of state variables, and u is the input voltage. The state-space averaged model (2.24) can be written in a more general form as

$$\frac{\mathrm{d}}{\mathrm{d}t}\langle x\rangle = g\big(\langle x\rangle, u\big), \tag{2.59}$$

where $\langle x\rangle$ represents the local average of x and

$$\begin{aligned} g\big(\langle x\rangle, u\big) &\triangleq \frac{1}{T_s}\int_t^{t+T_s} f\big(t, \langle x\rangle, u\big)\,\mathrm{d}t \\ &= \big[dA_0 + (1-d)A_1\big]\langle x\rangle + \big[dB_0 + (1-d)B_1\big]u. \end{aligned} \tag{2.60}$$

The KBM algorithm places the averaging method in a geometric framework by considering a change of variables [4]

$$x(t) = \big\langle x(t)\big\rangle + \Psi_1\big(t, \big\langle x(t)\big\rangle\big), \tag{2.61}$$

where $x(t)$ and $\langle x(t)\rangle$ are the solutions to (2.58) and (2.60), respectively, and $\Psi_1(t, \langle x(t)\rangle)$ is the ripple correction term to be obtained using the following procedure:

1. Take the difference between (2.58) and (2.60) and compute its indefinite integral:

$$\gamma_1(t,y) = \int f(t,y,u) - g(y,u)\,\mathrm{d}t. \tag{2.62}$$

2. Compute the time-average of γ_1 over one switching cycle and remove it from γ_1:

$$\Psi_1(t,y) = \gamma_1(t,y) - \frac{1}{T_s}\int_0^{T_s} \gamma_1(\tau, y)\,\mathrm{d}\tau. \tag{2.63}$$

The estimated ripple function Ψ_1 is piecewise linear because function f is piecewise linear, and is called the first-order ripple estimate. Generally, more accurate estimations can be obtained by considering the following change of variables [4]

$$x(t) = \langle x(t)\rangle + \Psi_1\big(t, \langle x(t)\rangle\big) + \Psi_2\big(t, \langle x(t)\rangle\big) + \Psi_3\big(t, \langle x(t)\rangle\big) + \cdots, \tag{2.64}$$

where Ψ_i is called the ith-order ripple estimation which are ith-order polynomials in t and can be determined sequentially using a procedure similar to that outlined above.

The first-order (linear) ripple current functions for the three basic converter topologies are given below, where v_{in} is the input voltage, v_O is the average of the output capacitor voltage, and T_s is the switching cycle:

$$\hat{i}_L(t) = \frac{v_{\mathrm{in}}}{2L}(d-1)(dT_s - 2t), \tag{2.65}$$

$$\hat{i}_L(t) = \frac{\langle v_O\rangle}{2L}(d-1)(dT_s - 2t), \tag{2.66}$$

$$\hat{i}_L(t) = \frac{v_{\mathrm{in}} + \langle v_O\rangle}{2L}(d-1)(dT_s - 2t) \tag{2.67}$$

for the buck, boost and buck–boost converters, respectively. These functions are valid for the interval when the switch is ON, that is, $t \in [0, dT_s]$. Similar expressions can be obtained for the OFF-interval of the switch and for other converter topologies.

2.4.3 Complete Averaged Models and Subharmonic Instability

As an example, consider again the boost DC–DC converter shown in Fig. 2.15b. The state-space averaged model of the converter can be written as follows when a resistive load is assumed:

$$\frac{\mathrm{d}}{\mathrm{d}t}\begin{pmatrix}\langle i_L\rangle\\ \langle v_O\rangle\end{pmatrix} = \begin{pmatrix}0 & \frac{d-1}{L}\\ \frac{1-d}{C} & -\frac{1}{CR}\end{pmatrix}\begin{pmatrix}\langle i_L\rangle\\ \langle v_O\rangle\end{pmatrix} + \begin{pmatrix}\frac{v_{\mathrm{in}}}{L}\\ 0\end{pmatrix}. \tag{2.68}$$

The instantaneous inductor current at the switch turn-off instant can be written as $\langle i_L\rangle + \hat{i}_L(dT_s)$ so that the following duty ratio constraint is obtained:

$$i_p - m_c dT_s = \langle i_L\rangle + \hat{i}_L(dT_s). \tag{2.69}$$

Based on the first-order ripple current function (2.66), this can be more specifically written as

$$i_p - m_c dT_s = \langle i_L\rangle + \frac{dT_s(1-d)\langle v_O\rangle}{2L}, \tag{2.70}$$

where the variables are as defined in Fig. 2.15b and Fig. 2.20. This combined with (2.68) defines a complete averaged model for boost converter under peak-current control.

The averaged model (2.68) and (2.70) can be used to study various characteristics of the converter. Of particular interest is the ability of the model to predict the subharmonic instability discussed in the previous subsection. Here we will examine this at three different levels: (a) the modulator gain, (b) the current loop gain, and (c) dynamics of the entire converter. In all cases, we will see that the averaged model can correctly predict the subharmonic instability [19].

The nonlinear duty ratio constraint (2.70) can be linearised to define a small-signal transfer function from the average inductor current to the duty ratio, as given below:

$$\frac{D(s)}{\langle I_L(s)\rangle} = \frac{1}{-[m_c + (1/2 - D)(M_1 + M_2)]T_s}. \tag{2.71}$$

This essentially defines the gain of the modulator from the inductor current to the duty ratio under peak-current control. Symbols in capital letters in the above equation (as well as throughout the rest of the section) represent steady-state values of the corresponding variables in lower cases. (For example, M_1 represents the steady-state value of m_1.) As can be seen, when no slope compensation is included (i.e. $m_c = 0$), the small-signal modulator gain (2.71) becomes positive when the steady-state duty ratio, D, exceeds 0.5. This corresponds to a current feedback control loop with a positive gain, hence indicating unstable operation.

To find the complete current loop gain, we note that the average model of the inductor current in a continuous conduction mode PWM converter can be written in general as

$$\frac{\mathrm{d}}{\mathrm{d}t}\langle i_L\rangle = dm_1 - (1-d)m_2.$$

Under the assumption that the converter input and output voltages are constant, the duty ratio to the average inductor current transfer function can be obtained from this model as

$$\frac{\langle I_L(s)\rangle}{D(s)} = \frac{M_1 + M_2}{s}. \tag{2.72}$$

Combining this with (2.71) yields the current loop gain of a boost converter under peak-current control:

$$I_c(s) = \frac{M_1 + M_2}{-[m_c + (1/2 - D)(M_1 + M_2)]T_s}\,\frac{1}{s}. \tag{2.73}$$

As can be seen, the loop gain is unstable for $D > 0.5$ when no slope compensation is used. This prediction is consistent with the sampled-data analysis results [26].

To examine the stability of the overall boost converter under peak-current control, note that the averaged model (2.68) is nonlinear but can be linearized around a given operation point to give the following linear model:

$$\frac{\mathrm{d}}{\mathrm{d}t}\begin{pmatrix}\widehat{\langle i_L\rangle}\\ \widehat{\langle v_O\rangle}\end{pmatrix}=\begin{pmatrix}0 & \frac{D-1}{L}\\ \frac{1-D}{C} & -\frac{1}{CR}\end{pmatrix}\begin{pmatrix}\widehat{\langle i_L\rangle}\\ \widehat{\langle v_O\rangle}\end{pmatrix}$$
$$+\frac{V_{\text{in}}}{1-D}\begin{pmatrix}\frac{1}{L}\\ -\frac{1}{CR(1-D)}\end{pmatrix}\hat{d}. \tag{2.74}$$

Meantime, linearisation of (2.70) gives the following small-signal duty-ratio constraint

$$\hat{d}=\frac{1-D}{V_x}\left[\begin{pmatrix}\frac{2L}{T_s}\\ D(D-1)\end{pmatrix}^{\top}\begin{pmatrix}\widehat{\langle i_L\rangle}\\ \widehat{\langle v_O\rangle}\end{pmatrix}+\frac{2L}{T_s}\hat{i}_p\right],$$

where $V_x=2(D-1)Lm_c+(2D-1)V_{\text{in}}$. A complete small-signal model of the converter is obtained by substituting the above expression into (2.74). The system matrix of the combined small-signal model is found to be

$$A=\begin{pmatrix}0 & \frac{D-1}{L}\\ \frac{1-D}{C} & -\frac{1}{CR}\end{pmatrix}+\frac{V_{\text{in}}}{V_x}\begin{pmatrix}\frac{2}{T_s} & \frac{D(D-1)}{L}\\ -\frac{2L}{RCT_s(1-D)} & \frac{D}{CR}\end{pmatrix}. \tag{2.75}$$

Stability of the converter depends on the eigenvalues of A, that is, the roots of the characteristic polynomial $\det(sI-A)$. In the case when no slope compensation is used, A can be written as

$$A=\begin{pmatrix}\frac{2}{T_s}\frac{1}{2D-1} & \frac{D-1}{L}\frac{3D-1}{2D-1}\\ \frac{1-D}{C}-\frac{2L}{RCT_s(1-D)(2D-1)} & \frac{1}{CR}\frac{1-D}{2D-1}\end{pmatrix}. \tag{2.76}$$

Hence

$$\det(sI-A)=s^2+\frac{s}{1-2D}\left(\frac{2}{T_s}+\frac{1-D}{RC}\right)+a_0. \tag{2.77}$$

Note that the coefficient of s in (2.77) is negative when $D>0.5$, in which case at least one root of the characteristic polynomial will be positive because the sum of the two roots is equal to the inverse of this coefficient. This confirms again the ability of (2.70) as the duty-ratio constraint to predict subharmonic instability under peak-current control.

2.5 Summary

Pulse-width modulation is the foundation of control in power electronics. Among various modulation methods, constant-frequency PWM using a sawtooth or triangle carrier is most common in practice. When the reference is a sine wave, the use of a triangle carrier has an additional benefit that odd-order sideband components of odd-order carrier harmonics as well as even-order sideband components of even-order carrier harmonics are eliminated. Closed-form spectral models developed using double Fourier analysis facilitated the analysis and comparison of these PWM methods.

The pulse-width modulator is also an integral part of the feedback control loop and needs to be properly modelled for control design. A constant-frequency modulator can be modelled by a constant gain, but such linear time-invariant model is only valid up to half the carrier frequency. Beyond this frequency, the modulator response to a perturbation in the reference may be affected by sideband components of the perturbation (when the perturbation frequency is a multiple of one half of the carrier frequency) and cannot be modelled by a constant gain. A variable-frequency modulator can also be modelled by a DC gain with a leading phase. The DC gain changes with the ON and OFF time, but the variation is small and can be ignored in practice. The leading phase adds to the phase margin of the control loop and is beneficial for control stability.

Multiple PWM converters can be connected in parallel or in series to form a modular design with scalable current or voltage capacities. In such modular systems, interleaving offers an opportunity to reduce the combined input or output harmonics through harmonic cancellation among different modules. The traditional symmetric interleaving method eliminates all but N-multiples of the carrier harmonic for N modules connected in parallel or in series. Additionally, the harmonic cancellation effect is such that, when a carrier harmonic is eliminated, all of its sideband harmonics are eliminated as well. Asymmetric interleaving makes it possible to selectively reduce different carrier harmonics and their sideband components to achieve different objectives, such as minimisation of EMI filter size.

Peak-current control is a special form of control in which there is no explicit PWM process. The current ripple is effectively used as a carrier signal in an equivalent PWM process. Averaged models can be used to study both steady-state and dynamic characteristics, including ripple instability, of peak-current controlled converters. A key step in the development of such averaged model is a duty ratio constraint that models the equivalent PWM process. Duty ratio constraints developed based on ripple estimation using the KBM method are provided for commonly used converters and are shown to correctly predict the ripple instability.

References

1. Beechner, T., Sun, J.: Asymmetric interleaving—a new approach to operating parallel converters. In: Proc. of the IEEE Energy Conversion Congress and Exposition, San Jose, California, USA, pp. 99–105 (2009)
2. Bennett, W.R.: New results in the calculation of modulation products. Bell Syst. Tech. J. **12**(4), 238–243 (1933)
3. Black, H.S.: Modulation Theory. Van Nostrand Reihnold, New York (1953)
4. Bogoliubov, N.N., Mitropolsky, Y.A.: Asymptotic Methods in the Theory of Nonlinear Oscillations. Hindustan Publishing Corporation, New Delhi (1961)
5. Bowes, S.R., Bird, B.M.: Novel approach to the analysis and synthesis of modulation processes in power convertors. IEE Proc. **122**(5), 507–513 (1975)

6. Cuk, S., Middlebrook, R.D.: A general unified approach to modeling switching power stages. In: Proc. of the IEEE Power Electronics Specialists Conference, Cleveland, Ohio, USA, pp. 18–31 (1976)
7. Dixon, L.H.: Average Current-Mode Control of Switching Power Supplies. Unitrode Power Supply Design Seminar Handbook, Merrimack, New Hampshire, USA (1990)
8. Erickson, R.W., Maksimovic, D.: Fundamentals of Power Electronics, 2nd edn. Kluwer Academic Publishers, Dordrecht (2001)
9. Holmes, D.G., Lipo, T.A.: Pulse Width Modulation for Power Converters—Principles and Practice, 1st edn. Wiley-IEEE Press, Piscataway (2003)
10. Hsu, S.P., Brown, A., Rensink, L., Middlebrook, R.D.: Modeling and analysis of switching dc-to-dc converters in constant-frequency current-programmed mode. In: Proc. of the IEEE Power Electronics Specialists Conference, San Diego, California, USA, pp. 284–301 (1979)
11. Maksimovic, D., Jang, Y., Erickson, R.: Nonlinear-carrier control for high-power-factor boost rectifiers. IEEE Trans. Power Electron. **11**(4), 578–584 (1996)
12. Ridley, R.B.: A new, continuous-time model for current-mode control. IEEE Trans. Power Electron. **6**(2), 271–280 (1991)
13. Sanchez, M., Popert, F.: Uber die Berechrung der Spektren modulierter Impulsfolgen. Arch. Elecktr. Ubertrag. **9**(10), 441–452 (1955)
14. Sun, J.: Small-signal modeling of variable-frequency pulse-width modulators. IEEE Trans. Aerosp. Electron. Syst. **38**(3), 1104–1108 (2002)
15. Sun, J.: Modeling and analysis of pulse-width modulation. In: Professional Education Seminar at IEEE Applied Power Electronics Conference, Austin, Texas, USA (2008)
16. Sun, J., Bass, R.M.: Automated ripple analysis of PWM DC/DC converters using the KBM method. In: Proc. of the IEEE International Conference on Industrial Electronics, Control, and Instrumentations, Atlanta, Georgia, USA, pp. 621–626 (1997)
17. Sun, J., Bass, R.M.: A new approach to averaged modeling of current-programmed PWM converters. In: Proc. of the IEEE International Conference on Industrial Electronics, Control, and Instrumentation, New Orleans, Louisiana, USA, pp. 599–604 (1997)
18. Sun, J., Chen, M.: Nonlinear average current control using partial current measurement. IEEE Trans. Power Electron. **23**(4), 1641–1648 (2008)
19. Sun, J., Choi, B.: Averaged modeling and switching instability prediction for peak current control. In: Proc. of the IEEE Power Electronics Specialists Conference, Recife, Brazil, pp. 2764–2770 (2005)
20. Sun, J., Grotstollen, H.: Optimal space vector modulation and regular-sampled PWM: a reexamination. In: Proc. of the IEEE Industry Applications Society Annual Meeting, San Diego, California, USA, pp. 956–963 (1996)
21. Sun, J., Grotstollen, H.: Symbolic analysis methods for averaged modeling of switching power converters. IEEE Trans. Power Electron. **12**(3), 537–546 (1997)
22. Sun, J., Fröhleke, N., Grotstollen, H.: Harmonic reduction techniques for single-switch three-phase boost rectifiers. In: Proc. of the IEEE Industry Applications Society Annual Meeting, San Diego, California, USA, pp. 1225–1232 (1996)
23. Sun, J., Mitchell, D.M., Greuel, M., Krein, P.T., Bass, R.M.: Averaged modeling of PWM converters operating in discontinuous conduction mode. IEEE Trans. Power Electron. **6**(4), 482–492 (2001)
24. Tolstov, G.P.: Fourier Series. Prentice-Hall, Englewood Cliffs (1962)
25. UCC38501 BiCMOS PFC/PWM Combination Controller datasheet. Texas Instruments, Dallas, Texas, USA (1999)
26. Verghese, G.C., Bruzos, C.A., Mahabir, K.N.: Averaged and sampled-data models for current mode control: A reexamination. In: Proc. of the IEEE Power Electronics Specialists Conference, Milwaukee, Wisconsin, USA, pp. 484–491 (1989)

Chapter 3
Commercial Power Electronics Controllers

Chin Chang

3.1 Introduction

In modern electronic equipment, power converters are widely used and play important role on enabling more critical features and functions even though some end consumers do not realise it or take it as granted. Power converters as a means of power conversion from source to load need to be properly controlled and regulated in order to achieve stability at all operating conditions and environment. With the increasing market demand of more efficient, lower cost, smaller size power converters, the circuit stability and functionality are not to be sacrificed. Even more, it is a prerequisite for any commercial electronic products that utilise power converters.

With the great advancement in electronics, semiconductor, and information technology, there are many different types of electronic equipment and devices in the market today, ranging from consumer electronics (e.g., smartphones, cameras, TVs), computers, office automation to medical devices, telecommunication equipment, lighting and industrial equipment. Power converters are used to process power with levels ranging from mW and kW to MW and take the forms of linear power supply and switching-mode power supply. By interfacing different types of sources and loads, power converters could be classified as AC–DC, DC–DC, DC–AC and AC–AC converters. For example, in battery-powered handheld devices, DC–DC converters are widely used. In offline power supplies, AC–DC converters are usually required. In industrial welding, motor drive and fluorescent lighting, DC–AC converters are commonly utilised. To control these variety of power converters with their specific applications, many different, yet often optimised controllers are used in the forms ranging from a simple, few pins analog integrated-circuit (IC) to digital signal processor (DSP) with hundreds of pins. While it is not the purpose of this chapter to enlist and analyse the vast families of commercial power electronic con-

C. Chang (✉)
Power Management, SEMTECH Corporation, 200 Flynn Road, Camarillo, CA 93012-8790, USA
e-mail: CChang@Semtech.com

F. Vasca, L. Iannelli (eds.), *Dynamics and Control of Switched Electronic Systems*, Advances in Industrial Control,
DOI 10.1007/978-1-4471-2885-4_3,

trollers in the marketplace, we select some typical cases, describe the problems and discuss the solutions that the controllers provided.

We focus on switching-mode power converters, as members of the switched electronic systems, with the state-of-the-art power switching devices metal-oxide-semiconductor field-effect transistors (MOSFETs), bipolar junction transistors (BJT), diodes and passive components (inductor and capacitors). When the converters operate at switching mode, the switching frequency could be either constant or variable. With constant switching frequency, the power converters are usually controlled or regulated via pulse-width modulation (PWM). In variable switching frequency power converters, the power converters could be controlled and regulated via different modulations schemes. For example, there are hysteretic mode, constant ON-time mode, constant OFF-time mode, adaptive ON-time mode, variable frequency with constant duty ratio, or the combination of them, etc. Most likely, these different modulation and regulation schemes lead to different dynamic behaviour of the power converters. Complete model and analysis of them have been a challenge and continuously draw interest and effort from both academic and industrial communities as shown in the other chapters of this book. To a broad perspective, the switching-mode power converter is a subset of the switched electronic systems and nonlinear in general. In linearized small-signal models, different modulation schemes may display different transfer functions from modulation variables to the output voltage or current. Different controller transfer functions are usually needed to compensate these. To help classify the different controllers, we extend the traditional current mode control and voltage mode control definition to a broad extent so that almost all of these controllers could be simply classified into two different categories. Within these two categories, we discuss some commonly used commercial power converter controllers and ICs. For example, we will discuss the controllers for DC–DC switching-mode power supplies in industrial, computer and networking power management applications. In addition to reveal the practical control methods for these DC–DC synchronous buck converters, some unique requirements for these applications are addressed. For example, multiple-phase controllers with interleaving, current sharing, current sinking and souring, and so on. Controllers for AC–DC converters will be discussed with boost converter as power factor correction (PFC) stage. Also, a simple controller for Flyback converter as a single-stage PFC is analysed along with some discussions on the modulation schemes and control implementations to shape the input current. The dynamic response limitation of the traditional control method is analysed. For commercial controllers in DC–AC converters, the half-bridge resonant inverter in electronic ballast application is considered as example. Along with the lighting industry, we will discuss various structures and controllers for the emerging light-emitting diode (LED) illumination market.

The development of commercial power electronic controllers are mainly application and market driven. Due to the fact that different loads, ranging from customer integrated circuit chips, motors, fluorescent lamps, HID lamps, LEDs, exhibit different electrical characteristics, there are different requirements to power and drive these devices. There is no single topology and single controller that could cover all

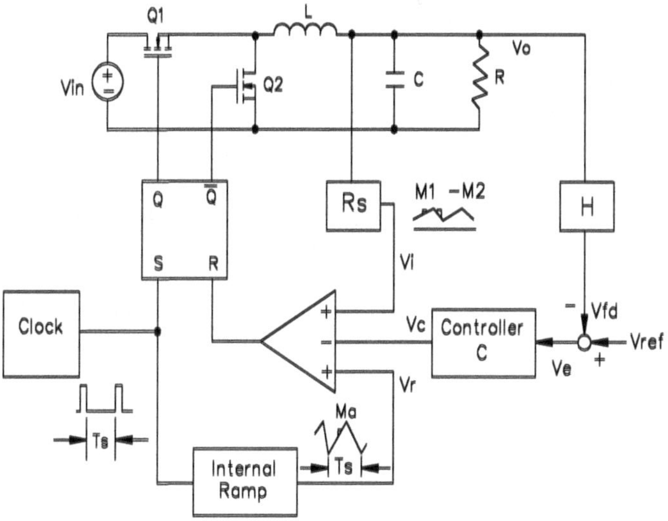

Fig. 3.1 A simple buck converter with control block diagram

the real-world applications. This makes the fields of switching-mode power converters and control enriching and fascinating [31].

3.2 Common Control Methods for Commercial Power Converters

In power converters, efficiency is very critical. High efficiency not only does it save energy, but also enables smaller and more reliable electrical equipment and devices. To achieve this goal, various switching-mode power converters are invented and developed in the past century to replace traditional linear regulators and power amplifiers. To reduce power losses in converters, the active semiconductor devices, such as transistors and diodes, are operated at switching modes (ON = saturation or OFF = cutoff), and only energy storage and transferring passive components (e.g. inductors, capacitors and transformers) are used in the power stage. To reduce the value and size of the passive components, the switches are commonly operated at higher frequencies such as from kHz to MHz. From electric circuit point of view, the operation of modern switching mode power converters is a sequence of topological switching among passive components or networks with loseless power switches at high frequency. The role of the power converter controller is to ensure high efficient power transfer and conversion with guaranteed stability and performance [23].

A simple power converter diagram is shown in Fig. 3.1, where V_{in} is the input power source voltage, Q_1 and Q_2 are the power switches, L and C are the output filtering inductor and capacitor, and R is the load resistance. Also shown in Fig. 3.1 are the control circuitry blocks used to regulate the output voltage V_o against possible input voltage variation, load resistance variation, and component value variations due to working environment changes.

The basic DC–DC power converters used today are the buck, boost and buck–boost converters and their corresponding isolated versions with the addition of transformer. In these basic converters, normally two power switches, e.g. MOSFET, insulated gate bipolar transistor (IGBT), bipolar transistor, or diode, and one inductor (or transformer) are used along with the input and output capacitors at the terminals of the converters. Via high-frequency switching implemented by the power switches, the input energy is normally installed in the inductor or magnetic field and then released to the output during each switching cycle T_s. In steady state, the magnetic flux (volt–second) in the inductor and the electric charge in the capacitors should be balanced, i.e.

$$\Phi_L(T_s) - \Phi_L(0) = L\left[i_L(T_s) - i_L(0)\right] = \int_0^{T_s} v_L(t)\,\mathrm{d}t = 0 \tag{3.1}$$

and

$$Q_C(T_s) - Q_C(0) = C\left[v_C(T_s) - v_C(0)\right] = \int_0^{T_s} i_C(t)\,\mathrm{d}t = 0. \tag{3.2}$$

These two equations govern the steady-state behaviour of the simple switching-mode buck, boost, buck–boost and their corresponding isolated power converters. The voltage and current conversion ratios of these converters could be derived from these two equations. If one further zooms into the circuit operation at each switching period, it is clear that the subcircuits are simple linear RLC (resistor–inductor–capacitor) circuitry with different excitation voltage/current, different load and initial conditions. The state variables, inductor current i_L and capacitor voltage v_C, in the subcircuits could be used to describe the dynamic behaviour of the converter in switching cycle. It is natural to control these state variables to achieve desired power converter behaviour.

Depending on the state variables used and controlled, one could generally classify the control methods in power converters as current mode control and voltage mode control. Specifically, if only the capacitor voltage is sensed and regulated, it is called voltage-mode control. If inductor current is also sensed and used to control the power converter output voltage or output current, it is called current-mode control. With these, Table 3.1 lists the common control methods in these two categories.

In general, the current-mode control involves two feedback loops to regulate the targeted voltage or current as shown in Fig. 3.1. In many topologies, it requires extra current sensing resistor, related filtering circuitry, error amplifier, slope compensation, and related control circuitry. Such added complexity is usually introduced to

Table 3.1 Some control methods in commercial power converters

Current mode control	Voltage mode control
Peak current mode	Voltage mode
Valley current mode	Hysteretic
Average current mode	Constant ON time
Charge control	Constant OFF time
Variable frequency control involving current information	Adaptive ON time
Mixed current–voltage mode	Variable ON–OFF time
et al.	et al.

improve the circuit dynamic performance, e.g. extended loop bandwidth, greater stability margin and/or reduced noise audio-susceptibility. Also, some extended functionality, e.g. parallel operation with load current sharing, pulse-by-pulse current limit, come naturally with the current-mode control.

Traditionally, the voltage-mode control and current-mode control methods were treated differently. But, in some simple cases such as a DC–DC synchronous buck converter under current-mode control shown in Fig. 3.1, we have the following observations.

In Fig. 3.1, H is the voltage feedback gain function, and R_s is the current feedback gain function. The PWM modulator function [8] is

$$g(t, v_C, I_o) = (v_C - R_s I_o) - (m_1 + m_a)t \tag{3.3}$$

with $m_1 = R_s(V_{\text{in}} - V_o)/L$, $-m_2 = -R_s V_o/L$ and m_a being the slope of internal ramp signal. The circuit operating frequency is constant $1/T_s$. The output load current is I_o. The three-input comparator follows the logic: output is "high" when $(V_r + V_i - V_C) > 0$, and output is "low" when $(V_r + V_i - V_C) < 0$. Depending on the parameters relative magnitude in the modulation function, the circuit in Fig. 3.1 could operate at different control modes. For example, when the current feedback gain R_s is set to 0, the circuit operates at voltage mode. When one wants to operate the circuit in current mode, R_s is usually set as a constant. Furthermore, to avoid sub-harmonic oscillation for all the duty ratios, say D, the internal ramp slope should satisfy

$$m_a \geq \frac{1}{2} m_2 = \frac{D}{2(1-D)} m_1. \tag{3.4}$$

The inequality in (3.4) provides the lower bound of the internal ramp slope for current loop stability. In practice, m_a is set to a larger value. In the extreme case that m_a goes to infinity, the inner current loop becomes unconditionally stable, and the control mode becomes voltage mode. Equivalently, R_s approaches 0.

As a result, it is possible to analyse both the voltage-mode control and current-mode control in an unified framework for this case.

Table 3.2 Some control methods classified by switching frequency

Constant frequency	Variable frequency
Peak current mode	Hysteretic
Valley current mode	Constant ON time
Average current mode	Constant OFF time
Charge control	Adaptive ON time
Voltage mode	Variable ON–OFF time
Mixed current–voltage mode	Variable-frequency control involving current information
et al.	et al.

In switching mode power converters, there is an important parameter that leads to another simple classification of control methods in commercial controllers. It is the switching frequency. As shown in Table 3.2, most commercial power electronic controllers fall into one of the two categories: constant-switching-frequency controllers or variable-switching-frequency controllers. Many control methods have their unique characteristics. It is up to the design engineers to use a suitable control method on a selected power convert topology to fulfil a specific application requirement. Apparently, there is no single control method on a single circuit topology that will cover all the applications. For example, in many DC–DC power supplies, one wants to precisely regulate the output voltage in the presence of dramatic change in input voltage and load current. In a typical AC–DC PFC circuit, one needs to control the output voltage and shape the input current in the presence of input voltage and load current variations. In designing an DC–AC electronic ballast, engineers want to ignite the lamp first and then regulate the output lamp current, both in shape and magnitude, in the presence of input voltage and load characteristic variations. Also, in developing AC–DC or DC–DC LED drivers, the LED current needs either to be regulated at constant DC for fixed light output, or to be controlled by a low frequency (e.g. 200 Hz) PWM signal for adjustable light output. Therefore, there are vast, quite different commercial power electronic controllers in the market to meet the wide spectrum of applications.

3.3 Controllers for DC–DC Switching-Mode Power Supplies

DC–DC switching-mode power supplies have been widely used in telecommunication, computers, consumer electronics, instruments, automotive, medical, military and aerospace areas.

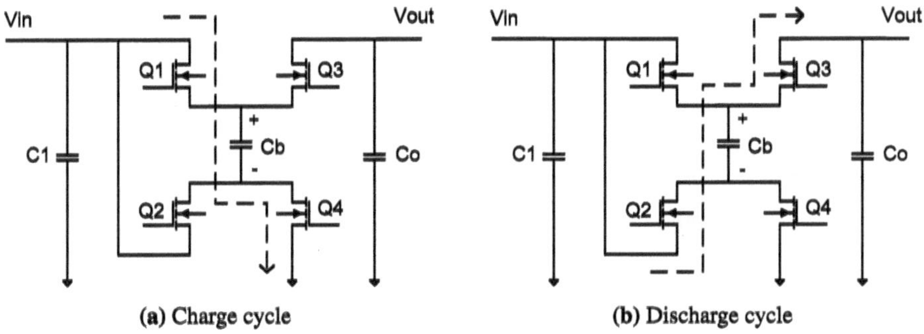

Fig. 3.2 A simple charge pump circuit diagram

3.3.1 Charge Pump Controller and Regulators in Portable Devices

In battery-operated portable electronic devices, the input power source voltage is low (e.g. 1.5–3.6 V for lithium battery cell). To regulate the relative lower power (<1 W) under the limitation of small size, low cost and simplicity, charge pump controller and regulators are the natural choices.

Figure 3.2 illustrates a simple charge pump circuit with the commonly called "flying capacitor" C_b. During each of the high-frequency charging cycle, where Q_1 and Q_4 are ON, and Q_2 and Q_3 are OFF, C_b is charged to V_{in}. At the following discharging cycle, where Q_1 and Q_4 are OFF, and Q_2 and Q_3 are ON, C_b is connected in series with V_{in} to deliver the power to output. Normally, the output voltage $V_o \approx 2V_{in}$. The circuit in Fig. 3.2 is also called a voltage doubler. The charge balance on the flying capacitor C_b at each switching cycle governs the steady-state operation of the circuit. In general, the input-to-output voltage gain (defined as $M = V_o/V_{in}$) of charge pump circuitry is discrete. Over the years, various topologies were developed to produce charge pumps with fraction gains (e.g. 1/2, 1/3, 1/4, etc.) and multiple gains (e.g., 1.5, 2, 2.5, 3, 4, 5, etc.). The Dickson's charge pump [17] and its extensions to multiple stages have found wide-range applications in nonvolatile memories, such as EEPROM (electrically erasable programmable read-only memory) and flash memories. For instance, in the mode of writing to the memory of RFID (radio frequency identification device), the charge pump circuit needs to convert the input voltage of 1.5 V to approximately 14 V for programming the EEPROM.

Due to the lower power and smaller size limitation, most charge pumps and their controllers are integrated inside an IC as shown in Fig. 3.3 [2]. The users only need to select proper input, output and flying capacitors to complete the design.

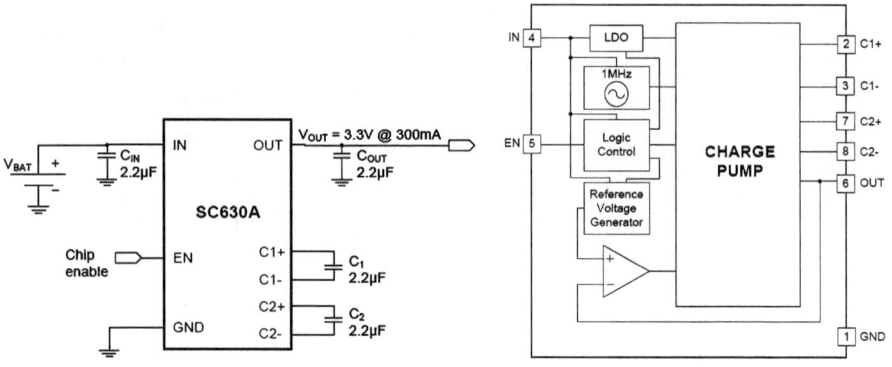

Fig. 3.3 A state-of-the-art charge pump with internal compensation

Another important application of the charge pump circuits in portable is LED backlight driver. We will discuss this in Sect. 3.6.

3.3.2 Inductor-Based DC–DC PWM Power Converters and Controllers

The three basic DC–DC PWM converters using inductor as energy transferring element are the buck, boost and buck–boost types as shown in Fig. 3.4. The power switch Q_1, could be MOSFET, BJT (bipolar junction transistor), IGBT or other types, is normally driven at high frequency (tens of kHz to MHz range) and used to achieve output voltage regulation in the presence of variations at input voltage V_{in}, load current I_o and other circuit components values.

With constant switching frequency, Fig. 3.5 illustrates the input-to-output voltage DC gain versus duty ratio D for the three basic DC–DC PWM converters. Among them, the curves are nonlinear for boost and buck–boost converters. To analyse the dynamics of these converters, the "state space averaging" method [37] is developed and used to model the power converters. Once the small-signal model is derived from the perturbation and linearisation of the averaged model at a quiescent operating point, the classic linear control design methods are used to derive the circuit compensation parameters, mostly linear resistor and capacitor values, for implementation.

In practice, the state-of-the-art control circuitry for switching-mode power converter are implemented by analog IC and digital IC. One of the most diversified commercial controllers is the MC34063 first incepted in the 1980s and its most recent improvements and derivatives [7, 45, 46].

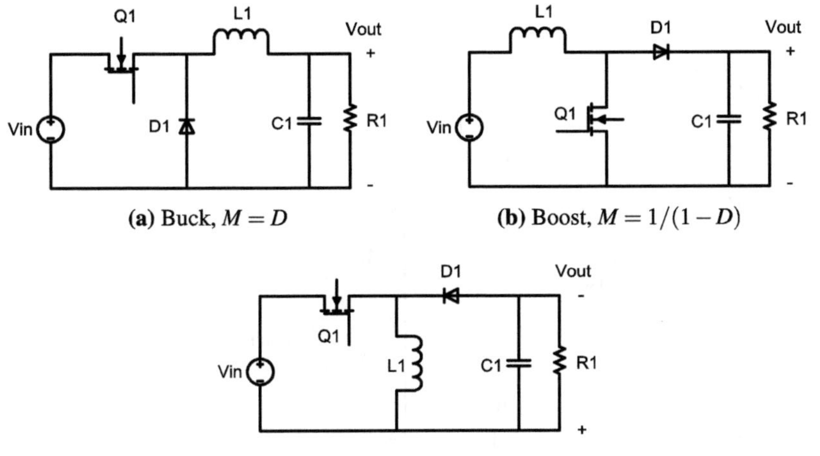

(a) Buck, $M = D$ **(b)** Boost, $M = 1/(1-D)$ **(c)** Buck–boost, $M = |D/(1-D)|$

Fig. 3.4 Three basic high-frequency DC–DC converters and their corresponding voltage gain M in continuous-conduction mode

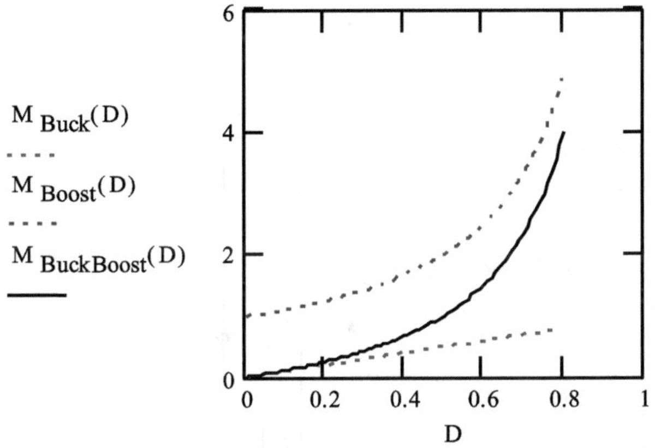

Fig. 3.5 DC voltage gain versus duty ratio for different DC–DC converters topologies

As shown in Fig. 3.6, the internal structure of MC34063 is very simple. It only contains a temperature-compensated voltage reference (1.25 V), comparator, controlled duty-cycle oscillator with an active peak current limit circuit, driver, and a high-current output switch Q_1. In low-power applications, Q_1 could be used as circuit power switch shown in Fig. 3.6. For higher-power applications, Q_1 could serve as driver to external power switch.

In MC34063, internal current sources are used to charge and discharge an external capacitor C_T connected from pin 3 to ground to program its oscillator frequency

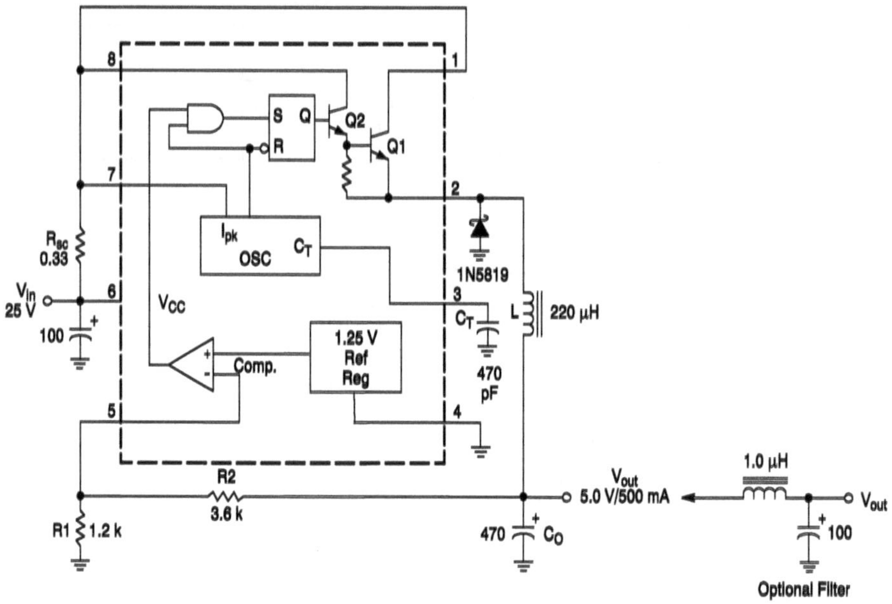

(a) Buck

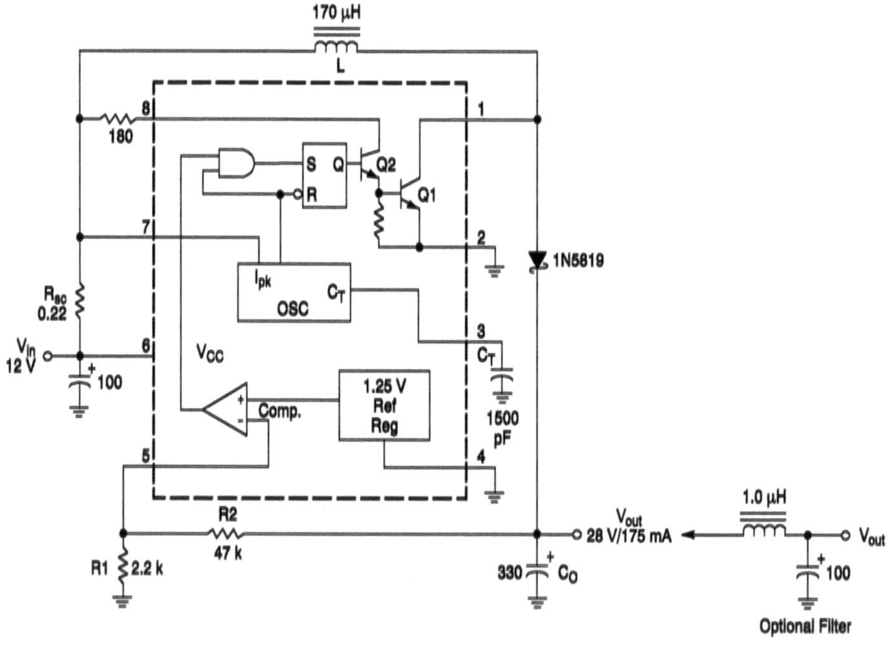

(b) Boost

Fig. 3.6 MC34063 and its applications in different DC–DC converters

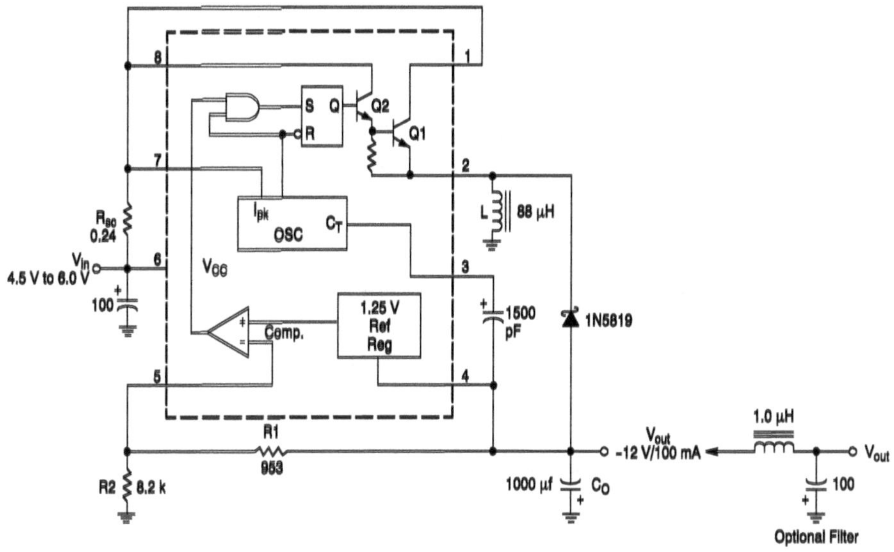

(c) Buck-boost

Fig. 3.6 (Continued)

f_{osc} via the following expression:

$$f_{osc} = \frac{200\ \mu A}{7 \times 0.5\ V} \frac{1}{C_T} (Hz). \quad (3.5)$$

But, the resulted triangular voltage waveform on C_T is not simply compared with a DC signal such as the output of an error amplifier to generate a PWM signal and then drive the power switches. Instead, it is gated by the comparator output as shown in Fig. 3.6. The comparator compares the internal reference voltage $V_{ref} = 1.25$ V with the output feedback voltage. When the output feedback voltage is less than the V_{ref}, the comparator output is "high". It allows the "set" and "reset" of the RS latch and therefore the switching of the power converter to increase the output voltage. The operating frequency is the programmed oscillating frequency. When the output voltage is greater than the V_{ref}, the comparator output is "low". Via the AND logic gate, it blocks the "set" function of the RS latch. The power switch will not turn on after its last turn off, and the circuit waits for the decay of converter output voltage towards the preset value. Typical operation waveforms are plotted in Fig. 3.7.

In practice, in order to avoid unwanted mistrigging, a small hysteresis (1.5 mV) is implemented in the comparator input of MC34063. This translates to the typical output voltage ripple of $1.5 V_{out}/V_{ref}$ (mV). Such ripple is required for the proper operation of the circuit.

In steady-state operation, when the output voltage ripple is small, the circuit operates at constant frequency set by (3.5). The circuit regulation is achieved by adjusting the leading edge of the PWM signal. It is similar to the constant frequency

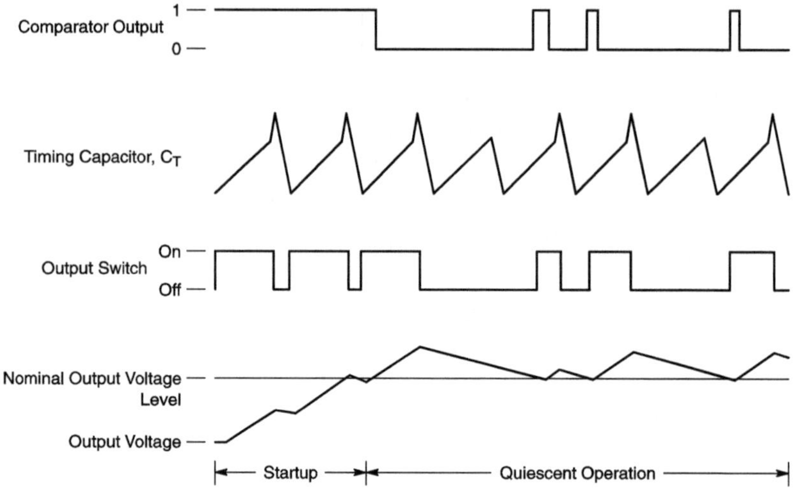

Fig. 3.7 Typical operation waveforms in MC34063

PWM controller. However, when there is a sudden change of the load current, the controller MC34063 responds within one switching cycle. If the load is added and leads to output voltage drop, the controller switches at maximum duty ratio with the same frequency. This is different from the traditional hysteresis control, where the power switch is indefinitely ON to help bring back the output voltage. It is also different from the constant ON-time and constant OFF-time control where the switching frequency is normally dramatically increased to help rise the output voltage. On the other hand, if the load is released abruptly, the output voltage will normally rise. The MC34063 controller will turn off the power switch and block all the consequent turn on pulses until the output voltage comes back to the targeted voltage set by V_{ref}. The action at this condition is similar to the behaviour of the traditional hysteretic, constant ON-time and constant OFF-time controllers.

Over the years, hysteretic control method and its various extensions and modifications are successfully used for power converter regulation. Some of their most common features are: variable switching frequency, noticeable output voltage ripple, fast transient response, free of compensation network. If designed and optimised for a specific condition, such control method usually leads to an elegant solution with simplicity and robustness. The dual edge modulated hysteretic control, modern constant ON-time, constant OFF-time and adaptive ON-time control are the examples of extended hysteretic control [43]. MC34063 could be considered as a new member of the hysteretic control family with its own characteristics.

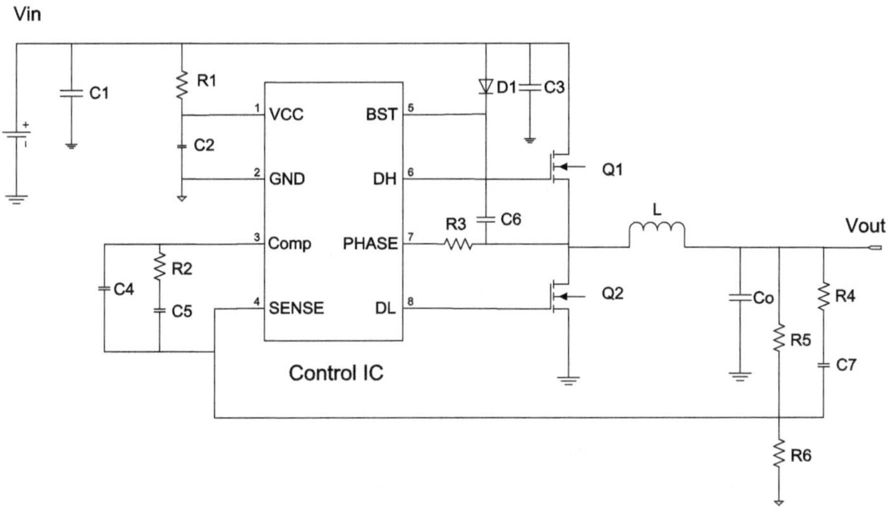

Fig. 3.8 Typical schematic of a buck converter for POL

3.3.3 DC–DC Buck Converters for Telecommunication and Networking

Telecommunication, networking equipment, data centre and servers are the back bone of today's internet, Ethernet and cloud computing. High efficiency, low noise and high reliability have been the major requirement of power solutions for these equipment. DC–DC buck converters have been widely used for point-of-load (POL) voltage regulation for the core processor, memory, logic and other interface devices. In the system, 12 V power rail is normally available for the POL power module input. The output voltages ranging from 5 V, 3.3 V, 2.5 V, 1.8 V, 1.2 V, 1 V to even sub-1 V depending on the load requirement. The current could range from a few amperes to even hundred of amperes.
To achieve high efficiency, especially for lower output voltage cases, synchronous rectification is deployed.

Figure 3.8 shows the schematic of a simple buck converter circuit, where the MOSFET Q_1 controls the duty ratio and MOSFET Q_2 functions as synchronous rectifier to reduce the conduction losses. Some typical waveforms in this circuit are shown in Fig. 3.9. Clearly both Q_1 and Q_2 turn on and turn off complementarily. To avoid possible shoot through from Q_1 to Q_2, certain amount of dead time (t_{d_1}) is inserted between the turn off of Q_1 and turn on of Q_2. Similar dead time (t_{d_2}) is also added between the turn off of Q_2 and turn on of Q_1. During the dead time, the inductor current goes through the body diode of Q_2. Normally, the body diode of the

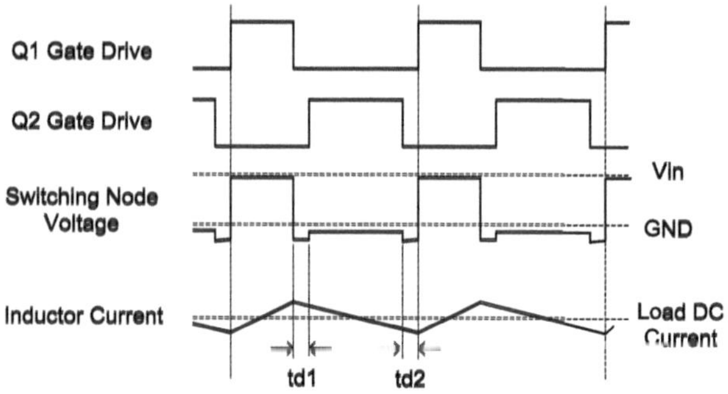

Fig. 3.9 Typical circuit waveforms of synchronous buck converter

MOSFET comes as a parasitic type, has higher V_f and larger reverse recovery current. Significant conduction losses could happen during t_{d_1} while both conduction and switching losses could occur during t_{d_2}. In practice, it is important to minimise the dead time while still prevent any shoot through for all the operating conditions. It was one of the key design challenges of control ICs, coupled with the required voltage level shift to drive the high side N-channel MOSFET Q_1 in Fig. 3.8.

The internal circuit block diagram of the control IC in Fig. 3.8 is shown in Fig. 3.10, where the control IC contains an oscillator with fixed frequency, an error amplifier with transconductance gain, a PWM modulator, and gate driver for both high-side and low-side MOSFETs. Clearly, it is a fixed-frequency voltage-mode controller with transconductance error amplifier. With it, the classic Type II (proportional–integral, PI) and Type III (proportional–integral–derivative, PID)

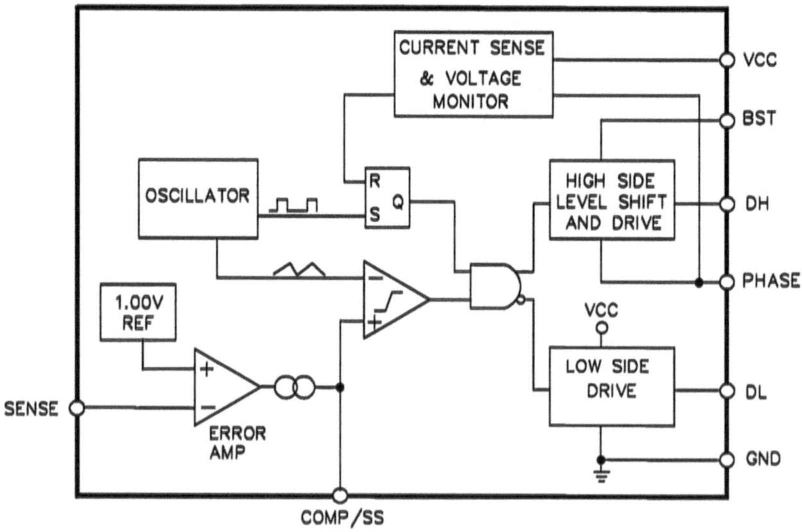

Fig. 3.10 Control IC internal structure

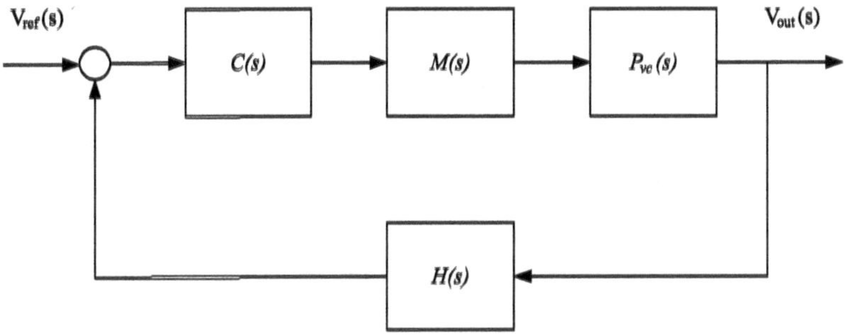

Fig. 3.11 A typical control block diagram of switching-mode power converter

compensation schemes are commonly used to achieve circuit stability and transient response performance.

To design a PI or PID controller for the power converter, a small-signal model of the power stage is needed. But, as we see in Figs. 3.8–3.10, the switching mode converter is not a simple linear system. Rather, it is a piece-wise linear system with very short time period (the inverse of the switching frequency). While it is possible to describe the circuit behaviour at each time interval, the circuit dynamics at large signal with longer time scale is important for the system stability. To model the circuit at longer time scale, various averaging methods have been developed. The state-space averaging method [36, 37] and the circuit averaging method [51, 52] are commonly utilised in switching-mode power converter analysis and design. By using these methods, a small-signal model of the synchronous buck converter power stage could be developed and its PWM control-to-output voltage transfer function is derived as

$$P_{vc}(s) = \frac{v_o(s)}{d(s)} = v_{in}\frac{1+\frac{s}{s_{z_1}}}{1+\frac{s}{\omega_0 Q}+\frac{s^2}{\omega_0^2}}, \tag{3.6}$$

where

$$s_{z_1} = \frac{1}{R_c C_o}, \qquad \omega_0 = \frac{1}{\sqrt{LC_o}}\sqrt{\frac{1+R_l/R_o}{1+R_c/R_o}},$$

$$Q = \frac{1}{\omega_0}\frac{1}{\frac{L}{R_l+R_o}+C_o\left(R_c+\frac{R_l R_o}{R_l+R_o}\right)}$$

with R_l being the equivalent series resistance (ESR) of the inductor L, R_c being the ESR of C_o, and R_o being the equivalent load resistance defined as V_{out}/I_o.

With the power stage PWM control-to-output voltage transfer function $P_{vc}(s)$, a control system diagram for the synchronous buck converter of Fig. 3.8 is shown in Fig. 3.11. Here, $M(s)$ is the PWM modulator gain. $H(s)$ is the output voltage sensing gain, and $C(s)$ is the compensator transfer function. For Type II compensation scheme, $H(s)$ is simply the constant gain of feedback voltage divider. For Type III compensation scheme, $H(s)$ contains a pole-zero pair for loop control along with

Table 3.3 Some capacitors used for the synchronous buck converter output

Type	Alum. Elec.	OS-CON	POSCAP	Tantalum	Ceramic	Ceramic
Series	MV-GX	SP	TPB	593D		
Package	10 round	10.5 round	7.3 × 4.3	7.3 × 4.3	1206	1206
Height	12.5 mm	10.5 mm	3.8 mm	4 mm		
Voltage	6.3 V	6.3 V	2.5 V	4 V	10 V	10 V
Value	820 μF	820 μF	680 μF	470 μF	10 μF	0.1 μF
ESR	85 mΩ	32 mΩ	40 mΩ	100 mΩ	4 mΩ	7 mΩ
ESL	5 nH	3.5 nH	5 nH	3.2 nH	8 nH	1 nH
Cost	low	high	high	medium	high	medium
Ripple Current	1.7 A	5 A	3 A	3.5 A		

Table 3.4 Suggested compensator types for different output capacitors

Compensator type	Location of zero crossover frequency f_o	Typical output capacitor
Type II (PI)	$f_{po} < f_{zo} < f_o < f_s/2$	electrolytic, tantalum
Type III (PID) – A	$f_{po} < f_o < f_{zo} < f_s/2$	tantalum, ceramic
Type III (PID) – B	$f_{po} < f_o < f_s/2 < f_{zo}$	ceramic

$C(s)$. The location of the poles and zeros in $C(s)H(s)$ design is determined by the targeted loop transfer function and the plant transfer function $P_{\mathrm{vc}}(s)$, which is related to the circuit parameters. For example, the zero in $P_{\mathrm{vc}}(s)$ is determined by the output capacitor (both capacitance value and its associated ESR value). For different capacitor types, the zero could land at different locations. In Table 3.3, various capacitor types along with their typical capacitance and ESR values are listed. Depending on the selected output capacitors, the desired loop crossover frequency and compensation type could be determined with the following Table 3.4 [41]. Once the compensator type and the loop crossover frequency are determined, the classical loop compensation methods could be used to finish the design of $C(s)$ and $H(s)$. A good summary of the design routine is in [41].

In many cases, commercial power converter controllers not only provide dynamic compensation at small signal level, they also interface with the power MOSFETs to execute the PWM control. In the controller and MOSFET interaction, one simple inequality needs to be met so that the MOSFETs could be turned on and turned off at any time as intended. Such inequality can be expressed as follows:

$$\int_0^{t_f} \min i_{\mathrm{drive}}\,\mathrm{d}t > \max Q_g, \tag{3.7}$$

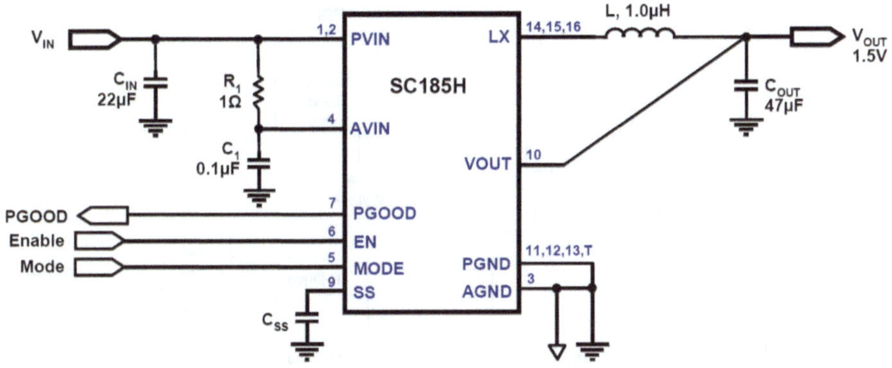

Fig. 3.12 A switching regulator with integrated MOSFETs and controller

where, i_{drive} is the gate drive current delivered by the control IC, Q_g is the MOSFET gate total charge, and t_f is the gate voltage transition time during switching. While the inequality is simple, in practice, it usually involves a fair amount of engineering effort to obtain the best and cost effective MOSFETs that meet both the power stage efficiency requirement and control IC's driving capability. Also, concrete circuit layout is required for best system electrical and thermal performance.

With the advancement of power semiconductor, circuit design, and electronic packaging technologies, it becomes possible to integrate the control IC with the power MOSFETs at lower input voltage (e.g. <30 V) and lower output current levels (e.g. <10 A).

One example is shown in Fig. 3.12, where the compensation network is also integrated within the IC. With this level of integration, the time-consuming process of optimally matching the MOSFETs and the controller driver is eliminated. Certainly, they are optimised for the best efficiency in the designated power range. Also, the internal compensation is tuned to provide decent circuit stability and transient response performance for selected output capacitor and inductor. Normally, this approach is best suited for high-volume applications with narrow input and output voltage ranges.

It has been a great research endeavour to develop required technologies to produce highly integrated power electronics modules as a basic building block or simply a component for power processing [32]. Such effort is aimed to enable unprecedented growth of power electronics industry by demonstrating

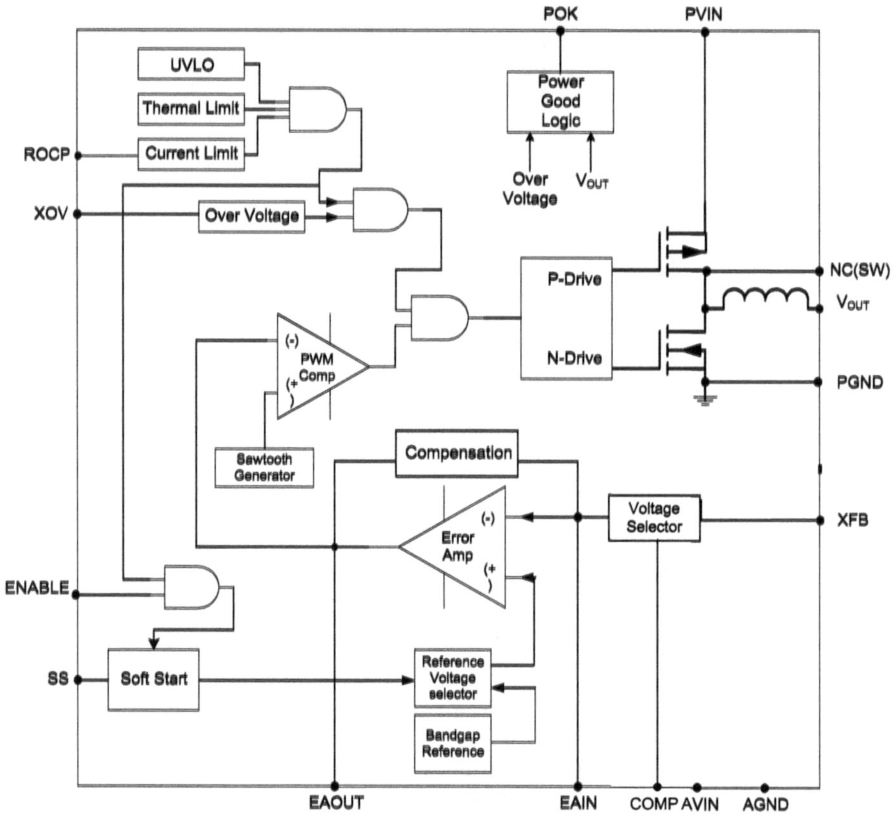

Fig. 3.13 A switching regulator with integrated MOSFETs, inductor and controller

10-fold improvements in quality, reliability and cost effectiveness of power electronics systems. A commercially available power supply on a chip (PwrSoC) 5 MHz DC–DC converter is shown in Fig. 3.13 [5], where the simple constant-frequency voltage-mode control is employed.

3.3.4 *DC-DC Buck Converter for Computer and Memory Applications*

Buck converter-based voltage regulator modules (VRM) have been successfully commercialised to power microprocessors, graphic cards and memory chips in servers, desktop and portable computers. Some of the key technical challenges in these applications are:

- high efficiency for low-voltage high-current (e.g. output voltage ranging 0.5–1.6 V, output current up to 150 A with 12 V input bus),
- fast load current transient (current step up to 120 A within 1 μs and maximal slew rate up to 300 A/μs),
- very small output overshoot voltage (within 50 mV) and very short settle time (within 25 μs) [50].

In the process of developing high power density VRMs, various control ICs are designed to support the multi-phase buck converter architecture to meet high efficiency, deliver high output current and achieve outstanding dynamic performance.

Figure 3.14 shows the internal circuit block diagram of a control IC [25] that drives up to four-phase buck converter. It is noted that, all four buck converters are put in parallel with common V_{in} and V_o, share single error amplifier and have the same switching frequency. Normally, the switching clock to the four phases are shifted with 90°. It is an example of the general interleaving scheme with equal phase delay among multiple phases [13]. Interleaving, even though simple, allows VRM to have much reduced input and output voltage ripple. Or, for a given input and output voltage ripple specification, it needs reduced number of input and output filter capacitors than the case without interleaving. Furthermore, interleaving effectively extends the loop crossover frequency and leads to much fast transient response. Many VRM controllers today support multi-phase interleaving.

In multiple phase structure, current balance among the phases is very important for minimal input, output voltage ripple, equal thermal distribution and ease of control.

Many commercial controllers implemented different schemes to get the current balance or load current sharing among the phases. The one in Fig. 3.14 first obtains the average phase current at each switching cycle and compares it with the sampled current at each channel. The resulted error signal is filtered and then added to the voltage loop error amplifier output to adjust the pulse width at corresponding channel and to correct any current unbalance. All the channels share one error amplifier which has a typical DC gain of 96 dB and unit gain bandwidth of 20 MHz [25]. Around such error amplifier, Type III compensation could be used in this voltage-mode controller to maintain the circuit stability and loop transient response. Normally, one chooses a loop crossover frequency high enough to meet the required dynamic performance, and low enough to stay away from the limit of half the PWM switching frequency.

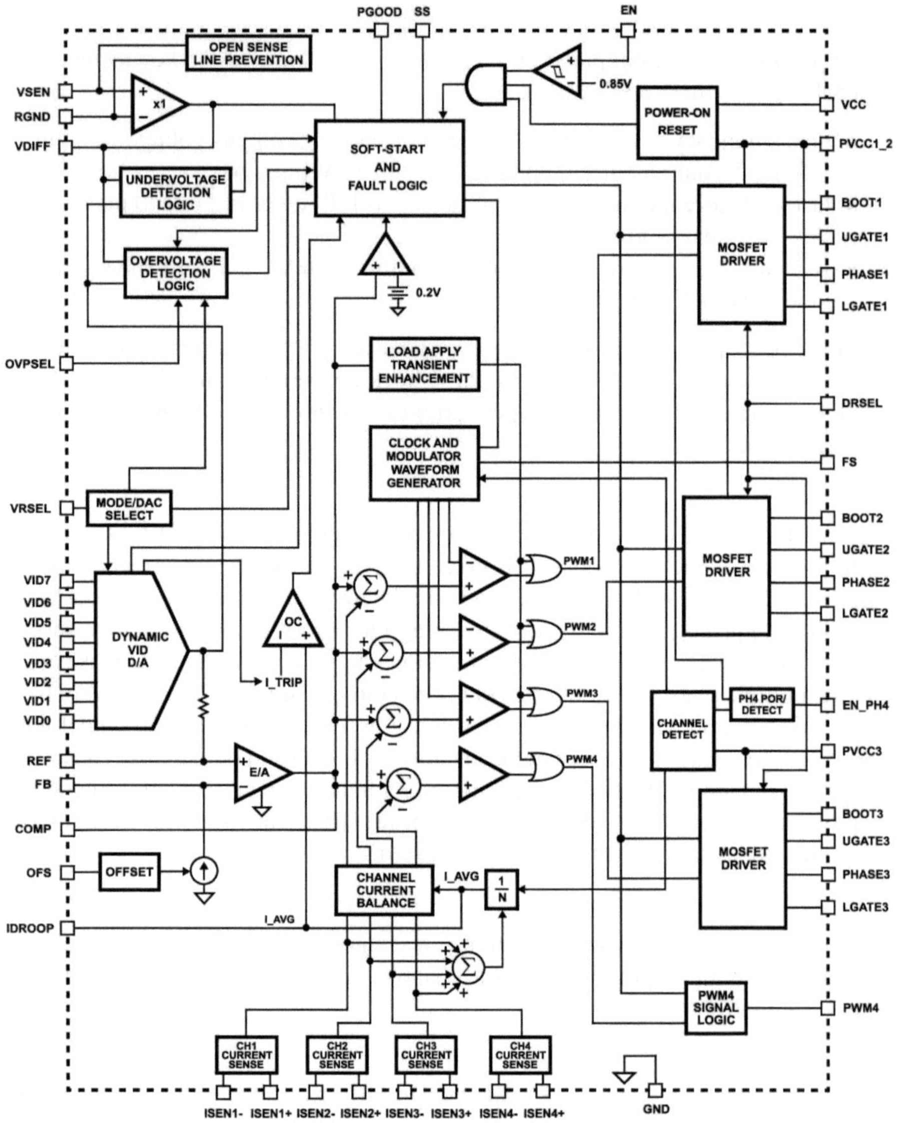

Fig. 3.14 A multiple phase buck PWM controller for VRM

In Fig. 3.14, the average phase current signal is made available to the external circuitry. By injecting this load current signal into the output voltage feedback loop, one could generate an output voltage droop that is related to the output current to meet the output load line regulation for most microprocessors. This simple modification, however, changed the dynamic behaviour of the simple voltage-mode control to a similar current-mode control. The double pole formed by the inductor and output capacitor is nicely split. In this case, simpler Type II compensation could be used to stabilise the system.

As we see, the multi-phase controller in Fig. 3.14 is actually a single controller with multiple phase shifted PWM modulators and MOSFET drivers. It was meant to deliver a single output voltage with large load current. Inside computers, there are other electronic devices that require two or more different power rail voltages. For example, the modern double data rate (DDR) synchronous dynamic random access memory (SDRAM) chips need two power rails for VTT and VDDQ. Furthermore, VTT is required to track the VDDQ voltage and have both current sink and source capability for load termination [40].

To meet this and other power rail requirement, a versatile dual buck converter controller such as shown in Fig. 3.15 is developed [22]. Basically, it consists of two almost identical current-mode PWM controllers with a common high-frequency oscillator. The PWM switching of both controllers are at the same constant frequency, but with 180° phase shift. In Fig. 3.16a, the two independent PWM controllers regulate their corresponding power stages for different output voltages and currents [21]. But, when they share the same input voltage V_{in}, due to the 180° switching clock phase shift, the effective total input current and voltage ripple could be reduced. When used to power DDR SDRAM modules, one output is connected to VDDQ, and the other is connected to VTT. To set that VTT $= 1/2$ VDDQ, the error amplifier reference voltage REFIN for the VTT channel is simply made accessible and connected to VDDQ via a voltage divider as shown in Fig. 3.16a. As a result, VTT follows VDDQ both statically and dynamically. With regarding to VTT sink and source current capability, it is interesting to observe that such function is naturally buried in the synchronous buck converter topology no matter it is under voltage-mode control or current-mode control. For the current-mode PWM controller as in Fig. 3.15, proper conditioning of the sensed current signal (both positive and negative) is implemented so that the output voltage could be regulated at both current sink and source modes. Furthermore, current limits are installed for both modes.

As shown in Fig. 3.16b, the controller in Fig. 3.15 could also be conveniently configured to control two buck converters connected in parallel with single output voltage. In this case, the second channel controller is disabled by pulling its error amplifier negative feedback pin IN2 to "high" (above 1.8 V). This action also connects the first channel error amplifier output to the second channel current loop input. As a result, both channels have the same inductor peak current and consequently the same average current because they operate at the same frequency and same duty ratio with common input and output voltages. When the inductor current sensing circuitry and gain are matched, the two channels will equally share the average load current.

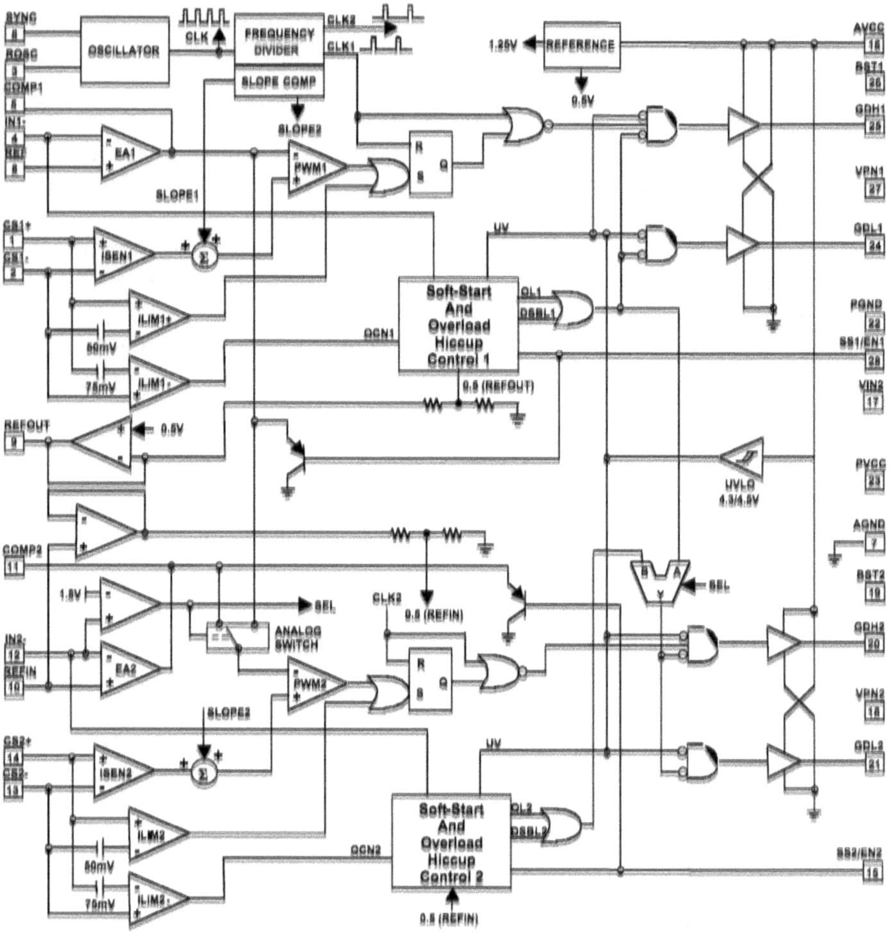

Fig. 3.15 A versatile dual buck PWM controller for DDR memory and other applications

3.3.5 Digital Control of DC–DC Converters

Digital control, due to its advantages of programmability, flexibility, and more possible built-in intelligence over analog control, has been explored for power converters. Over the years, it has taken different forms of microprocessor based, field programmable gate array (FPGA) based, DSP based and more recently application specific IC based. Continuous endeavour on power converter digital control has resulted in numerous publications and commercial control ICs.

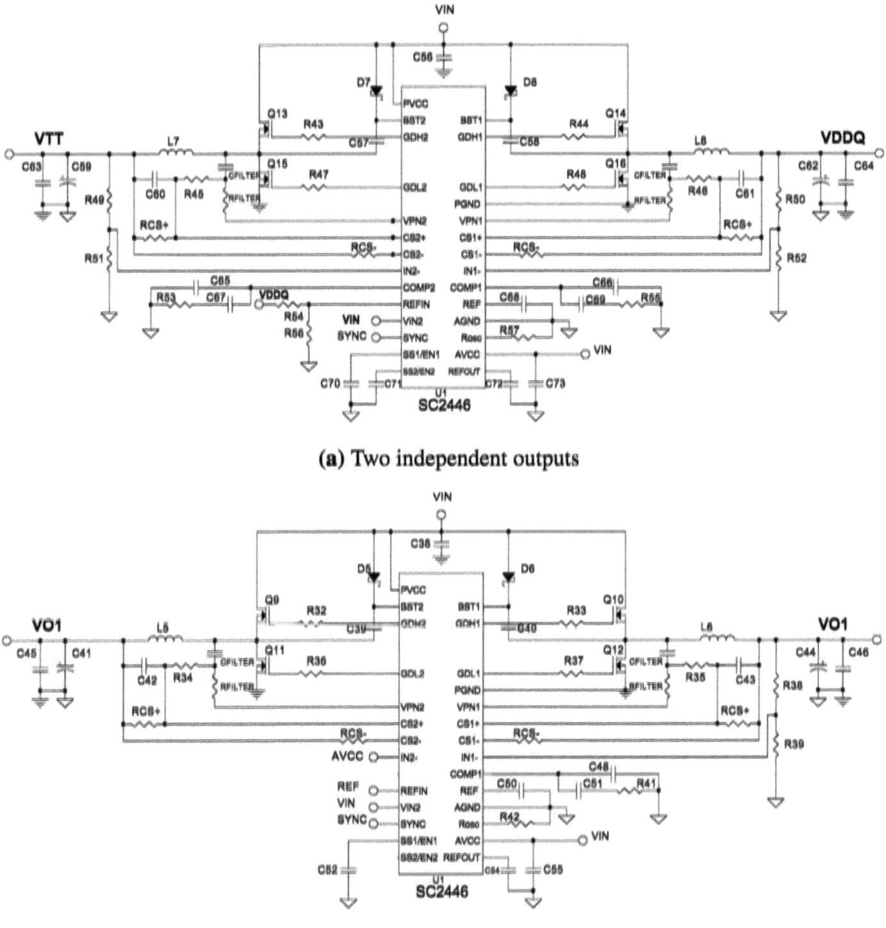

(a) Two independent outputs

(b) Single output with dual phase

Fig. 3.16 Two typical configurations of a dual buck PWM controller

One PWM controller shown in Fig. 3.17 [18] offers comprehensive features enabled by digital control. For example, to optimise the circuit operation and performance, the controller conducts fine calibration of output voltage, output current and temperature measurements via PMBus. To obtain the best control algorithm for a specific design and circuit operating condition, the controller also conducts automatic system identification online and generates the transfer function of the converter power stage. To accomplish this, the IC internally synthesises a sine wave signal and injects it, as a small signal disturbance, into the loop at the voltage reference input. Then, the IC measures the response of the disturbance at the output of the digital compensator and calculates the power-stage transfer function. A simple closed-loop digital control block diagram is shown in Fig. 3.18, where the blocks of DPWM, compensator $G_c(z)$, gain scheduling, and ADC all reside in the digital PWM con-

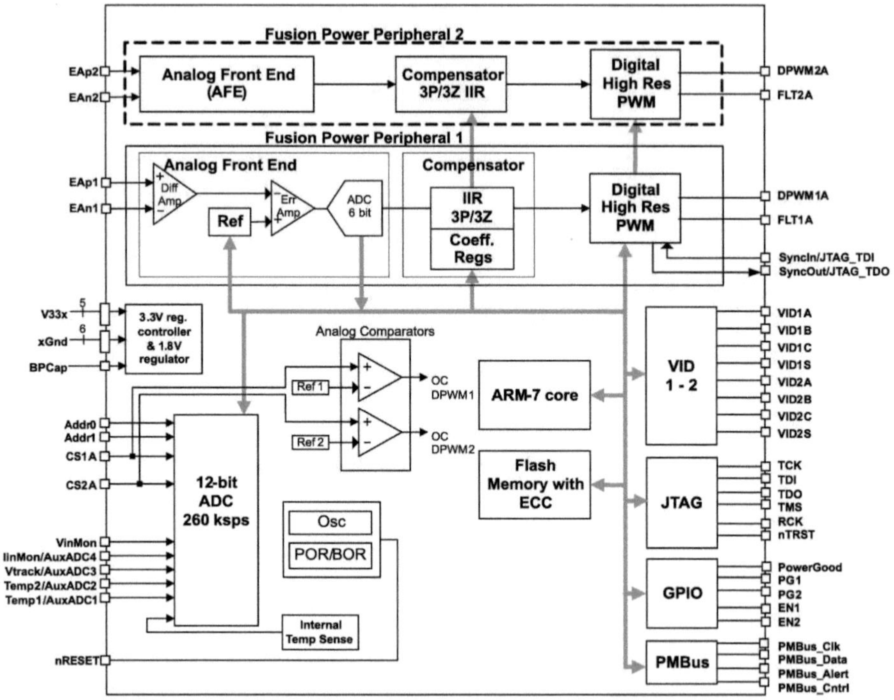

Fig. 3.17 Digital PWM controller for dual buck converters

Fig. 3.18 A simplified block diagram of a digital controlled PWM converter

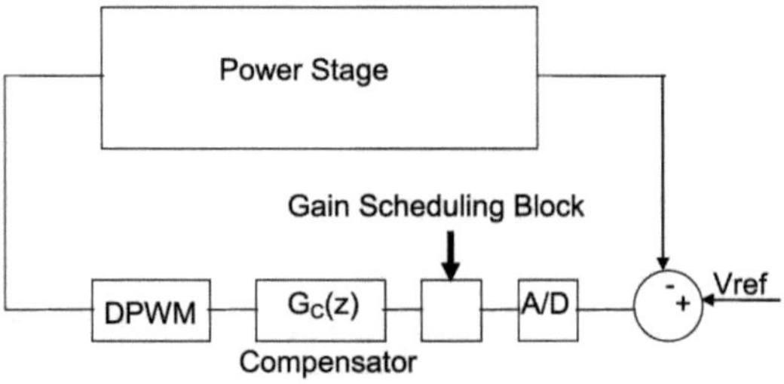

troller. While there are so many different options on the structure and implementation of the compensator $G_c(z)$, the commercial digital controller in Fig. 3.17 [18] takes the form of second-order infinite impulse response (IIR) filter cascaded with a first-order IIR filter. The coefficients of the filters determine the poles and zeros of the digital controller. Their design is based on the converter power-stage operating condition and targeted closed-loop gain, phase margin and crossover frequency.

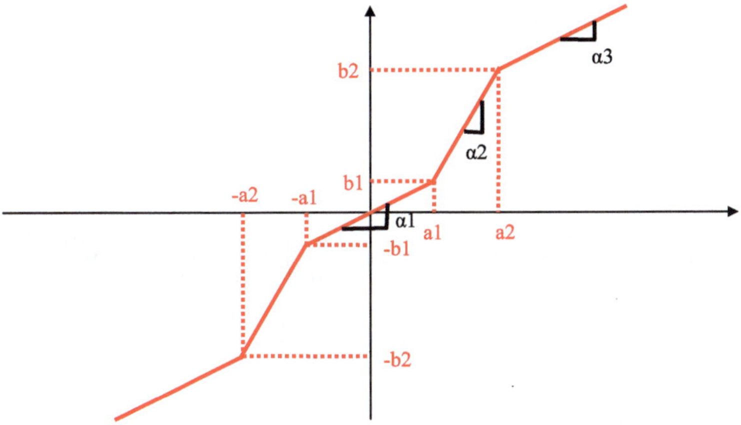

Fig. 3.19 Typical input–output characteristics of the gain scheduling block

Among the benefits offered by digital control, one could conveniently develop nonlinear control scheme to improve converter transient response performance while still maintain solid system stability. A simple, yet effective nonlinear digital control approach, gain scheduling [38], is used to achieve this goal as shown Fig. 3.18. In this approach, the nonlinearity is introduced by selectively modifying the output of the linear ADC. When the ADC error signal is small, a lower gain is applied. It makes the closed-loop system less sensitive to noise. When the ADC error signal is greater than a certain threshold value, a higher gain is applied. It provides larger corrective action during the load transient.

A typical characteristic of the gain scheduling block is shown in Fig. 3.19, and its characteristics are expressed in the following equations:

$$e(n) = f(n) = \begin{cases} \alpha_1 e, & |e| < a_1, \\ b_1 + \alpha_2(e - a_1), & a_1 \le e \le a_2, \\ -b_1 + \alpha_2(e + a_1), & -a_2 \le e \le -a_1, \\ b_2 + \alpha_3(e - a_2), & e \ge a_2, \\ -b_2 + \alpha_3(e + a_2), & e \le -a_2. \end{cases} \tag{3.8}$$

When the digital control loop contains nonlinear component such as in Fig. 3.19, the simple describing function could be used to study its effect on the loop stability. Recalling that, with sinusoidal input $x(t) = a\sin(\omega t)$ to the nonlinear component, the Fourier expansion of the output $y(t)$ is $y(t) = b_0 + b_1\sin(\omega t) + b_2\sin(2\omega t) + \cdots$, the gain G of the nonlinear components could be described as $G = b_1/a$. This gain could be amplitude or frequency dependent. Applying the describing function method to the nonlinear gain characteristics shown in Fig. 3.19, one could derive the following interesting gain factors:

$$G = \alpha_1 \tag{3.9}$$

for $a < a_1$,

$$\begin{aligned} G = \alpha_1 &+ \frac{\alpha_2 - \alpha_1}{2}\left[2 - \frac{4}{\pi}\arcsin\left(\frac{a_1}{a}\right) + \frac{2}{\pi}\sin\left(2\arcsin\frac{a_1}{a}\right)\right] \\ &- (\alpha_2 - \alpha_1)\frac{a_1}{a}\frac{4}{\pi}\cos\left(\arcsin\frac{a_1}{a}\right) \end{aligned} \tag{3.10}$$

for $a_1 < a < a_2$, and

$$\begin{aligned} G = \alpha_1 &+ \frac{\alpha_2 - \alpha_1}{2}\left[2 - \frac{4}{\pi}\arcsin\left(\frac{a_1}{a}\right) + \frac{2}{\pi}\sin\left(2\arcsin\frac{a_1}{a}\right)\right] \\ &- (\alpha_2 - \alpha_1)\frac{a_1}{a}\frac{4}{\pi}\cos\left(\arcsin\frac{a_1}{a}\right) \\ &+ \frac{\alpha_3 - \alpha_2}{2}\left[2 - \frac{4}{\pi}\arcsin\left(\frac{a_2}{a}\right) + \frac{2}{\pi}\sin\left(2\arcsin\frac{a_2}{a}\right)\right] \\ &- (\alpha_3 - \alpha_2)\frac{a_2}{a}\frac{4}{\pi}\cos\left(\arcsin\frac{a_2}{a}\right) \end{aligned} \tag{3.11}$$

for $a > a_2$.

With the above-derived nonlinear gains, one could proceed with loop stability analysis and design [38]. In practice, it should be noted that zero-error bin and saturation effects always exist due to the resolution of the A/D and the existence of the saturation on the A/D and DPWM. When designing a digital compensator for the power stage, one should be aware that the gain could be significantly smaller than the designed value because of the above-mentioned nonlinear effects.

3.4 Controllers for AC–DC Power Factor Correction

In today's common AC power distribution systems, AC–DC power conversion is needed for majority of electronic devices and equipment ranging from household electronics, personal computers, medical equipment to telecommunication base stations. With the growing number of these electronic apparatus directly connected to the power grid, it is important to efficiently utilise the power system capacity, maintain power line quality and preserve the system stability.

One important parameter to describe the power loading effect to the power grid is the power factor of the electronic devices. In general term, the power factor (PF) is defined as the ratio of the real power (P_r) it consumes and the apparent power (P_a) it receives from the power grid.

Here, the real power P_r is the average power calculated by

$$P_r = \frac{1}{T}\int_0^T v_{\text{in}}(t) i_{\text{in}}(t)\, \mathrm{d}t, \tag{3.12}$$

and the apparent power P_a is calculated by

$$P_a = I_{\text{in,rms}} \times V_{\text{in,rms}} = \sqrt{\frac{1}{T}\int_0^T i_{\text{in}}(t)^2\, \mathrm{d}t} \times \sqrt{\frac{1}{T}\int_0^T v_{\text{in}}(t)^2\, \mathrm{d}t} \tag{3.13}$$

for periodical waveforms $v_{\text{in}}(t)$ and $i_{\text{in}}(t)$ with the same period T. It could be shown that, if the electronic device input impedance exhibits resistive characteristics with equivalent resistance R_{in}, the power factor $PF = 1$. Note that this statement is true no matter the waveforms $v_{\text{in}}(t)$ and $i_{\text{in}}(t)$ are sinusoidal or not, as long as one signal (either $v_{\text{in}}(t)$ or $i_{\text{in}}(t)$) follows the other signal with a linear ratio R_{in}. In today's power grid, the voltage $v_{\text{in}}(t)$ is commonly sinusoidal.

The goal of power factor correction in modern AC–DC power converters is to sharp the input current $i_{\text{in}}(t)$ to follow $v_{\text{in}}(t)$ as close as possible. To simply describe the closeness of the current $i_{\text{in}}(t)$ waveform to sinusoidal signal, the total harmonic distortion (THD) is defined and introduced as

$$THD = \frac{\sqrt{\sum_{k=2}^{+\infty} I_k^2}}{I_1}, \tag{3.14}$$

where I_1 is the magnitude of the fundamental frequency component, and I_k is the magnitude of kth harmonic frequency current.

Clearly, if $i_{\text{in}}(t)$ is a sinusoidal waveform with line frequency $\omega_1 = 2\pi f_1$, its THD will be zero. In practice, however, the input current $i_{\text{in}}(t)$ may not be pure sinusoidal. The harmonic current injected to the power grid will cause power losses and electromagnetic interference (EMI). To maintain the quality and efficiency of the power distribution system, Commission Electrotechnique Internationale established and continuously update electromagnetic compatibility (EMC) standard, e.g. IEC 61000, to regulate the industry practice. Specifically, IEC 61000-3-2 clearly specifies the harmonic current content limits for various electronic equipment with input current less than 16 A per phase and further classification from Class A to

Class D [30]. Over the years, such standard guideline is gradually adopted in many regions of the world. To meet this stringent requirement, active power factor correction is commonly required.

The circuit topologies that are widely used for active power factor correction are boost and flyback converters. They are based on the two basic converter topologies in Fig. 3.4 that exhibit voltage conversion gain greater than 1. With rectified sinusoidal input voltage $v_{\text{in}}(t) = V_{\text{pk}}|\sin(\omega t)|$, these converters are capable of maintaining a constant output voltage V_o. Also, they have the freedom to sharp the input current $i_{\text{in}}(t)$ while a required average output load current I_o is delivered.

Assuming that the targeted input current $i_{\text{in}}(t) = I_{\text{pk}}|\sin(\omega t)|$ and neglecting the power conversion losses, one could derive that the converter equivalent input resistance

$$R_{\text{in}} = \frac{V_{\text{pk}}^2}{2V_o I_o} \tag{3.15}$$

and the converter instantaneous output current

$$i_o(t) = \frac{V_{\text{pk}}^2}{2V_o R_{\text{in}}}\big(1 - \cos(2\omega t)\big). \tag{3.16}$$

By substituting (3.15) to (3.16), it is further shown that

$$i_o(t) = I_o\big(1 - \cos(2\omega t)\big). \tag{3.17}$$

The difference between the converter instantaneous output current $i_o(t)$ and average output current I_o is supplied by the output capacitor. A typical circuit structure of AC–DC converter with PFC is shown in Fig. 3.20. Basically, the control IC takes a DC voltage signal (normally from the IC internal reference) as output voltage reference, uses rectified V_{in} sinusoidal waveform as input current reference, synthesises a proper PWM control signal to the AC–DC power stage, so that the output voltage is regulated at a preset DC voltage and input current follows a sinusoidal shape. At the same time, the average input current is determined by the output load current I_o via the equation

$$\langle i_{\text{in}} \rangle = \frac{4}{\pi}\frac{V_o}{V_{\text{pk}}} I_o. \tag{3.18}$$

There are many different control methods and commercially available control ICs for PFC. Here, we discuss a few commonly used approaches.

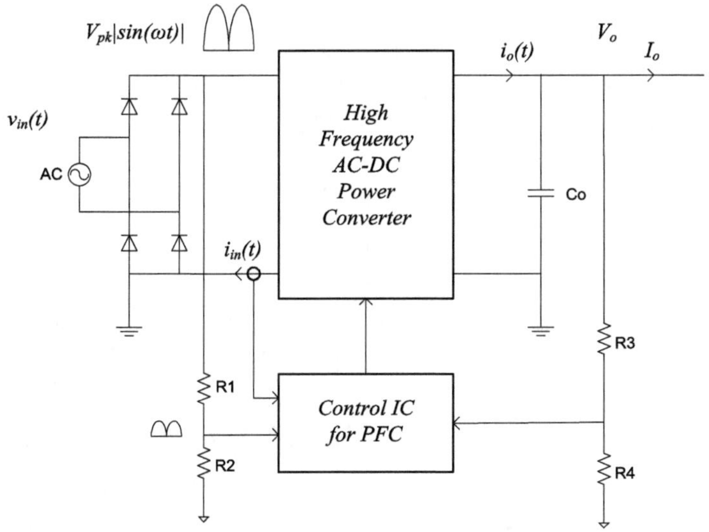

Fig. 3.20 A typical AC–DC converter and control IC structure with PFC

3.4.1 Average Current-Mode Control-Based PFC Controller

UC3854 [29] shown in Fig. 3.21 is one of the early PFC controllers used in the industry. It operates at a constant switching frequency programmed by the external resistor and capacitor connected to pins RSET and CT respectively. It has a voltage error amplifier and a current error amplifier readily being configured with average current-mode control structure for output voltage regulation. In order to shape the input current, an analog multiplier/divider is inserted between the two error amplifiers.

This block senses the rectified sinusoidal waveform from $v_{in}(t)$, multiplies it by the voltage error amplifier output (VA Out) and sends the modulated signal (at pin 5, Mult Out) to the current error amplifier as reference. By designing a fast average current loop with external compensation components, the input current $i_{in}(t)$ is then effectively controlled to follow the shape of $v_{in}(t)$ while its amplitude is adjusted to deliver the required load current according to (3.18). From (3.18) it is clear that, for a given output power, the input current amplitude is reduced when the input voltage is increased. This relationship is conveniently implemented by the division function inside the block and the square function block entering the terminal C. From the voltage divider and the RC filters, the VRMS (pin 8) terminal senses a scaled input voltage RMS value $\alpha_1 V_{in}$ with constant coefficient α_1. Such signal is subsequently squared and fed to the terminal C of the block. While obtaining the rectified sinusoidal reference waveform from $v_{in}(t)$, a scaled input voltage RMS value $\alpha_2 V_{in}$ (with constant coefficient α_2) also enters IAC (pin 6) and further appears at terminal B of the block. Along with all the inputs at terminals A, B and C, the

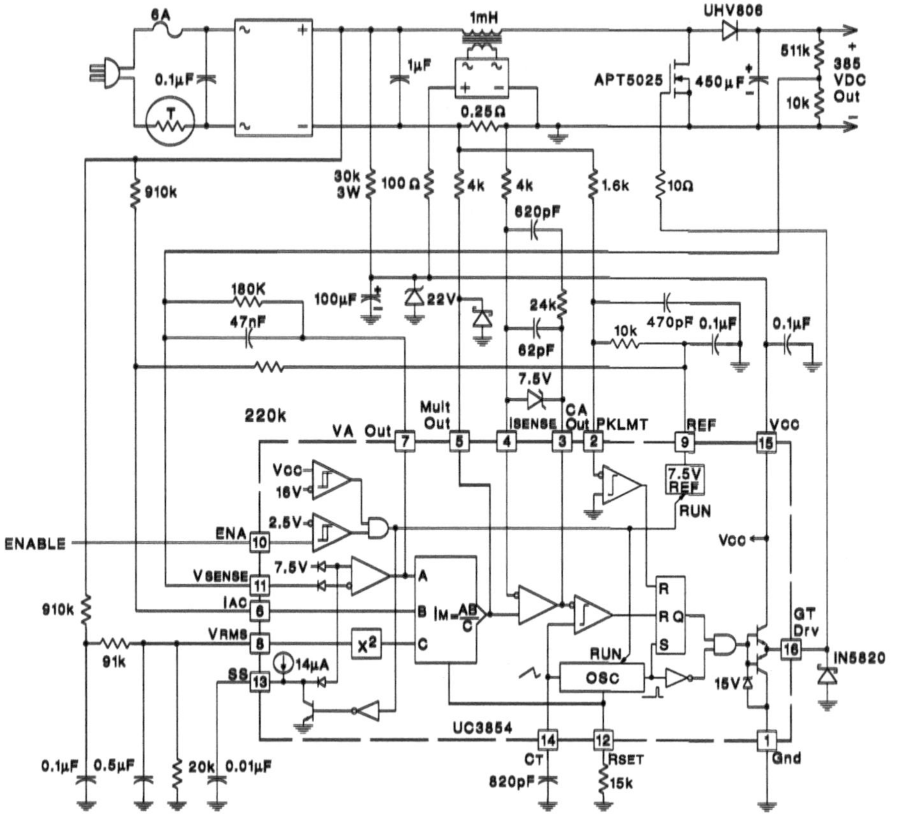

Courtesy Texas Instruments

Fig. 3.21 An universal line boost based PFC with average current mode control

output of the block is

$$Mult_{\text{Out}} = \frac{VA_{\text{Out}}\,\alpha_2 V_{\text{in}} \sin(\omega t)}{(\alpha_1 V_{\text{in}})^2} = \frac{\alpha_2 \sin(\omega t)}{\alpha_1^2 V_{\text{in}}} VA_{\text{Out}}. \tag{3.19}$$

As a consequence, the amplitude of reference signal to input current $i_{\text{in}}(t)$ is automatically scaled by the input voltage magnitude in inverse proportion. The voltage error amplifier output VA_{Out} is then mainly used to regulate the output voltage and output power.

As discussed in [20], the average current-mode control is a general control method for DC–DC switching power supplies. Compared with the peak current-mode control, it does not need slope compensation, has better noise immunity, and has better average current control accuracy. Also, it applies to both DCM and CCM. These are very important features for AC–DC converter with power factor correction. For example, to achieve unity power factor, the input line average current needs to be controlled to follow the input line voltage very closely. In many practical designs, the AC–DC converter operates at DCM and CCM at different portion of each input line voltage cycle. A common control method for both DCM and CCM is needed to achieve high-quality PFC. Yes, some design constraints and optimisation consideration call for the mixed DCM and CCM operation for PFC circuit. For example, we know that the input voltage to a PFC circuit is sinusoidal, while the output voltage is normally regulated to a constant value. With a fixed switching frequency, maintaining CCM for all load current conditions requires an inductor with larger volume, higher cost and excessive losses. On the other hand, if one wants to use a very small inductor so that the circuit always operates at DCM for all the line and load conditions, it usually leads to huge inductor current and associated circuitry losses, especially at lower input voltage and higher load current condition. An engineering compromise among circuit cost, losses and size requires a reasonable inductor value that usually leads a mix of CCM and DCM operation of the PFC circuit over the period of each input voltage line cycle. But, the PWM switching frequency is maintained constant as some engineers prefer.

3.4.2 Hysteretic Control-Based PFC Controller

In contrast to directly controlling the input average current and shaping it to sinusoidal as in average current-mode control, the hysteretic approach controls the envelop of the input current sawtooth waveform so that its average value is sinusoidal.

A simple, but very effective control IC for PFC boost converter is shown in Fig. 3.22 [39], where a sinusoidal waveform scaled from input line voltage is multiplied by the voltage loop error amplifier output to form the upper hysteresis limit of the input line current. Here, the rising edge of the input current sawtooth is conveniently sensed from the MOSFET branch. Once the input current crosses the upper hysteresis limit, the boost switch Q_1 is turned off. As a result, the inductor current (also the input current) ramps down with negative slope $(V_{pk}\sin(\omega t) - V_o)/L$ as shown in Fig. 3.23. When it hits the lower hysteresis limit—zero—the MOSFET Q_1 is turned back ON, and the inductor current ramps up with a positive slope $V_{pk}\sin(\omega t)/L$. As shown in Fig. 3.22, the sense of input current zero crossing point is from a winding coupled to the inductor. It simply used the fact that the voltage

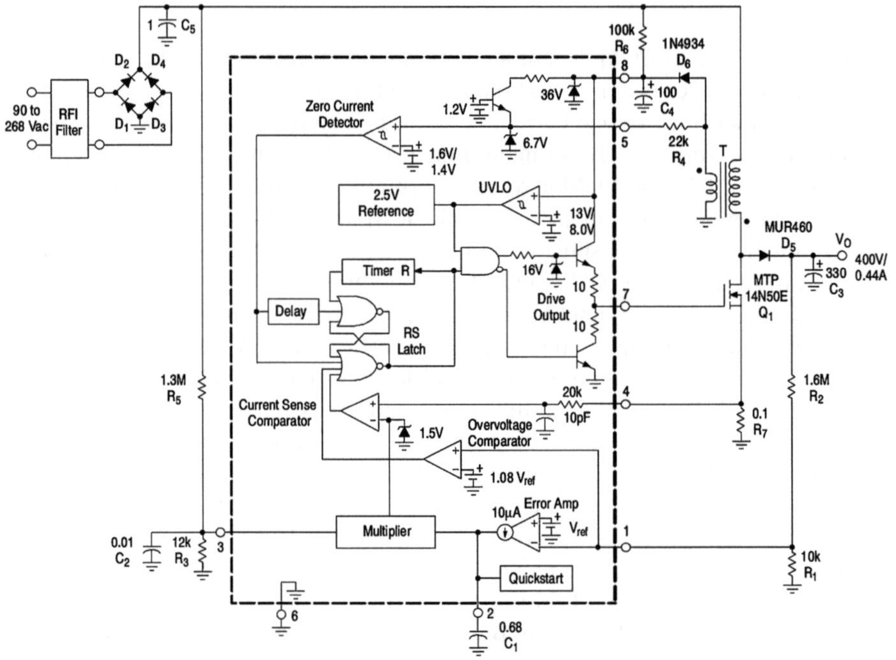

Fig. 3.22 An universal line boost-based PFC with hysteretic control

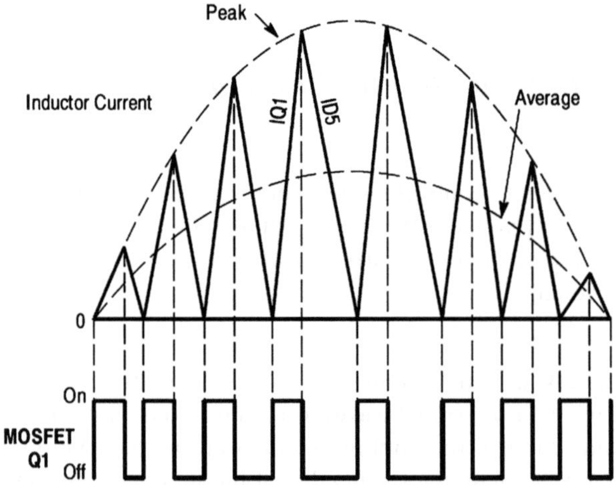

Fig. 3.23 Typical inductor and MOSFET gate waveforms of boost PFC circuit

across an inductor is proportional to its current change rate. When the inductor current ramps down and reaches zero, the current slew rate changes abruptly.

Compared with simple current magnitude measurement and detection, such approach has much higher signal-to-noise ratio at zero crossing point and leads to crisp turn on of the MOSFET for the next switching cycle. As a consequence, the boost PFC circuit operates at boundary conduction mode (BCM) with typical inductor current and MOSFET gate drive waveforms as shown in Fig. 3.23. Clearly, over each input voltage line cycle, the PWM switching frequency is variant. But, the PWM switching ON time seems invariant. It is indeed the case and could be proven as follows.

Let $i_{Q_1}(t)$ be the drain current of the MOSFET switch Q_1, which is equal to the inductor current $i_L(t)$ and the input line current $i_{in}(t)$ during the PWM ON time t_{ON}. Such current is sensed through the current sense resistor R_7 and compared with the output of the Multiplier at the end of t_{ON} as shown in Fig. 3.22. This leads to a simple equation

$$R_7 i_L(t_{ON}) = Mult_{Out}. \tag{3.20}$$

From Fig. 3.23 we know that

$$i_L(t_{ON}) = \frac{t_{ON}}{L} V_{pk} \sin(\omega t), \tag{3.21}$$

and from Fig. 3.22 that

$$Mult_{Out} = VA_{Out}\, \alpha_3 V_{pk} \sin(\omega t) \tag{3.22}$$

at time t. Here, VA_{Out} is the voltage error amplifier output at Pin 2, and α_3 is a constant covering the input voltage sense divider (R_3 and R_5) ratio and the multiplier gain. When the PWM switching frequency f_s (normally hundreds of kHz), is much greater than the input voltage line frequency f_l (typical 50–60 Hz), one could get the following equation by substituting (3.21) and (3.22) into (3.20) and making the coefficients equal:

$$t_{ON} = \alpha_3 L VA_{Out}. \tag{3.23}$$

Clearly, the boundary conduction mode forced by the control method in this IC for PFC automatically warrants constant ON time operation of the boost converter. On the other hand, Eq. (3.21) tells that, if one could simply maintain a constant ON time switching, the input line PFC would be automatically achieved without sensing the inductor current and the multiplier!

When one wants to regulate the output power and output voltage, the error amplifier output VA_{Out} is adjusted, and so is the ON time of the PWM switching according to (3.23). As long as VA_{Out} signal is maintained flat, the input line current follows a sinusoidal waveform as determined by (3.21). In the boost converter operating at BCM, the power switch Q_1 in Fig. 3.22 operates with zero current turn on, and the output rectifier diode has minimal reverse

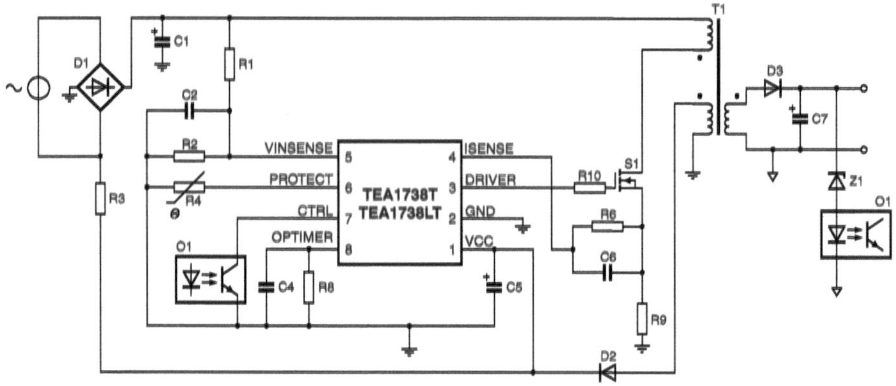

Fig. 3.24 A simple PWM controller for flyback PFC at DCM

recovery current. The reduced power stage losses and the simple control IC structure have made the solution in Fig. 3.22 widely accepted for PFC in low to medium power ranges. Furthermore, the BCM scheme has been extended to flyback and SEPIC circuit topologies with acceptable PFC performance. But, due to the variable operating frequency nature of this control method, sizeable input EMI filter is commonly deployed to meet the EMC standards.

3.4.3 Simple PFC Controller for Flyback Converter in Discontinuous Conduction Mode

As we know, to achieve PFC with a constant switching frequency, average current-mode control could be readily used. But, for the widely used simple flyback circuit topology, PFC under constant switching frequency comes naturally when the circuit operates at DCM. Figure 3.24 shows a typical flyback circuit controlled by a PWM controller with internal fixed constant switching frequency f_s [48]. For the input line voltage $V_{pk}\sin(\omega t)$, output voltage V_o and load current I_o, it is shown [24] that, if the magnetising inductance (L_M) of the transformer T_1 is designed such that the inequality

$$L_M < \frac{n^2 V_o}{4 f_s I_o (1 + n V_o / V_{pk})^2} \tag{3.24}$$

is satisfied, the flyback circuit operates at DCM over the entire input voltage line cycle.

The typical switch current is shown in Fig. 3.25. From it, the average switch current, which is also the input current i_{in} in flyback converter, over one switching

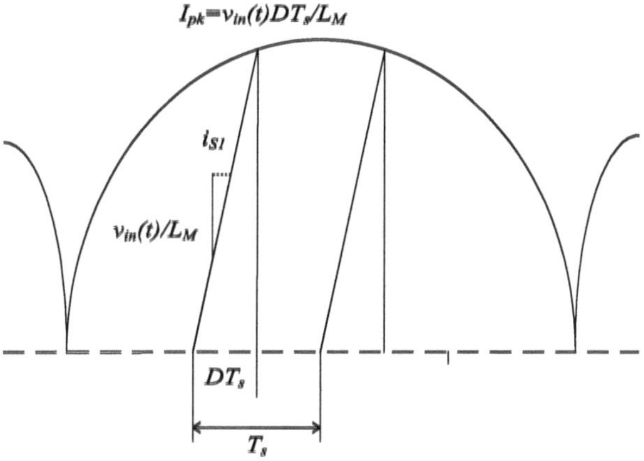

Fig. 3.25 Typical switch current of flyback PFC circuit at DCM

period T_s is calculated as

$$\langle i_{\text{in}}(t)\rangle = \langle i_{S_1}(t)\rangle = \frac{1}{T_s}\int_0^{T_s} i_{S_1}(t)\,\mathrm{d}t = \frac{1}{T_s}\int_0^{T_s} \frac{v_{\text{in}}}{L_M}t\,\mathrm{d}t = \frac{D^2T_s}{2L_M}v_{\text{in}}. \quad (3.25)$$

Clearly, with constant duty ratio D and constant switching frequency f_s, the average input current naturally follows the input voltage with magnitude $(D^2T_sV_{\text{pk}})/(2L_M)$. In Fig. 3.24, the CTRL (pin 7) signal generates proper duty ratio D to regulate the output voltage and output power.

3.4.4 Some Notes

With the growing demand of lower noise, more efficient electronic equipment and devices connected to the power grid, PFC becomes a necessity even at power less than 75 W. While the single-stage PFC schemes as discussed in this section fit to this category effectively, there is one fundamental trade off that should be noted. It is the output load transient response performance and PFC performance. As we know, under load current transient, the voltage error amplifier is responsible for acting fast and generating proper modulation signal to the power stage so that the output voltage has minimal deviation from a preset value. In typical DC–DC power converter, the voltage error amplifier is designed with wide bandwidth for fast load regulation. But, for the single stage AC–DC converter with PFC, the converter output current $i_o(t)$ carries a significant amount of low frequency (2 times input line frequency) component as shown in (3.17) and Fig. 3.20. A large output capacitor C_o is normally used to balance the converter output current $i_o(t)$ and the required load DC current I_o.

The resulted voltage ripple on C_o is calculated as

$$\tilde{v}_o(t) = \int_0^t \frac{i_C(t)}{C_o}\,\mathrm{d}t = \int_0^t \frac{I_o}{C_o}\cos(2\omega t)\,\mathrm{d}t = \frac{I_o}{2\omega C_o}\sin(2\omega t). \tag{3.26}$$

Such voltage ripple, scaled by the feedback voltage divider (e.g. R_3 and R_4 in Fig. 3.20) ratio, appears at the input of voltage error amplifier. If this low-frequency ripple (at $2f_l$) is not effectively filtered by the compensation network and further appears at the voltage error amplifier output, it will inject the corresponding low-frequency harmonic to the input line current, cause distortion and reduce the power factor. Equations (3.19), (3.23) and (3.25) indicated the importance of preventing any low-frequency components at voltage error amplifier output for high-quality PFC with average current-mode control, hysteretic control and constant duty ratio based PFC methods, respectively. In order to filter out the low-frequency ripple at $2f_l$, the voltage error amplifier along with the compensation network must have very limited bandwidth. Many engineering designs choose the bandwidth around $1/5$–$1/10$ of $2f_l$ to achieve a high-quality power factor. The penalty is then the deteriorated load transient response performance.

Over the past few decades, more and more PFC controllers are available in the market [42, 44] to work with different circuit topologies and control methods. Some research activities had also been done to achieve a high power factor at single-stage converters with excellent load transient response performance. Also the two-stage power conversion approach—PFC front stage cascaded with DC–DC converter—is used in many commercial applications to meet the high power factor and fast load transient response requirement [26].

3.5 Controllers for DC–AC Electronic Ballasts

DC–AC power converters have been used in motor drives, inductive heating, electronic ballast, and many other medical and industrial equipment. The commercial controllers range from microcontroller and DSP-based digital control, to single analogue control ICs with a few pins depending on the complexity of the load and required system performance. In contrast to most DC output power converters, the AC output system and circuit performance is mostly related to the load characteristics. Solid understanding of the load behaviour is imperative for proper circuit and controller design with targeted system performance.

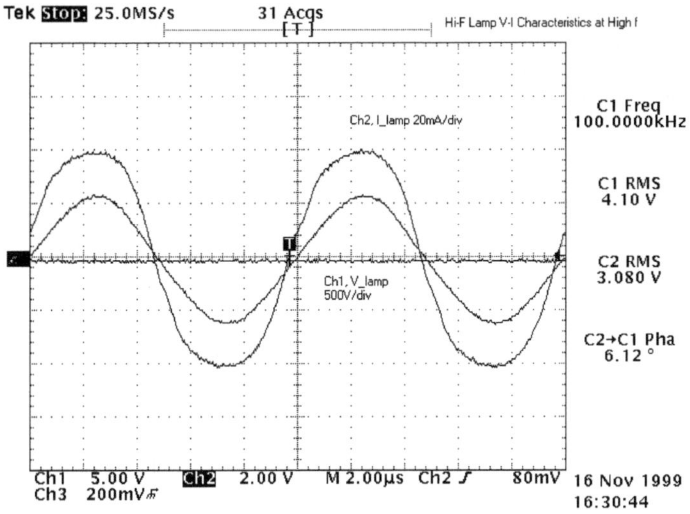

Fig. 3.26 Measured fluorescent lamp terminal voltage and current waveforms at high-frequency operation

In this section, we focus on the high-frequency electronic ballast for fluorescent lamps widely used in office buildings, factories, stores and industrial sites.

3.5.1 Lamp Characteristics and Ballast Interaction

For most gas discharge-based lamps, e.g. fluorescent, ignition is a prerequisite of the normal operation. The ignition voltage V_{ig} could range from a few hundred volts to tens of kilovolts depending on the type of lamps and the auxiliary devices. Normally, the duration of such ignition voltage is in milliseconds range. After ignition, the lamp enters normal operation with much reduced terminal voltage and increased current throughput. Figure 3.26 shows the measured lamp voltage $v_l(t)$ and lamp current $i_l(t)$ waveforms operating at high frequency 50 kHz. It is observed that the waveforms $v_l(t)$ and $i_l(t)$ are in phase even though they are not perfectly sinusoidal. By drawing them in v–i phase plane as in Fig. 3.27, it is clear that the fluorescent lamp terminal characteristics behave as a nonlinear resistor when the excursion frequency of the v–i curve is high, e.g. 50 kHz. Even though the exact equivalent resistance value at each time instant of the switching period is not a constant, it is commonly approximated by a DC resistance as $R_l = V_l/I_l$, where V_l and I_l are the root mean square (rms) values of $v_l(t)$ and $i_l(t)$, respectively. Interestingly, when one further characterises the lamp electrical behaviour at different

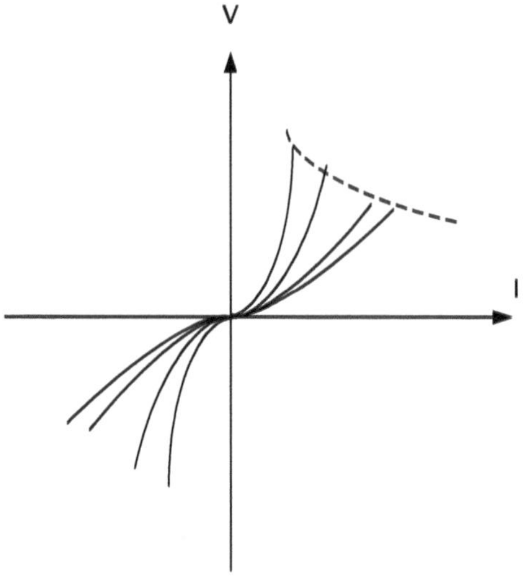

Fig. 3.27 Fluorescent lamp terminal voltage and current waveforms at high-frequency operation with different rms current values

lamp current, it is found that the v–i curve follows different slope as shown in Fig. 3.27 [9]. This implies that the above equivalent DC resistance R_l is a function of I_l.

There have been many different ways to capture this relationship. Here, we simply use curve fitting of empirical data and arrive at the following equation:

$$V_l(I_l) = a_1 e^{a_2 \sqrt[5]{I_l}} I_l + a_3, \tag{3.27}$$

where I_l is the lamp current in mA, and a_i with $i = 1, 2, 3$ are constant real numbers.

By differentiating (3.27), we further get the equivalent dynamic resistance $r_{ac}(I_l)$ of the lamp as

$$r_{ac}(I_l) = a_1 e^{a_2 \sqrt[5]{I_l}} \left(1 + \frac{a_2}{5} \sqrt[5]{I_l}\right). \tag{3.28}$$

Clearly, the lamp dynamic impedance is a strong function of the lamp current, especially at lower lamp current levels. For a typical fluorescent lamp TLD 36 W, the parameters are $a_1 = 750000$, $a_2 = -12$ and $a_3 = 80$.

It should be emphasised that the negative dynamic impedance $r_{ac}(I_l)$ only describes the lamp low-frequency dynamic behaviour with zero order. A first-order transfer function with right half-plane zero could be derived to describe the lamp dynamic behaviour [16] at low frequency.

Specifically, the lamp dynamic impedance is modelled as

$$Z_l(s) = K_p \frac{1 + s/z_1}{1 + s/p_1}, \tag{3.29}$$

where $K_p = r_{ac}(I_l) < 0$, $p_1 = 1/\tau(I_l)$, and

$$z_1 = \frac{r_{ac}(I_l)}{\tau(I_l)(R_l(I_l) + \delta R(I_l))}.$$

Here, the lamp time constant τ is also a function of the lamp current, which is approximated as

$$\tau(I_l) = b_1 e^{b_2 I_l} + b_3 I_l + b_4 \tag{3.30}$$

with constants b_i for $i = 1, 2, 3, 4$. The corresponding set of parameters for TLD 36 W is $b_1 = 4 \times 10^{-5}$, $b_2 = -50$, $b_3 = 3 \times 10^{-5}$, $b_4 = 3.77 \times 10^{-5}$. It is noted that the lamp dynamic model in (3.29) satisfies the following interpolation conditions:

$$\lim_{s \to +\infty} Z_l(s) = R_l(I_l) + \delta R(I_l), \tag{3.31}$$

which is the lamp impedance at switching frequency, and

$$\lim_{s \to 0} Z_l(s) = r_{ac}(I_l), \tag{3.32}$$

which is the dynamic lamp impedance at zero frequency.

Due to the negative dynamic impedance of the gas discharge lamps, it is difficult to drive the lamp with a voltage source, such as the one directly from the power grid. Various electronic ballasts have been developed to drive the fluorescent lamps, and they become an integral part of the lighting industry.

In discussing lamp–ballast interaction and their stability, one port circuit diagram as in Fig. 3.28 is often used. From it one could derive a *sufficient* (but not necessary) condition for the circuit stability as

$$|Z_o(s)| > |Z_l(s)|, \tag{3.33}$$

where $Z_l(s)$ is the lamp dynamic impedance as specified in (3.29), and $Z_o(s)$ is the ballast output dynamic impedance. Here, one should be cautious that the ballast output dynamic impedance is different from the typical output impedance calculated at switching frequency f_s. As a matter of fact, the typical output impedance calculated at switching frequency f_s is only $|Z_o(0)|$. The comparison between the lamp curve slope and the ballast load line slope to determine the circuit stability argument exactly falls into this category. Further treatment on the lamp–ballast interaction stability analysis can be found in [16].

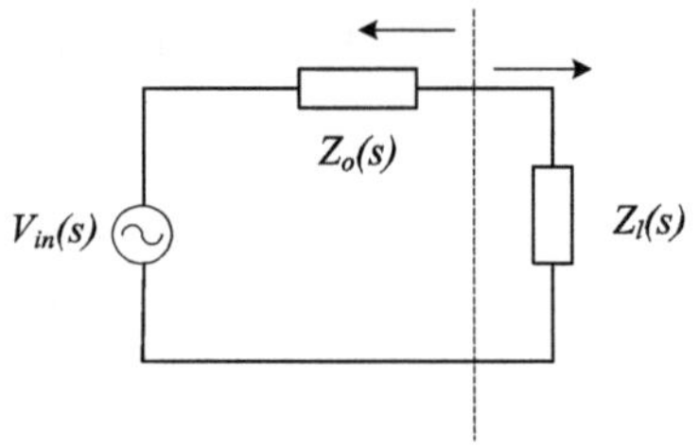

Fig. 3.28 One port circuit block diagram for stability analysis

3.5.2 Half Bridge LC Resonant Converter for Single or Compact Fluorescent Lamp

In developing an electronic ballast for gas discharge-based fluorescent lamps, the circuit is required to generate high voltage V_{ig} for ignition and provide lamp current at normal operating voltage V_{lp}. The ratio of these two voltages, defined here as voltage index $U = V_{ig}/V_{lp}$, not only describes the lamp characteristics, but also relates to the ballast design.

For example, for a fluorescent lamp TL70 F32T8, $V_{ig} = 550\,\mathrm{V}$ and $V_{lp} = 140\,\mathrm{V}$, the voltage index of this lamp is then about $U = 3.92$. For another fluorescent lamp with filament preheating, V_{ig} is reduced to 350 V. With the sample operating voltage $V_{lp} = 140\,\mathrm{V}$, the voltage index of this lamp is about $U = 2.5$.

Comparing with the equivalent dynamic impedance of the lamp, the voltage index is less sensitive to the operating point of the lamp. In some applications, it is convenient to use the voltage index for ballast design as well.

Figure 3.29 shows a high-frequency electronic ballast circuit diagram for a fluorescent lamp [3]. The power stage is a half bridge resonant inverter with two power MOSFETs T_1, T_2 and passive resonant components L_2 and C_5. The lamp is loaded in parallel to the resonant capacitor C_5. In typical design as in Fig. 3.29, the DC blocking capacitors C_3 and C_4 are much greater than C_5 [10]. In steady state, the voltage at the middle of the capacitor divider C_3 and C_4, denoted as V_a, is simply $V_{in}/2$. The voltage at the other side of resonant tank input is the switching node of the half bridge, denoted as $V_b(t)$. When the MOSFETs T_1 and T_2 switch complementarily with 50% duty ratio (minus a small non-overlap time caused phase angle θ_{dead}), the input voltage to the resonant tank $v_{ab}(t)$ $(= V_a(t) - V_b(t))$ is a square wave with peak-to-peak magnitude V_{in} as shown in Fig. 3.30. Correspondingly, its fundamental harmonic component amplitude is $V_{S1} = 2V_{in}/\pi$.

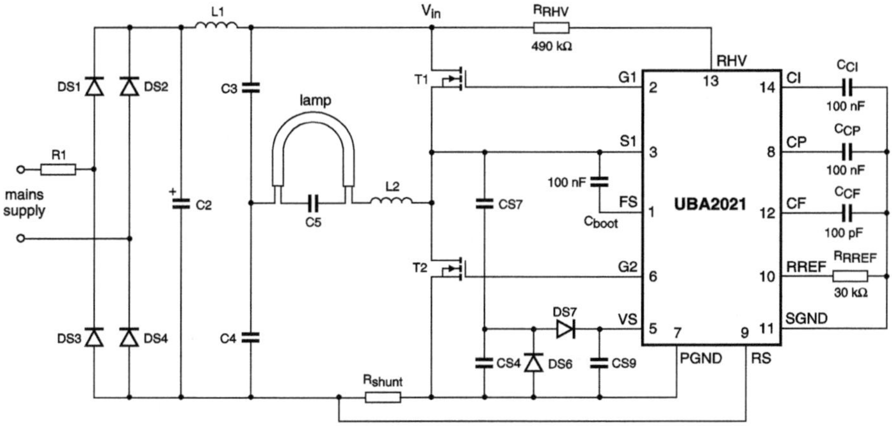

Fig. 3.29 An electronic ballast circuit schematics with IC controller

When the resonant tank consisting of L_2, C_5 and the lamp is excited by the square wave voltage $v_{ab}(t)$ with frequency f, the voltage across the lamp is approximately

$$V_l = |A_v| V_{S1} = \frac{1}{|1 - \Omega^2 + j\Omega/Q|} V_{S1} \tag{3.34}$$

under high-Q condition, where, $\Omega = \omega/\omega_r$, $\omega = 2\pi f$, $\omega_r = 1/\sqrt{L_2 C_5}$ and $Q = R_l/(\omega_r L_2)$. As we know, the lamp equivalent resistance R_l changes dramatically before and after the lamp ignition. For a 13 W compact fluorescent lamp (CFL) with operating voltage $V_l = 75$ V, operating current $I_l = 170$ mA and ignition voltage $V_{ig} = 460$ V, its equivalent impedance R_l could range from 441.2 Ω in steady state to over 1 MΩ before ignition. In a design example [10] with $L_2 = 3.1$ mH, $C_5 = 3.9$ nF and $V_{in} = 320$ V, the voltage across the lamp V_l as a function of the half bridge switching frequency f, as expressed in (3.34), is graphed in Fig. 3.31a for nominal lamp impedance $R_l = 441.2\ \Omega$ and large lamp impedance $R_l = 1$ MΩ before ignition. Clearly, the voltage across the lamp could be conveniently controlled by modulating the switching frequency f, thus pulse frequency modulation (PFM). It is different from the popular PWM method in DC–DC and AC–DC converters.

When adjusting the switching frequency to control the resonant tank output voltage and the lamp power, both frequency regions, above resonant frequency ($f > f_r$) and below resonant frequency ($f < f_r$), could be used. But, in practice, it is beneficial to operate the circuit at above the resonant frequency region. As we see from Fig. 3.31b, when $f > f_r$, the phase angle of the resonant tank input impedance

$$\varphi_{in} = \arctan\left\{\Omega\left[\frac{1}{Q} - Q\left(1 - \Omega^2\right)\right]\right\} \tag{3.35}$$

is positive. This means that inductor current is lagging the voltage $v_{ab}(t)$ as shown in Fig. 3.30. As a consequence, both MOSFETs T_1 and T_2 operate at zero voltage switching (ZVS) condition and lead to much reduced switching losses, especially

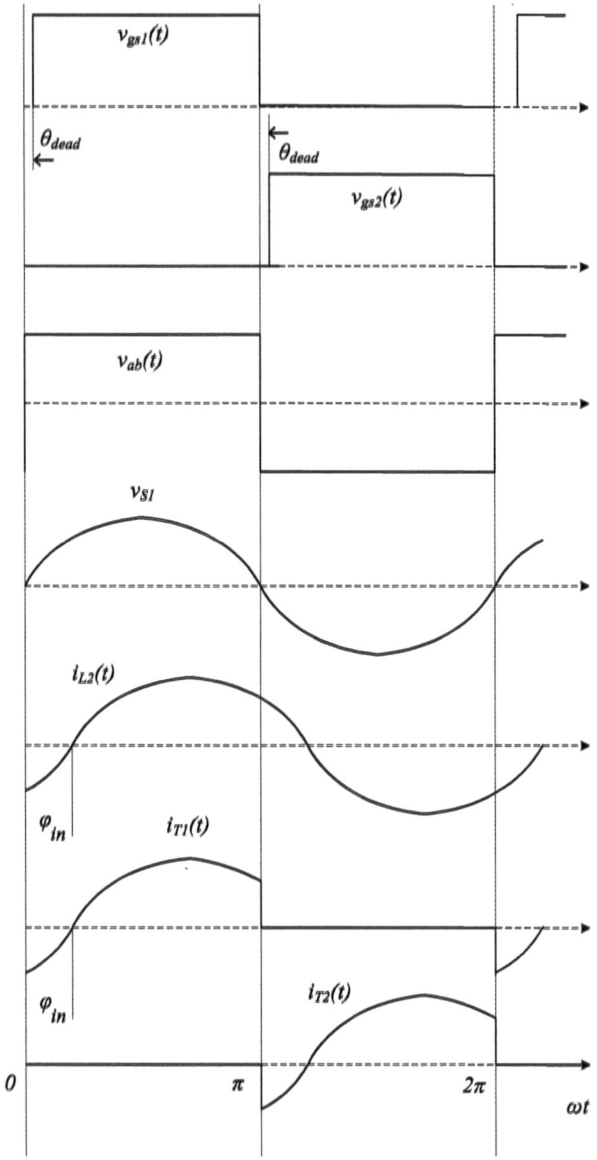

Fig. 3.30 Typical waveforms in an electronic ballast circuit

at high operating frequencies. This feature is widely deployed in commercial high-frequency electronic ballasts.

> In order to extend lamp life and reduce circuit stress, it is common to preheat the filament before ignition.

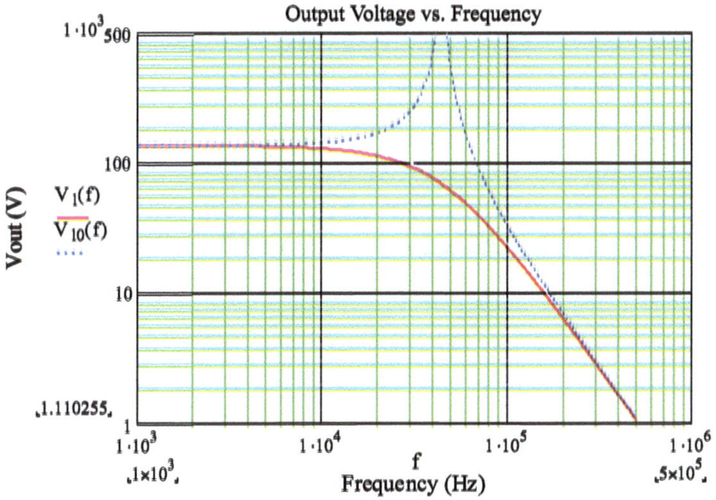

(a) Lamp voltage input impedance

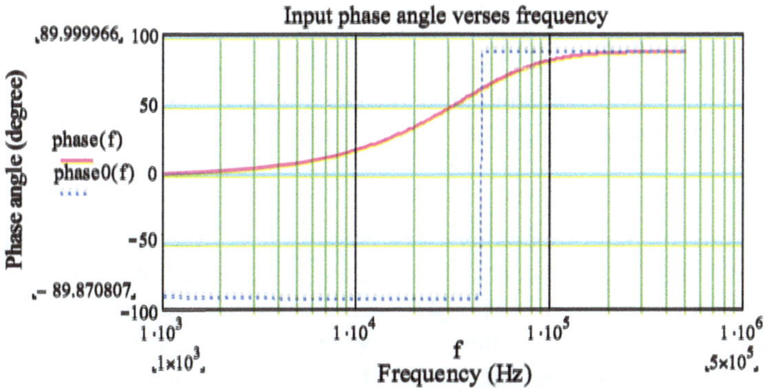

(b) Resonant tank input impedance

Fig. 3.31 Generated input impedances of an electronic ballast circuit design

The control IC in [3] integrated variable-frequency oscillator, adjustable timer and adjustable current settings to complete proper start up, preheat, ignition and normal operation for optimal lamp and ballast performance. Specifically, the control IC starts the circuit operation at frequency f_{start} much higher than the resonant frequency f_r. Correspondingly, the voltage at lamp is very low as shown in Fig. 3.31a. Then, the control IC quickly sweeps frequency down to f_{heat} at time t_1 as shown in Fig. 3.32. At this time, the *LC* resonant tank is seriesly loaded by the lamp filament. The inductor current heats up the filament for a pre-determined time duration $(t_2 - t_1)$. Because the filament resistance is usually small, the voltage across the lamp still follows the resonant tank voltage gain curve in Fig. 3.31a. From time t_2, the control IC further sweeps the frequency down towards f_{ig}. As the voltage across the lamp rises to V_{ig} at frequency f_{ig}, the lamp is ignited. Instantly, the *LC* resonant

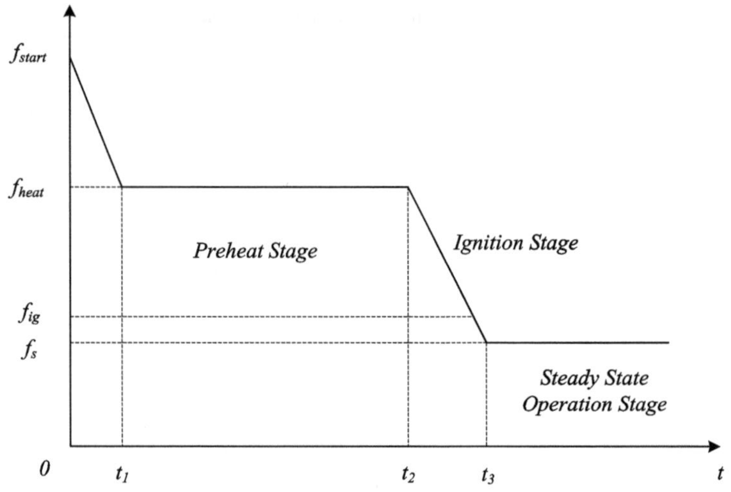

Fig. 3.32 Lamp ballast operation timing diagram

tank is loaded by R_l, and lamp voltage gain follows the lower curve in Fig. 3.31a. After quick adjustment, the lamp enters steady-state operation stage at time t_3 with frequency f_s and rated lamp voltage V_l and current I_l.

> In selecting the lamp operating frequency, some factors need to be considered. For example, it is beneficial to operate the lamp above 20 kHz for lamp efficacy improvement (by 15%) [27]. Also, one needs to avoid possible interference with infrared communication systems (32–36 kHz) commonly used in households and stores. But, it is ideal to keep the frequency below 50 kHz so that its third-harmonic component has least impact on circuit EMC performance.

In the circuit design of Fig. 3.29, the lamp operating frequency f_s is selected at 43 kHz, while the ignition frequency f_{ig} is selected at 50 kHz. With the selected lamp steady-state operating voltage V_l, current I_l, and frequency f_s, lamp ignition voltage V_{ig} and frequency f_{ig}, and the given DC input voltage V_{in}, one could design the resonant inductor L_2 and resonant capacitor C_5 via the following equations:

$$L_2 = \frac{R_l}{2\pi f_s}\sqrt{\left(\frac{\sqrt{2}V_{\mathrm{in}}}{\pi V_l}\right)^2 - \left[1 - \frac{f_s^2}{f_{\mathrm{ig}}^2}\left(1 + \frac{\sqrt{2}V_{\mathrm{in}}}{\pi V_{\mathrm{ig}}}\right)\right]^2}, \tag{3.36}$$

$$C_5 = \frac{1 + \frac{\sqrt{2}V_{\mathrm{in}}}{\pi V_{\mathrm{ig}}}}{(2\pi f_{\mathrm{ig}})^2 L_2}. \tag{3.37}$$

Once the lamp enters steady-state operation, a simple feedback loop could be used to regulate the lamp current and lamp power.

In [3], an even simpler approach, feedforward, is used to adjust the operating frequency so that the lamp power is limited within the maximum allowed value when the input voltage increases. From the control system point of view, it is an open-loop system with nonlinear feedforward. The lamp ballast circuit stability is governed by inequality (3.33).

3.5.3 Half-Bridge LLC Resonant Converter for Multiple Fluorescent Lamps

In office buildings, factories, stores and warehouses, multiple fluorescent lamps are commonly used. To save the cost, it is natural to use a single ballast to gear a group of lamps. This demand leads to the standard electronic ballasts for 2-lamp, 3-lamp and 4-lamp operations in the marketplace. For multiple (e.g. 3–4) lamp applications, the independent lamp operation (ILO) is an ideal feature. To the end users the meaning of multiple lamp ILO could be simply stated as follows

- If a lamp is added, it should be ignited and kept operation without disturbing the operation of other working lamps.
- If a lamp is removed from the socket, other lamps should be kept operation without interruption.

Under the assumption that only a single resonant converter is used to drive these multiple lamps (e.g. the lamp number $m = 4$), the inter-connections of these lamps must be determined so that the above conditions can be met.

For the multiple fluorescent lamp ILO, series connection of the lamps is not suitable because the second condition listed above is not satisfied. If the lamps are directly connected in parallel, the first condition will be violated because the voltage index U of the lamps is not equal to 1. As it turned out, the parallel connection shown in Fig. 3.33 via ballasting capacitors C_1–C_4 at each branch is a viable simple configuration. In this configuration, the output voltage V_o is usually designed to be V_{ig} such that the inserted lamps can be properly ignited. As a result, the voltage index U is related to the power factor $\cos\theta$ of each lamp branch via

$$\frac{1}{\cos\theta} = U. \tag{3.38}$$

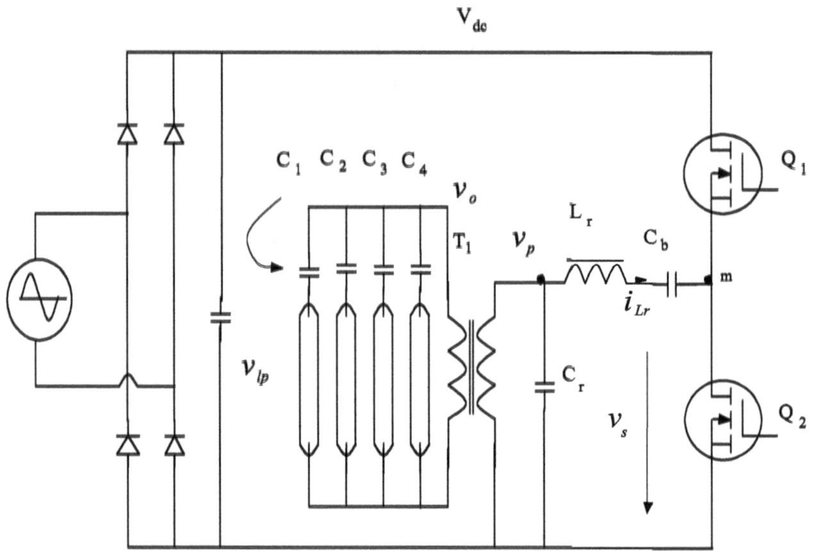

Fig. 3.33 A simplified circuit schematic of the 4-lamp ILO ballast

Here, it is important to point out that for multiple lamp ILO, the output voltage V_o should be kept constant regardless of the number of lamps. However, due to the strong dependence of the resonant tank gain on the lamp load, it is not a trivial task for the voltage-fed half-bridge resonant converter.

In the circuit shown in Fig. 3.33, the LLC resonant tank consists of the resonant inductor L_r, resonant capacitor C_r and the magnetising inductance L_m of the output transformer T_1.

Let n be the secondary side to primary side turns ratio of transformer T_1, R_{lp} be the equivalent lamp impedance, and f_4 be the ballast operating frequency at 4-lamp (full) load. If we assume that the ballasting capacitors $C_1 = C_2 = C_3 = C_4 = C_{\text{lp}}$, the required ballasting capacitor can be calculated by

$$C_{\text{lp}} = \frac{1}{2\pi f_4 R_{\text{lp}} \sqrt{U^2 - 1}}. \tag{3.39}$$

With this, we can find the reflected equivalent capacitance to the primary side of T_1 as

$$C_e = mn^2 \left(1 - \frac{1}{U^2}\right) C_{\text{lp}} \tag{3.40}$$

and the reflected equivalent resistance as

$$R_e = \frac{U^2}{mn^2} R_{\text{lp}}. \tag{3.41}$$

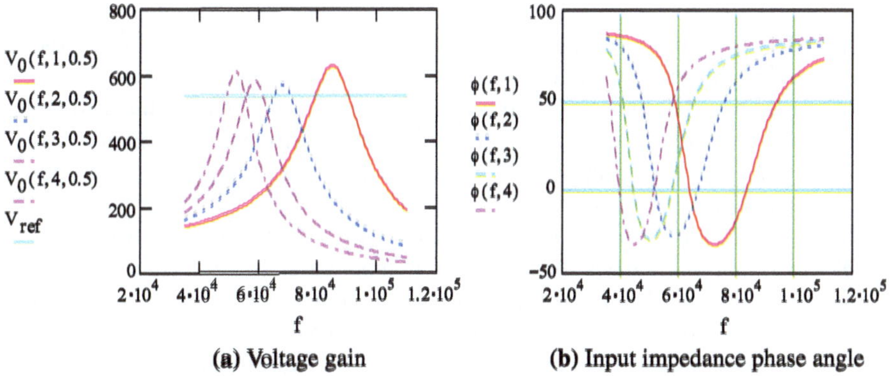

Fig. 3.34 Characteristics of the LLC resonant tank with different number of lamp loads

By using the fundamental component approximation approach, the static voltage gain of the LLC resonant tank loaded with lamps is derived as

$$A_v \triangleq \left|\frac{V_p}{V_s}\right| = \frac{1}{\frac{L_r}{L_s}\sqrt{[1-(\frac{\omega_s}{\omega_r})^2]^2+(\frac{\omega_s}{Q\omega_r})^2}}, \tag{3.42}$$

where $V_p = V_o/n$, $V_s = \sqrt{2}V_{dc}/\pi$, $L_s = L_m L_r/(L_m + L_r)$, $\omega_r = 1/\sqrt{L_s(C_e + C_r)}$ and $Q = R_e/(\omega_r L_s)$. Furthermore, if we define

$$\omega_{r_2} \triangleq \frac{1}{\sqrt{L_m(C_e + C_r)}}, \qquad Q_2 \triangleq R_e/(\omega_{r_2} L_m),$$

the input impedance of the LLC resonant converter can be found as

$$Z_{in}(j\omega) = j\omega(L_r + L_m)\frac{1-(\frac{\omega}{\omega_r})^2 + j\frac{\omega}{Q\omega_r}}{1-(\frac{\omega}{\omega_{r_2}})^2 + j\frac{\omega}{Q_2\omega_{r_2}}}. \tag{3.43}$$

Therefore, the phase angle of the resonant tank input impedance becomes

$$\phi_{in}(j\omega) = \frac{\pi}{2} + \arctan\frac{\frac{\omega}{Q\omega_r}}{1-(\frac{\omega}{\omega_r})^2} - \arctan\frac{\frac{\omega}{Q_2\omega_{r_2}}}{1-(\frac{\omega}{\omega_{r_2}})^2}. \tag{3.44}$$

With the lamp number m as a parameter, the plots of the voltage gain and the input impedance phase angle versus frequency are shown in Fig. 3.34. As we can see from Fig. 3.34a, the voltage gain curve shifts horizontally as the lamp number changes. In other words, the resonant frequency of the resonant tank loaded with parallel lamps is strongly load dependent.

In order to keep a constant output voltage, say $V_o = V_{\text{ref}} = 550\,\text{V}$, as the dotted line shown in Fig. 3.34a for different lamp numbers, the circuit operating frequency should be adjusted accordingly both at the steady-state and the transient processes. In particular, to prevent the extinction of other lamps during lamp addition or removal process, a fast frequency change to the proper operating point is required.

From Fig. 3.34a, it is noted that for a given $V_{\text{ref}} = 500\,\text{V}$, there are two possible operating frequencies for each lamp number case. The ZVS requirement, together with Fig. 3.34b, excludes the operation at the curve on the left-hand side of the peak point. It is common to use a voltage-controlled oscillator (VCO)-based simple frequency control in the above resonant frequency region. The control logic would be: If the output voltage V_o drops, the switching frequency should be reduced so that the output voltage can be pulled up. If the output voltage V_o jumps up, the switching frequency should be increased so that the output voltage can be pulled down. While this control scheme works well for single lamp operation as in Fig. 3.29 and for the lamp addition process in Fig. 3.34 with fast frequency sweeping, it fails in the lamp removal process.

The reason is explained as follows. Referring to Fig. 3.34a, let us assume that three lamps are operated by the converter and the operating point on the corresponding $m = 3$ curve is in the above resonant frequency region. If one lamp is removed from one parallel branch, the resonant tank gain curve instantly moves to the corresponding $m = 2$ gain curve. At this moment, the circuit operating frequency could be unchanged due to the limited controller and VCO bandwidth. Consequently, the new operating point on the two-lamp gain curve is on the left side of the resonant peak. With the reduced output voltage V_o as the lamp removal seen in Fig. 3.34a, the VCO-based controller will reduce the switching frequency in the hope of increasing the output voltage. However, because the resonant tank operates on the wrong side of the curve, this unfortunately leads to further decrease of V_o until all the lamps are completely extinguished. This malfunction prevents the simple VCO-based frequency control method from applying to the relatively complex multiple lamp parallel ILO case.

In the process of investigating proper control methods for voltage-fed half-bridge resonant converter for multiple lamp ILO, we have the following observation.

If combining Fig. 3.34a and Fig. 3.34b in the sense that relating the resonant tank voltage gain to the input phase angle with the switching frequency as a parameter for each lamp number case, we have the plot in Fig. 3.35. It is interesting to observe that the widely spread gain and phase curves for different lamp numbers in Fig. 3.34 are "merged" in Fig. 3.35. It tells us that if a constant phase angle is kept for different m, the output voltage V_o is maintained within a small limit band. For example, if we keep 25° phase angle for different m, the output voltage V_o is between $550\,\text{V}$ to $645\,\text{V}$ for $m = 1, 2, 3, 4$. As the phase angle increases, the limit band becomes smaller. On the other hand, if one wants to keep a constant $V_o = 550\,\text{V}$ for all m,

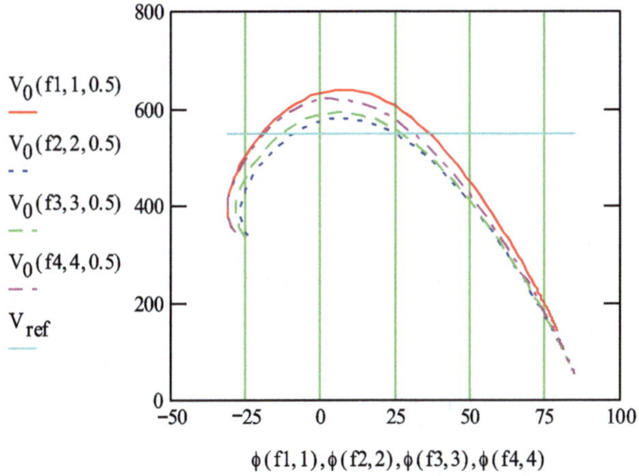

Fig. 3.35 Voltage gain vs. input phase angle for different lamp number cases

the required phase angle region would be just between 25° to 37.5°. By controlling phase angles, the output voltage can be adjusted to fulfil different requirements.

It is this basic observation that leads to the so-called phase-shift-based control method [11, 12]. Here, it should be pointed out that, for a given phase angle, even though the curves for different m in Fig. 3.35 are close, the corresponding frequencies for each m are still far apart.

As we know, the phase angle ϕ_{in} determined by (3.44) is actually the phase difference between $v_{ab}(t)$ and $i_{L_r}(t)$. Normally, when the MOSFETs Q_1 and Q_2 are switched with certain frequency, the resulted $v_{ab}(t)$ will drive the resonant tank and lead to $i_{L_r}(t)$ with phase lagging ϕ_{in}. If reverse the above process from a control point of view, we will have the following procedures:

1. Sense the resonant tank current signal $i_{L_r}(t)$;
2. Shift the sensed signal with leading phase angle ϕ_{in};
3. Drive Q_1 and Q_2 with an in-phase logic.

As a consequence, the phase angle between $v_{ab}(t)$ and $i_{L_r}(t)$ (say, $\phi_{in} = 25°$) is determined in the control process. In this case, the circuit operating frequency is forced to a point where the resonant tank produces the exact 25° phase lag solved from (3.44).

By regulating the pre-fixed phase angle value in the control process, one can indirectly control the circuit operating frequency and therefore the voltage gain. Figure 3.36 shows a functional block diagram of the phase-shift-based control scheme.

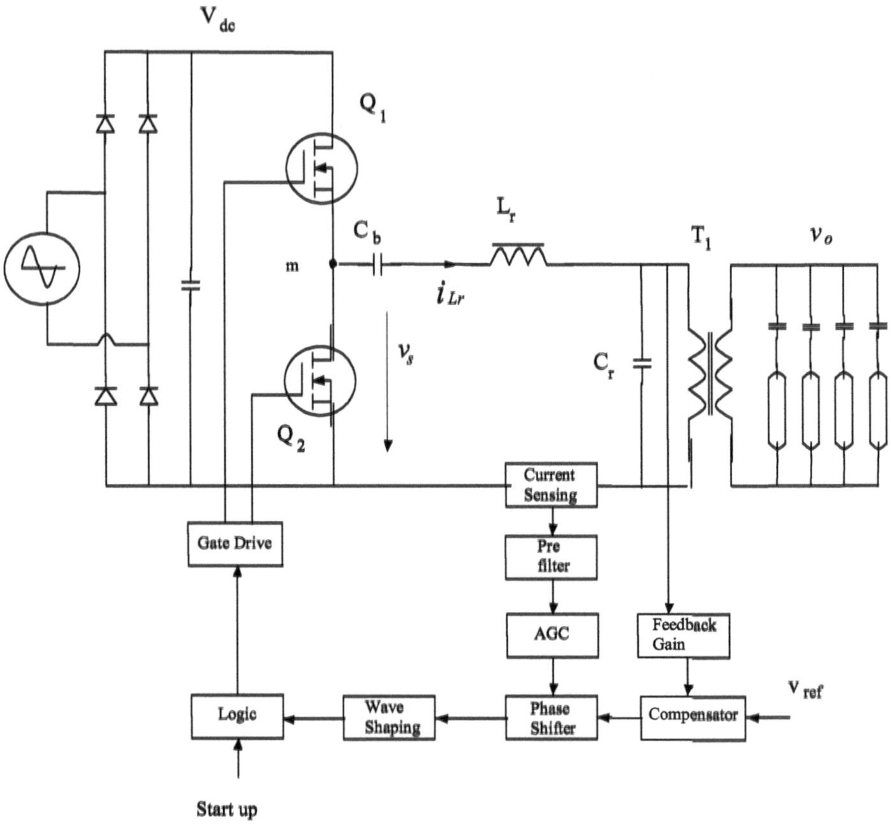

Fig. 3.36 A functional block diagram of 4-lamp ILO using phase-shift-based control

In applying the phase-shift-based control method to the multiple lamp ILO, one sets a reference signal to phase angle, and the control scheme rapidly changes the operating frequency such that the resulted phase angle is maintained almost constant. Thus, the resonant tank output voltage only has small variations during the lamp addition and removal transient processes. As a result, the multiple lamp ILO is achieved.

A commercial control IC is shown in Fig. 3.37 [19], where the resonant inductor current $i_{L_r}(t)$ is sensed from the current sense resistor RCS. The IC has an internal VCO with its frequency controlled by the voltage at VCO pin. At normal operation, the voltage at VCO pin is constant (controlled by the DIM pin) that generates a proper switching frequency and leads to a preset phase angle. The phase angle reference signal v_{ref} is shown in Fig. 3.38, while the sensed current signal from RCS is translated to a logic signal v_{fb} via internal comparator. The inductor current zero across timing information is carried in v_{fb}. When there is a timing different between

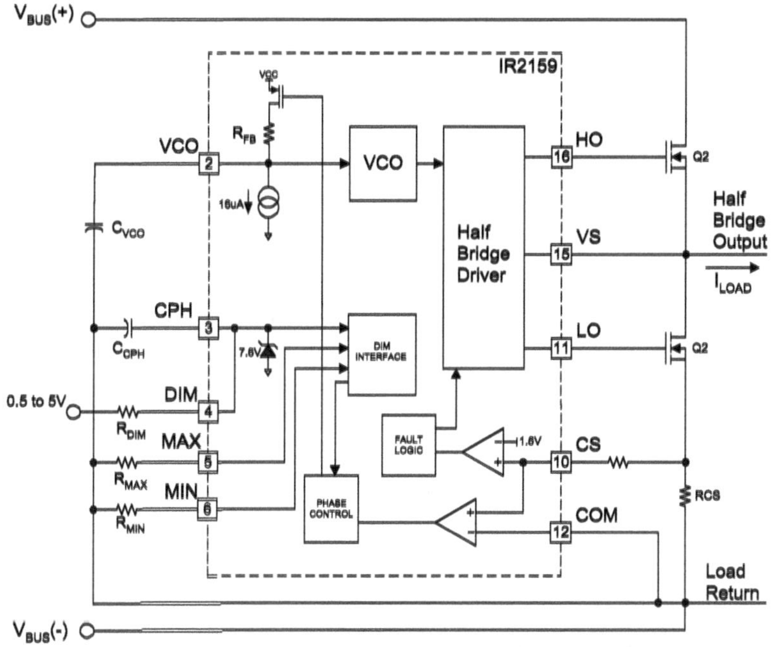

Fig. 3.37 A commercial resonant inverter control IC with phase control

v_{ref} and v_{fb}, an error pulse v_{err} is generated as depicted in Fig. 3.38. Such error pulse is then used to turn on the control PMOS and allow VCC to charge C_{vco} via the resistor R_{fb}. This immediately changes the VCO pin voltage and therefore the inverter switching frequency. Consequently, the input phase angle is adjusted so that it follows the falling edge of v_{ref}. This approach could be used to control dimmable fluorescent lamp as the IC was originally intended to.

> The phase-shift idea is general and is readily applicable in other resonant converter applications, for example, switching-mode power supplies [12].

3.6 Controllers for Light-Emitting Diode Drivers

Light-Emitting Diodes (LED) have been invented and used for many years. Recent technology progress expanded their applications from simple signal indicators to efficient light sources for illumination.

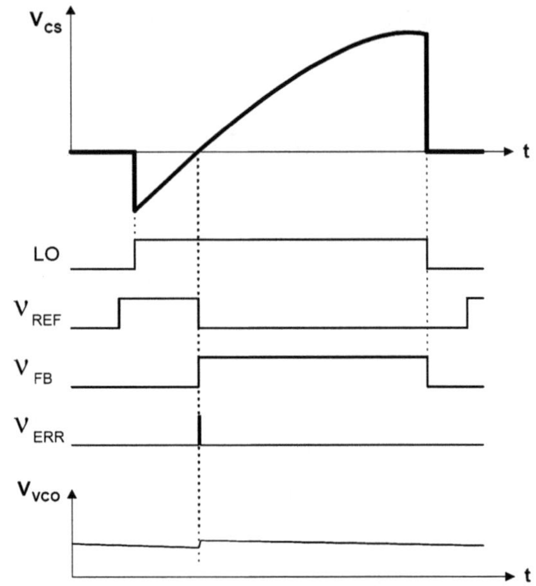

Fig. 3.38 Timing diagram of the phase control

3.6.1 LED Driver Requirement

Today, most white, green, cyan, blue and royal blue colour LEDs are based on InGaN materials while most red, red-orange and amber colour LEDs are made with AllanGap materials. The two different materials have slightly different v–i characteristics as shown in Fig. 3.39 [47]. Furthermore, the LED voltage exhibits negative temperature coefficient ($\approx -2\,\text{mV/°C}$ for both InGaN and AllnGap materials) as shown in Fig. 3.40.

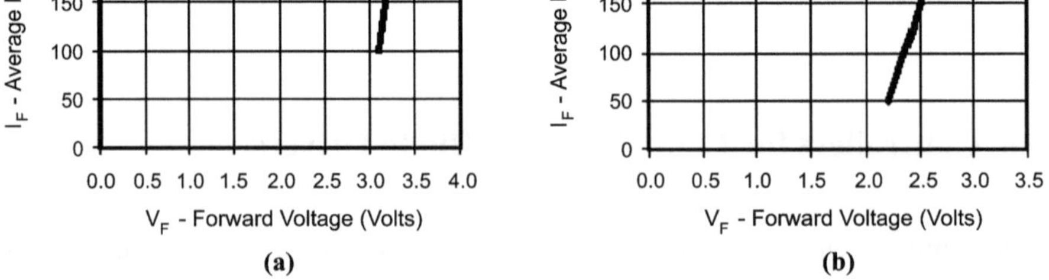

Fig. 3.39 Typical v–i curves of LED with InGaN (a) and AllanGap (b) materials

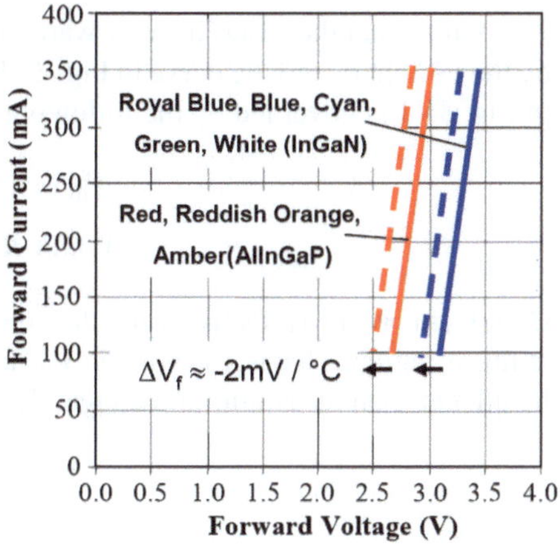

Fig. 3.40 Typical *v*–*i* curves of LED over the temperature

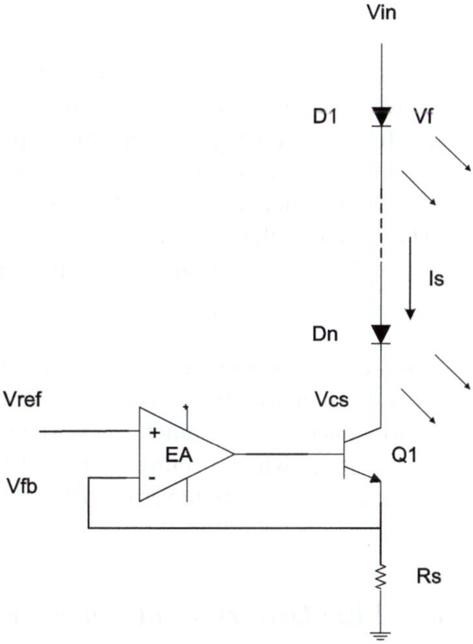

Fig. 3.41 A simple current source to drive an LED string

Due to the negative temperature coefficient of the LED forward voltage, it is not suggested to simply drive LEDs with voltage sources or connect LEDs in parallel without ballasting resistors. In practice, LEDs are usually driven with current sources. Figure 3.41 shows a simple current source structure and its connection to drive a string of n LEDs in series with current I_s.

Assume that the LEDs have forward voltage V_f at given current I_s as determined by the v–i characteristic curve in Fig. 3.39 and Fig. 3.40. The operation of the circuit in Fig. 3.41 is governed by the following basic equations:

$$V_{\text{in}} = nV_f + V_{\text{cs}}, \tag{3.45}$$

$$V_{\text{ref}} = R_s I_s. \tag{3.46}$$

Clearly, in order to effectively regulate the LED current, transistor Q_1 should operate at linear region. That is, the voltage V_{ce} across the collector and emitter of Q_1 should be greater than its saturation voltage $V_{\text{ce,sat}}$, i.e.

$$V_{\text{ce}} > V_{\text{ce,sat}}. \tag{3.47}$$

Based on (3.45) and (3.46), inequality (3.47) is further expressed as

$$V_{\text{in}} - nV_f - R_s I_s > V_{\text{ce,sat}}. \tag{3.48}$$

The inequality in (3.48) is a basic LED driver requirement, which guarantees a proper regulation of LED current I_s. Because the requirement is an inequality, for a given type and number of LEDs and transistor Q_1, any input voltage V_{in} higher than $nV_f + R_s I_s + V_{\text{ce,sat}}$ will meet the requirement. But, in engineering practice, the minimal V_{in} that satisfies (3.48) is the optimal solution. The reason is that any excessive V_{in} voltage over the LED required voltage nV_f will become power losses in the circuit.

Such excessive voltage is V_{cs} in Fig. 3.41 and in (3.45). It is clear that $V_{\text{cs}} = V_{\text{ce}} + V_{\text{ref}}$. In many LED driver controllers, V_{ref} is internally set at sub-1 V range in order to reduce V_{cs}. The other excessive voltage drop is then put on Q_1. Minimising the losses in Q_1 while meeting the basic LED driver requirement in (3.48) is a basic requirement in developing various LED drivers and controllers.

3.6.2 LED Drivers and Controllers

Based on the simple current source structure in Fig. 3.41, one can derive its topological equivalent structure as in Fig. 3.42 by interchanging the position of the serially connected LEDs and the transistor Q_1.
It is interesting to observe that the diagram in Fig. 3.42 reveals a LED driver circuit based on a simple low dropout (LDO) linear regulator with passing transistor Q_1 as marked with the block PS. Furthermore, it should be pointed

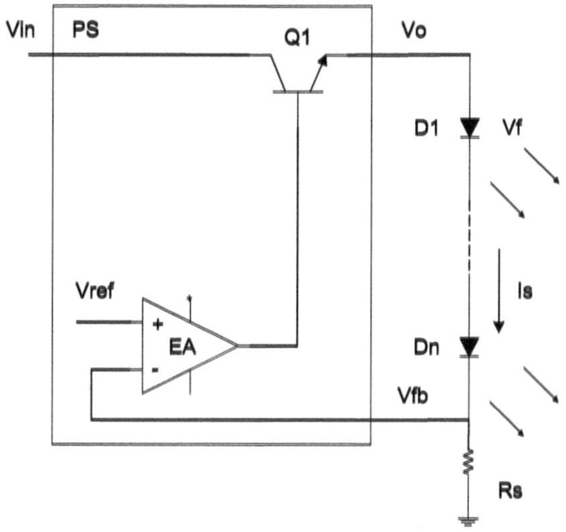

Fig. 3.42 A basic structure to drive an LED string

out that, even though very simple, the basic structure in Fig. 3.42 is shared in many commercial LED drivers and controllers for different applications. For example, in battery-powered portable electronics applications, V_{in} could be less than the total LED voltage drops at V_o. Switched capacitor-based charge pumps and inductor-based boost converters are then used in the PS section. In off-line based LED general illumination applications, flyback and half-bridge converters are commonly used in the PS section. In the following, we will discuss some typical LED drivers and controllers in different categories.

3.6.2.1 $V_{\text{in}} > V_o$, LDO Regulator Based

Due to efficiency reason, LDO is mostly limited to applications where V_{in} is slightly greater than V_o. One common application area is driving LED banks for LCD backlight in battery-powered handheld devices. Today, a single-cell Lithium Ion battery has typical voltage 3.6 V [34]. The emerging high-efficiency LED has typical forward voltage 2.9 V [33]. LDO-based LED driver becomes a natural choice with its reasonable efficiency, simplicity, smaller size and lower cost. Figure 3.43 show two commercial integrated LED drivers in this application. It is noted that the solution in Fig. 3.43a [35] used the LDO structure as in Fig. 3.42, while the solution in Fig. 3.43b [4] used the original current source structure as in Fig. 3.41. Even though they are topologically equivalent, there are some implementation differences in these two solutions.

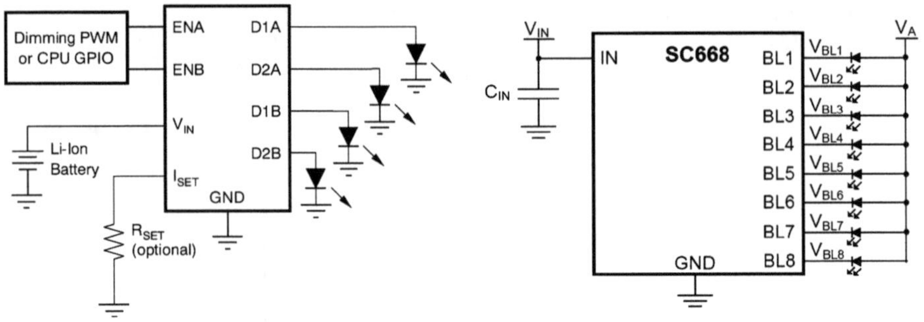

Courtesy Texas Instruments
(a)

Courtesy Semtech
(b)

Fig. 3.43 Two types of LDO-based LED driver

3.6.2.2 $V_{in} < V_o$, Charge Pump Converter Based

With the improvement of high-brightness LED technologies, LEDs are replacing xenon lamps as flash light in smart phone, camera and other portable electronic devices. During the flash, the LED draws substantial amount of current from the battery, which could lead to input voltage sag and cause system abnormal operation. To guarantee proper operation at all cases, a charge pump is normally used to provide sufficient energy and headroom for such pulsing application.

One early charge pump LED driver is shown in Fig. 3.44 [14], while many advanced solutions could be found in the marketplace. The charge pump solution is based on switched capacitor converter governed by the charge balance principle. Its output voltage usually has a fixed relation with its input voltage. For variable output voltages, multiple modes such as $\times 1$, $\times 1.33$, $\times 1.5$, $\times 2$ modes are implemented as in [15] and switched over to achieve corresponding output voltages. If more granularity of the charge pump output voltage is needed for more accurate LED current regulation, a simple way of cascading an LDO with a charge pump may be used. Due to the limitation of output power throughput, charge pump-based LEDs drivers are mainly deployed in portable device applications.

3.6.2.3 $V_{in} > V_o$, Buck Converter Based

When V_{in} is much greater than nV_f and the required LED current is higher, the LDO structure PS in Fig. 3.42 is commonly replaced by inductor-based switching-mode buck converters as shown in Fig. 3.45 [6].

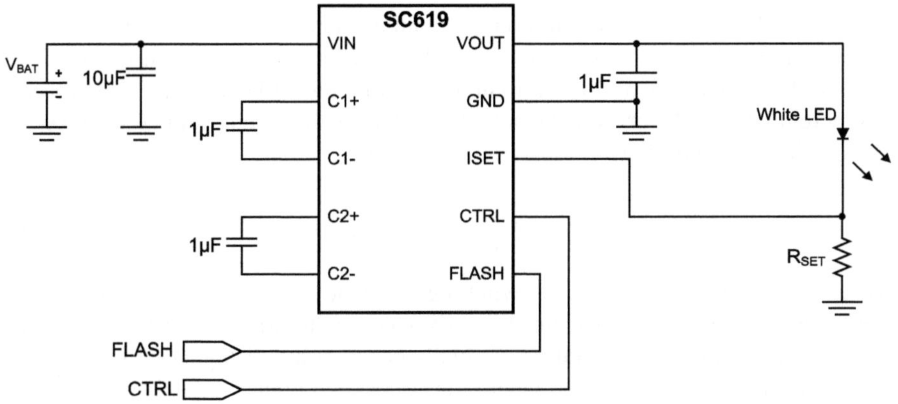

Fig. 3.44 A charge pump-based LED driver

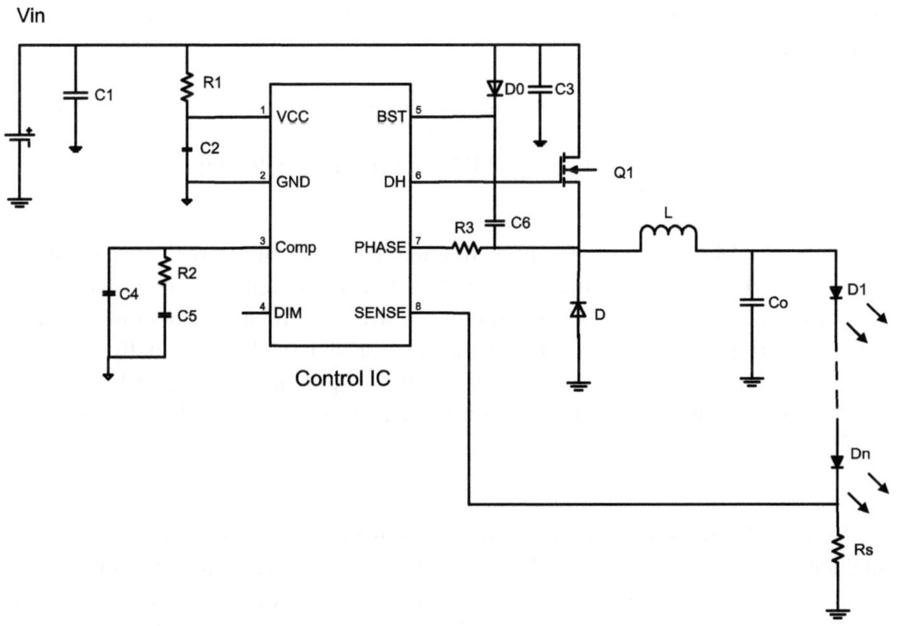

Fig. 3.45 Buck converter-based LED driver

In this type of converters, there are various configurations and control methods to achieve LED current regulation. One straightforward method is to simply adopt a DC–DC voltage regulator and convert it to a current source. By replacing the traditional voltage feedback path with the structure in Fig. 3.45, the output voltage is adjusted accordingly to regulate the LED current. The LED current is programmed by the current sensing resistor via $I_{\text{LED}} = V_{\text{SENSE}}/R_s = V_{\text{ref}}/R_s$ as in (3.46). In many cases, the internal control schemes (e.g. constant-frequency voltage mode, current mode or average current mode) remain unchanged when converting the constant output voltage regulation to constant output current regulation in Fig. 3.45. One of the advantages of this configuration is that the LED current could be accurately regulated with very small ripple. But, such small current ripple is achieved by using a large output filtering capacitor with value

$$C_o \gg \frac{1}{2\pi f_s(nr_{\mathrm{ac}} + R_s)}, \tag{3.49}$$

where f_s is the power converter switching frequency, n is the number of LEDs in series, r_{ac} is the LED equivalent AC resistance. Such large output filter capacitor coupled with higher voltage rating gradually becomes a limiting factor for the commercialisation of LED lighting, where lower cost and higher reliability is required. In many popular DC–DC converter topologies, such as boost, buck–boost, SEPIC and flyback converters, the output currents are pulsing type. The output filter capacitor is critical to reduce the output voltage ripple. In contrast, in buck converter, the inductor is connected to the output terminal. The output current ripple is simply the inductor current ripple

$$\Delta i_L = \frac{V_{\mathrm{in}} - V_o}{L} t_{\mathrm{ON}}, \tag{3.50}$$

where t_{ON} is the ON-time of the top MOSFET at each switching period.

When the LED allows some high-frequency current ripple without negative visible effect and performance degradation, it is possible to reduce the output filter capacitor size or even remove it. Recent investigation showed that current ripple up to ±20% of its average current is allowed for the LEDs tested [49]. This leads to the removal of the output capacitor C_o in Fig. 3.45 and the use of the output inductor to directly drive the LED string as in Fig. 3.46.

As a result, the sensed voltage across the current sensing resistor is $R_s i_L$ which includes a ripple component $R_s \Delta i_L$. With the sensed ripple signal, the ripple-based control methods [43] could be used to regulate the LED average current. The commercial controller in Fig. 3.46 [6] used a valley current constant ON-time control method to achieve LED average current regulation. A simplified control circuit block diagram is shown in Fig. 3.47. Here, the buck converter operates at CCM with voltage gain

$$\frac{V_o}{V_{\mathrm{in}}} = D = t_{\mathrm{ON}} f_s. \tag{3.51}$$

When the sensed signal $R_s i_L$ is less than the internal reference signal V_{ref}, the output of the comparator 1 goes "high" and sets the RS latch. The MOSFET Q_1 turns on and inductor current ramps up. At the same time, the switch SW is "open", and a current source with magnitude αV_{in} charges an internal timing capacitor C_t from 0 V.

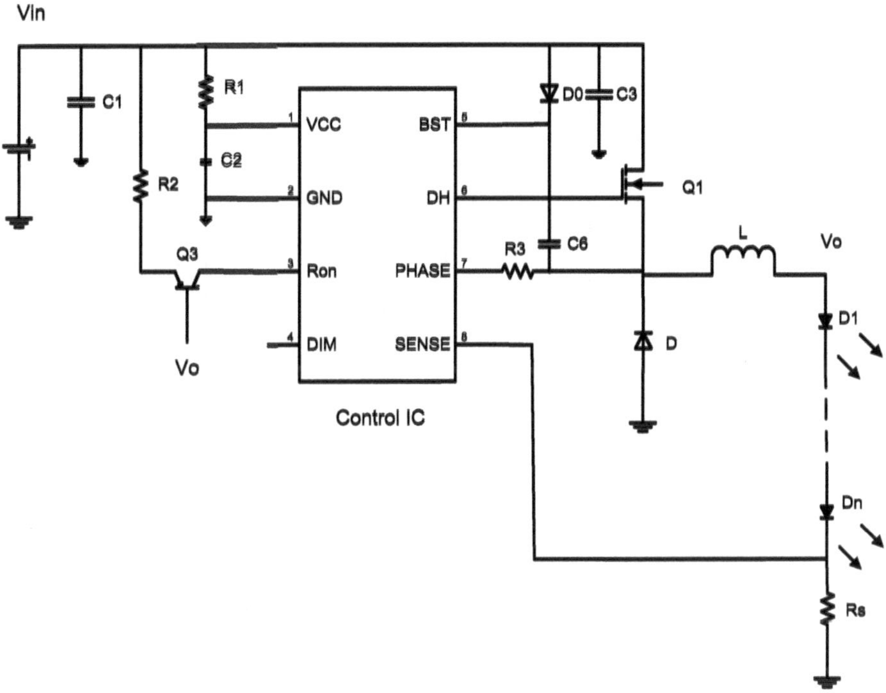

Fig. 3.46 Buck-based LED driver without output capacitor

When the voltage across C_t reaches a reference voltage level V_{th}, the comparator output turns "high", resets the RS latch and turns off Q_1. The Q_1 conduction time duration is determined as

$$t_{\text{ON}} = \frac{C_t V_{\text{th}}}{\alpha V_{\text{in}}}. \tag{3.52}$$

Substituting (3.52) into (3.51), the circuit operating frequency is then found as

$$f_s = \frac{\alpha V_o}{C_t V_{\text{th}}}. \tag{3.53}$$

Clearly, when the LED current is regulated and the converter output voltage V_o is maintained constant, (3.53) shows that the circuit operates at a constant frequency. If one wants to further make the frequency be independent of the converter output voltage, one could use a sensed output voltage V_o signal as V_{th}. The commercial controller in Fig. 3.46 uses a simple external PNP transistor Q_1 to complete this action. As a result, under the CCM, the circuit in Fig. 3.46 operates at constant frequency, while its ON time could be slowly modulated by V_{in} and V_o to regulate the LED current. It is noted that, with the LED current being the inductor current having ripple component, the control circuit in Fig. 3.46 regulates the LED current at valley point. It is maintained at $V_{\text{SENSE}}/R_s = V_{\text{ref}}/R_s$. Along with the current

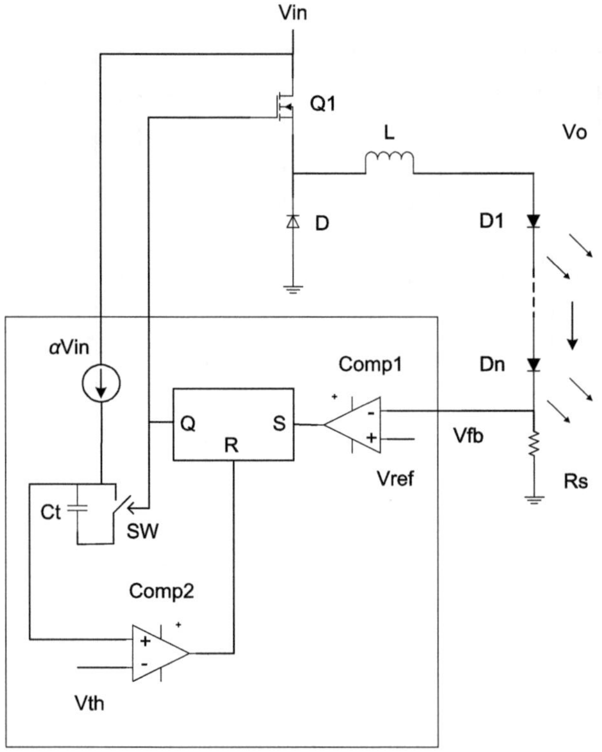

Fig. 3.47 Simple control circuit block diagram of buck-based LED driver

ripple in LEDs, the average LED current is regulated at

$$I_{avg} = \frac{V_{ref}}{R_s} + \frac{\Delta i_L}{2} = \frac{V_{ref}}{R_s} + \frac{C_t V_{th}}{2\alpha L}\left(1 - \frac{V_o}{V_{in}}\right). \quad (3.54)$$

Clearly, the current ripple is a function of the inductor value, input voltage V_{in} and output voltage V_o. So is the LED average current. This is one drawback of this simple control scheme where no external compensation components are needed. In practice, it is common to allow $\Delta_{i_L}/2$ to be about $\pm 20\%$ of I_{avg} in order to use reasonable inductor size and operate at reasonable frequency. In [6], detailed circuit design steps and tradeoffs are provided.

In adjusting LED light output, average LED current needs to change accordingly. Some ways to adjust the average LED current are amplitude modulation, pulse-width modulation with low frequency and the combination of both. Amplitude modulation is simple and straightforward. However, due to the possible colour shift and narrower adjustable range, this method is getting less usage. In contrast, the low frequency (e.g. 0.2–1 kHz) based PWM dimming method becomes more popular. It allows wide dimming range, uniform light output for multiple LEDs and better circuit efficacy.

In commercial controllers, a TTL/MOS-compatible logic input pin is usually used to receive the PWM dimming signal. In the controller of Fig. 3.46, it is the DIM pin. When the DIM signal is "high", the buck converter switches at a fixed frequency and regulates the LED current. When the DIM signal is "low", the buck converter stops switching, and LED current drops to zero as the inductor current diminishes. During this period, only the power stage switching stops. Other supporting circuitry inside IC (e.g. bandgap, V_{cc}, driver, etc.) are still active so that the buck converter could instantly wake up and enter to steady-state operation when the DIM signal goes back to "high" again. The LED current rise time, fall time and the delay time from DIM to the MOSFET ON/OFF determine the dimming range (100%/min D_{PWM}) at given PWM dimming frequency f_{PWM}. Usually, f_{PWM} is a few orders of magnitude lower than f_s.

As we know, for the buck converter, the switching MOSFET Q_1 is floating above the ground. Its proper switching requires a bootstrap circuitry with voltage greater than V_{in}. For lower input voltage applications, e.g. <28 V as in Fig. 3.46, it is sustainable. When a large number of high-power LED strings needs to be driven from off-line for general illumination, the voltage stress on the control IC will render to higher-cost and even impractical solutions. To solve this issue, an equivalent circuit topology to Fig. 3.46 is developed in Fig. 3.48. Here, the MOSFET Q_1 and its associated driver are referenced to ground while the LEDs are floated. The control structure, method and their performance remain close to the case in Fig. 3.46. Similar discussions also apply to the cases with and without the output capacitor C_o.

3.6.2.4 $V_{in} < V_o$, Boost Converter Based

When more LEDs are used for illumination, series connection is preferable due to its nature current sharing characteristics. As the number of LEDs in series increases for higher light lumen output, the total string LED forward voltage nV_f could be greater than the input voltage V_{in}. In this case, a boost converter such as in Fig. 3.49 is needed.

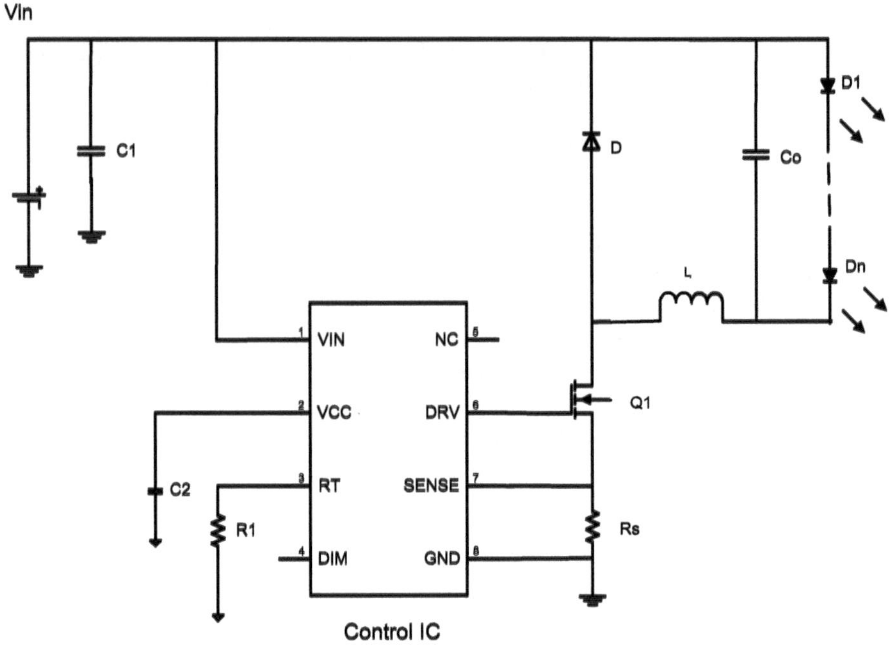

Fig. 3.48 Buck based LED driver with MOSFET referenced to ground

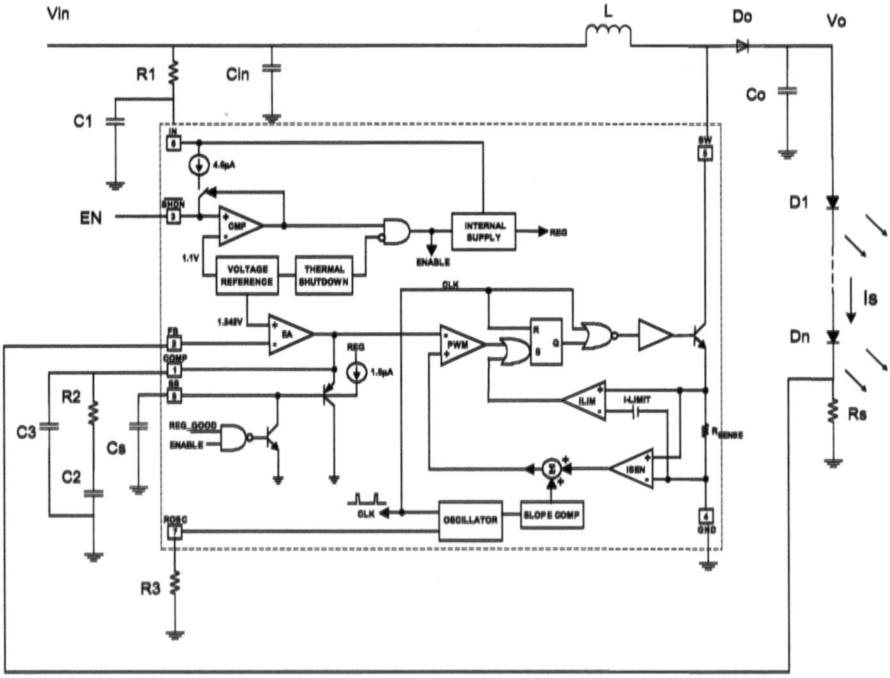

Fig. 3.49 A simple boost-based LED driver

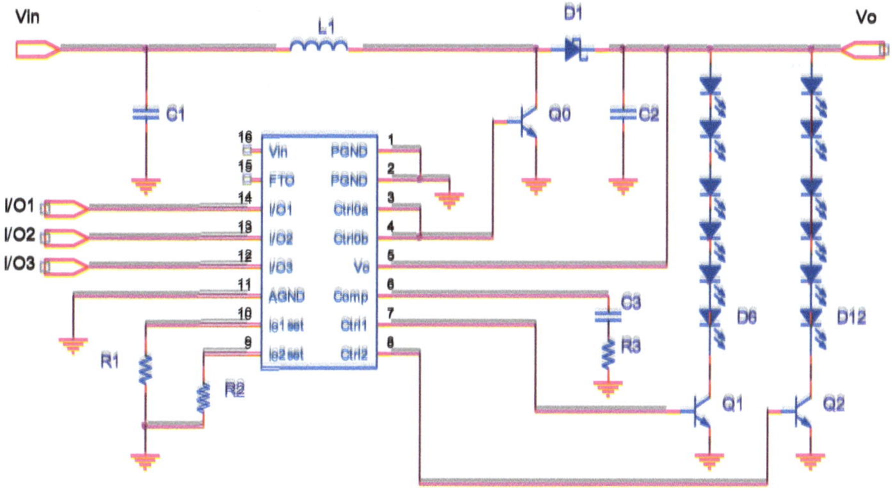

Fig. 3.50 A boost-based LED driver with PWM dimming switches

As we know, the output current of the boost converter is pulsing type even if it operates at CCM. An output capacitor C_o is necessary to filter the pulsing current and provide steady current to the LED string with negligible ripple. Consequently, the sensed LED current signal via R_s in Fig. 3.49 is almost DC type. The constant frequency voltage mode or current mode PWM controllers used in DC–DC converters could be readily applied for LED drivers. The LED driver in Fig. 3.49 is based on a constant-frequency current-mode PWM regulator [1]. In further developing commercial controllers for boost converter-based LED drivers, some special considerations are needed. First, as we see from Fig. 3.49, the LED current is regulated at V_{ref}/R_s. The associated power loss in the current sensing resistor is V_{ref}^2/R_s. It is clear that smaller reference voltage value V_{ref} leads to less power losses in the current sensing resistor. But, V_{ref} should not be too small to prevent reliable LED current sensing due to reduced signal-to-noise ratio. Many commercial controllers set the value of V_{ref} in the range 250–750 mV.

Secondly, in the boost converter topology, there is a DC path from V_{in} to V_o. Even if the controller disables the PWM switching, V_o terminal still has a voltage close to V_{in}, which is then applied to the anode of the LED string. In the cases that $V_{in} \ll n V_{knee}$, the LED could be considered completely OFF. However, when $V_{in} \approx n V_{knee}$, leakage current in the LED string gradually develops. Even with a few microamps leakage current, the LED string could be seen to glow with a faint light at certain ambient light condition. To prevent this from happening, one could insert a switch in the path from V_{in} to V_o and to ground path so that the DC path could be removed whenever needed. A simple way is to add a switch from the cathode of LED string to ground as shown in Fig. 3.50. Interestingly, the addition of the switch allows a convenient implementation for PWM dimming of the LED strings. When operating the switches Q_1 and Q_2 at certain PWM dimming frequency f_{PWM} with duty ratio D_{PWM}, the LED current will become a sequence of square waves. When

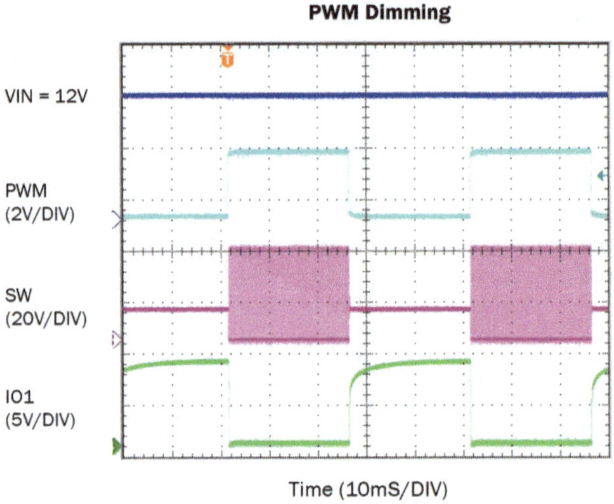

Fig. 3.51 Circuit waveforms of boost-based LED driver with PWM dimming

Q_1 and Q_2 are OFF during the PWM dimming OFF time, the LED string current is zero. If the boost converter continues to switch during this time, the output voltage will be over boosted and may hit over voltage protection (OVP) threshold. So, it is necessary to turn off the boost converter PWM switch Q_0 when Q_1 and Q_2 are OFF. This synchronisation is implemented in most commercial controllers. Some typical waveforms are shown in Fig. 3.51 with $V_{\text{in}} = 12\,\text{V}$, $V_{\text{out}} = 36\,\text{V}$ for 10 of 50 mA LEDs in series.

The state-of-the-art highly integrated LED drivers further use the series switches to develop local current sources for each string. That is, the switches Q_1 and Q_2 in Fig. 3.50 are used to regulate the corresponding string LED current during the PWM dimming ON time and turn the corresponding LED string OFF during the PWM dimming OFF time. This basically enables multi-string independent operation with the driving structure in Fig. 3.50.

Figure 3.52 depicts a commercial high-efficiency integrated LED driver for up to four strings of 150 mA LEDs [28]. Depending on the LED number and LED array configuration, the user could select one to four current sources I_{o_1}–I_{o_4}. The current sources could be programmed from a fraction of miniamps to 150 mA with the resistor connect from I_{oset} pin to ground. The current matching among the current sources are guaranteed to be ±2%. If more than 150 mA current source is need to drive higher-power LEDs or LED banks, the four current sources could be simply connected in parallel. Among them, the current sharing accuracy is just their current matching accuracy.

Figure 3.53 shows the internal circuitry block diagram of such commercial IC [28]. Clearly, the boost converter controller has an inner current loop with slope compensation as in traditional current-mode control. But, its outer loop is not to regulate the boost output voltage. Instead, it regulates the LED string current to its preset value while keeping the voltages at I_{o_1}–I_{o_4} pins to minimal. Surely, there

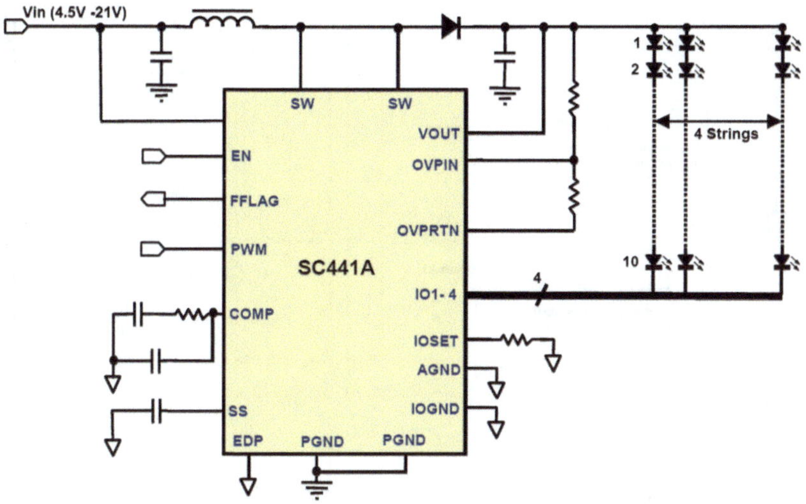

Fig. 3.52 A boost-type LED driver for multiple LED strings

are some variations of the V_f in LEDs, and the voltages at I_{o1}–I_{o4} may not be exactly the same. The control method in [28] then minimises the lowest voltage at I_{o1}–I_{o4} pins with type II compensation via the RC network connected from COMP pin to ground. In this way, the power losses associated with the current sources are minimised, and the total system efficiency is increased. To further minimise power losses during PWM dimming OFF time, where the current sources are turned off, a simple switch is integrated in SC441A from OVPRTN to ground. The timing of this simple switch is synchronised with the PWM dimming signal at the PWM pin. During PWM dimming ON time, the switch is closed, and OVPRTN pin is connected to ground. Along with the external voltage divider and OVPIN, the circuit provides OVP for the boost converter output. During PWM dimming OFF time, the boost converter power switch is turned off as shown in Fig. 3.51. As a consequence, the boost converter output voltage will not increase. On the other hand, one does not want the output voltage to sag either during this period of time. Otherwise, excessive inrush current may occur at the beginning of the next ON period. The small internal switch at OVPRTN just disconnects the OVP feedback resistor divider from the ground and helps reduce the output capacitor discharge current during the PWM OFF time. Also, the associated power losses at the feedback resistors are removed.

Multiple LED strings have been used for LCD backlight and illumination in notebook computer, smart phones, tablet, car information display, spot lamps and other electronic devices. Many of commercial LED drivers and controllers for multiple LED strings share similar structure as in Fig. 3.50. For different applications and specifications, the boost converter power switch and/or the current sources may take integrated form or discrete form.

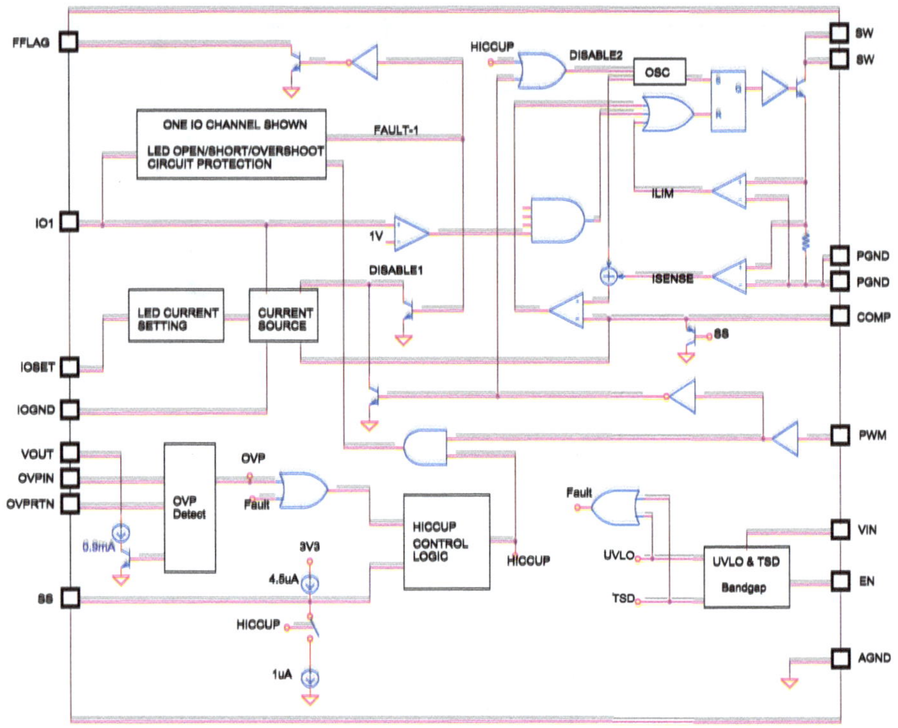

Fig. 3.53 A commercial IC controller for boost-type LED driver

3.7 Summary

In this chapter, we selected a few typical commercial power electronic controllers to illustrate their development and application in solving real-world problems. It is clear that, for different applications, the circuit topologies and controllers take different forms. Even for the same application with the same circuit topology, different controllers and methods are still used to achieve different performance and characteristics.

In addition to making the power switching circuitry and system work, engineers are striving to make them operate more efficiently, perform more elegantly, last more extensively and cost more economically. To reach this goal, deep understanding of the real-world problems is prerequisite, and a set of powerful engineering tools and solid analytic skills are needed. The basic buck, boost and buck–boost switching power converter topologies along with their variances and derivatives provided the starting points to meet different power requirements. The developed state-space averaging and various switching average-based modelling methods helped engineers to understand the power converters behaviour and dynamics. The traditional PID compensation methods empowered engineers to successfully solve many engineering problems and issues. However, with higher goals and more challenges from the changing world, significant research and engineering efforts are still required to understand and develop switch-mode power converters, a subset of switched electronic systems, from circuit topology, devices, packaging, modelling, simulation, and control perspectives. Novel power electronic controllers with the state-of-the-art control technologies will surely help meet the new requirement and find success in the marketplace.

References

1. 2 Amp, 2 MHz step-up switching regulator with soft-start, SC4501 datasheet. Semtech, Camarillo, California, USA (2011). URL http://www.semtech.com/images/datasheet/sc4501.pdf
2. 1 MHz fixed 3.3V output charge pump regulator, SC630A datasheet. Semtech, Camarillo, California, USA (2011). URL http://www.semtech.com/images/datasheet/sc630a.pdf
3. 630V driver IC for CFL and TL lamps, UBA2021, datasheet. NXP, Eindhoven, The Netherlands (2008). URL http://www.nxp.com/documents/data_sheet/UBA2021.pdf
4. 8 LED Light Management Unit, SC668 datasheet. Semtech, Camarillo, California, USA (2011). URL http://www.semtech.com/images/datasheet/sc668.pdf
5. 9A synchronous buck PWM DC–DC converter with integrated inductor external output voltage programming, EN5396QI datasheet. Enpirion, Hampton, New Jersey, USA (2009). URL http://www.enpirion.com/products-en5396qi.htm
6. A constant current buck regulator for driving high power LEDs, LM3404/04HV datasheet. Analog Devices, Norwood, Massachusetts, USA (2010). URL http://www.national.com/ds/LM/LM3404.pdf
7. Alberkrack, J.: Theory and applications of the MC34063 and μA78S40 switching regulator control circuits, AN920/D. Intusoft, San Pedro, California, USA (2002). URL http://www.intusoft.com/onsemipdfs/AN920-D.pdf
8. Arbetter, B., Maksimovic, D.: Feedforward pulse width modulators for switching power converters. IEEE Trans. Power Electron. **12**(2), 361–368 (1997)
9. Ben-Yaakov, S., Shvartsas, M., Glozman, S.: Statics and dynamics of fluorescent lamps operating at high frequency: Modeling and simulation. In: Proc. of the IEEE Applied Power Electronics Conference, Dallas, Texas, USA, pp. 792–801 (1999)
10. CFL 13W demo PCB with UBA2021 for integrated lamp-ballast designs, Application note. NXP, Eindhoven, The Netherlands (2008). URL http://www.nxp.com/documents/application_note/AN98091.pdf
11. Chang, C., Bruning, G.: High frequency electronic ballast for multiple lamp independent operation. US patent 6326740 (2001)
12. Chang, C., Chang, J.: Switching scheme for power supply having a voltage-fed inverter. US patent 5781418 (1998)
13. Chang, C., Knights, M.: Interleaving technique in distributed power conversion systems. IEEE Trans. Circuits Syst. **42**(5), 245–251 (1995)
14. Charge pump flash LED driver with safety timer, SC619 datasheet. Semtech, Camarillo, California, USA (2008). URL http://www.semtech.com/images/datasheet/sc619.pdf
15. Charge Pump, 7-Channel Smart LED driver with I2C interface, ADP8860 datasheet. Analog Devices, Norwood, Massachusetts, USA (2009). URL http://www.analog.com/static/imported-files/data_sheets/ADP8860.pdf
16. Deng, E., Cuk, S.: Negative incremental impedance and stability of fluorescent lamp. In: Proc. of the IEEE Applied Power Electronics Conference, Atlanta, Georgia, USA, pp. 1050–1056 (1997)
17. Dickson, J.F.: On-chip high-voltage generation in NMOS integrated circuits using an improved voltage multiplier technique. IEEE J. Solid-State Circuits **11**(3), 374–378 (1976)
18. Digital PWM system controller with 4-bit, 6-bit, or 8-bit VID support, UCD9222 datasheet. Texas Instruments, Dallas, Texas, USA (2011). URL http://focus.ti.com/lit/ds/symlink/ucd9222.pdf

19. Dimming ballast control IC, IR21592, datasheet. International Rectifier, El Segundo, California, USA (2005). URL http://www.irf.com/product-info/datasheets/data/ir21592.pdf
20. Dixon, L.: Average current mode control of switching power supplies, Application note U-140. Texas Instruments, Dallas, Texas, USA (1999). URL http://focus.ti.com/lit/an/slua079/slua079.pdf
21. Dual-phase, single or two output synchronous step-down controllers, SC2446 datasheet. Semtech, Camarillo, California, USA (2004). URL http://www.semtech.com/images/datasheet/sc2446.pdf
22. Dual-phase, single or dual output synchronous step-down controller, SC2446A datasheet. Semtech, Camarillo, California, USA (2005). URL http://www.semtech.com/images/datasheet/sc2446a.pdf
23. Erickson, R.W., Maksimovic, D.: Fundamentals of Power Electronics, 2nd edn. Kluwer Academic Publishers, Dordrecht (2001)
24. Erickson, R., Madigan, M., Singer, S.: Design of a simple high power factor rectifier based on the flyback converter. In: Proc. of the IEEE Applied Power Electronics Conference, pp. 792–801 (1990)
25. Four-phase Buck PWM controller with integrated MOSFET drivers for Intel VR10, VR11, and AMD applications, ISL6312 datasheet. Intersil, Milpitas, California, USA (2011). URL http://www.intersil.com/data/fn/fn9289.pdf
26. Greenchip III SMPS control IC, TEA1750, datasheet. NXP, Eindhoven, The Netherlands (2008). URL http://www.nxp.com/documents/data_sheet/TEA1750.pdf
27. Hammer, E.: Fluorescent lamp operating characteristics at high frequency. J. Illum. Eng. Soc. **Oct.**, 211–224 (1984)
28. High efficiency integrated driver for 4-strings of 150 mA LEDs, SC441A datasheet. Semtech, Camarillo, California, USA (2010). URL http://www.semtech.com/images/datasheet/sc441a.pdf
29. High power factor preregulator, UC3854 datasheet. Texas Instruments, Dallas, Texas, USA (2009). URL http://focus.ti.com/docs/prod/folders/print/uc3854.html
30. IEC 61000-3-2. Commission Electrotechnique Internationale, Geneva, Switzerland (2001). URL http://std.iec.ch/
31. Lee, F., et al.: Virginia Power Electronics Center Publication Series, vols. 1–8. Virginia Tech, Blackburg (1995)
32. Lee, F.C., van Wyk, J.D., Boroyevich, D., Lu, G., Liang, Z., Barbosa, P.: Technology trends toward a system-in-a-module in power electronics. IEEE Circuits Syst. Mag. **2**(4), 4–22 (2002)
33. Light Emitting Diode, NHSW046A datasheet. Nichia, Tokushima, Japan (2011). URL http://www.nichia.co.jp/en/product/led.html
34. Linden, D., Reddy, T.B.: Handbook of Batteries. McGraw-Hill, New York (2002)
35. Low dropout, two-bank LED driver with PWM brightness control, TPS7510x datasheet. Texas Instruments, Dallas, Texas, USA (2010). URL http://www.ti.com/lit/ds/sbvs080h/sbvs080h.pdf
36. Middlebrook, R.D., Cuk, S.: A general unified approach to modeling switching converter power stages. Int. J. Electron. **42**(6), 521–550 (1977)
37. Middlebrook, R.D., Cuk, S.: Advances in Switched-mode Power Conversion. TESLAco, Irvine (1983)
38. Peng, H., Chang, C.: A simple nonlinear gain scheduling method in digital PWM converter control. In: Proc. of IEEE 6th International Power Electronics and Motion Control Conference, Wuhan, China, pp. 331–336 (2009)
39. Power factor controllers, MC34262/MC33262 datasheet. Onsemi, Phoenix, Arizona, USA (2010). URL http://www.onsemi.com/PowerSolutions/product.do?id=MC34262
40. Stub series terminated logic for 1.8V (SSTL-18), JESD8-15A, Addendum 15 to JESD8 series (revision of JESD8-15). JEDEC solid state technology association, Arlington, Virginia, USA (2003). URL http://www.jedec.org/sites/default/files/docs/jesd8-15a.pdf
41. Qiao, M., Parto, P., Amirani, R.: Stabilize the Buck converter with transconductance amplifier, AN-1043. International Rectifier, El Segundo, California, USA (2002). URL http://www.irf.com/technical-info/appnotes/an-1043.pdf

42. Redl, R.: Power-factor correction in single-phase switching-mode power supplies—An overview. Int. J. Electron. **77**(5), 555–582 (1994)
43. Redl, R., Sun, J.: Ripple-based control of switching regulators—An overview. IEEE Trans. Power Electron. **24**(12), 2669–2680 (2009)
44. Rossetto, L., Spiazzi, G., Tenti, P.: Control techniques for power factor correction converters. In: Proc. of the Power Electronics, Motion Control Conference, Warsaw, Poland, pp. 1310–1318 (1994)
45. Step-Up/Down/Inverting switching regulators, MC34063A, MC33063A, NCV33063A datasheet. Onsemi, Phoenix, Arizona, USA (2004). URL http://www.onsemi.com/pub_link/Collateral/MC34063A-D.pdf
46. Step-Up/Down/Inverting switching regulators, NCP3063, NCP3063B, NCV3063 datasheet. Onsemi, Phoenix, Arizona, USA (2009). URL http://www.onsemi.com/pub_link/Collateral/NCP3063-D.pdf
47. Swennen, E., van der Meulen, P.: LED characteristics. Lumileds, Sweden (2008). URL http://www.riksutstallningar.se/upload/Semiarier/pdf/led.pdf
48. TEA1738T datasheet. GreenChip SMPS control IC (2011). URL http://www.nxp.com/documents/data_sheet/TEA1738T_TEA1738LT.pdf
49. Van der Broeck, H., Sauerlander, G., Wendt, M.: Power driver topologies and control scheme for LEDs. In: Proc. of the IEEE Applied Power Electronics Conference, Anaheim, California, USA, pp. 1319–1325 (2007)
50. Voltage regulator module (VRM) and enterprise voltage regulator-down (EVRD) 11.1, Design Guidelines. Intel (2009). URL http://www.intel.com/Assets/PDF/designguide/321736.pdf
51. Vorperian, V.: Simplified analysis of PWM converters using the model of the PWM switch. IEEE Trans. Aerosp. Electron. Syst. **26**(3), 490–505 (1990)
52. Vorperian, V., Tymerski, R., Lee, F.: Equivalent circuit models for resonant and PWM switches. IEEE Trans. Power Electron. **4**(2), 205–214 (1989)

Chapter 4
Grid Synchronisation Based on Frequency-Locked Loop Schemes

Gerardo Escobar, Ngai-Man Ho, and Sami Pettersson

4.1 Introduction

The use of switched power converters connected to the grid is an application that has become very common nowadays. On the one hand, there is a growing integration of renewable energy sources into the grid, and on the other hand, there is the need to improve the quality of the power delivered to the grid. A crucial aspect in the control of *grid-connected power converters* is the detection of the fundamental component of the utility voltage and perhaps the fundamental frequency or the phase angle of such a component [6]. Moreover, in a three-phase connection under unbalanced operation, it makes more sense to detect the positive-sequence of the fundamental component. This detection of the utility voltage component at fundamental frequency is essential for the control of distributed generation and storage systems, flexible ac transmission systems (FACTS) [5], power line conditioners and uninterruptible power supplies (UPS) [2], grid-connected inverters for alternative energy sources and other grid-connected power conditioning equipments. The detected information of such a fundamental component either in magnitude and phase angle or as a grid voltage vector and fundamental frequency is used for the synchronisation of the converter output variables, power flux calculations, or for the transformation of state variables into rotating reference frames [10]. In all these applications, both amplitude and phase must be obtained as fast and as accurate as possible. This issue becomes very challenging when the system operates under harmonic distortion, and, in the case of the three-phase systems, under unbalanced conditions.

G. Escobar (✉) · N.-M. Ho · S. Pettersson
ABB Switzerland Ltd, Segelhofstrasse 1K, 5405 Baden-Dattwil, Switzerland
e-mail: gerardo.escobar@ch.abb.com

N.-M. Ho
e-mail: carl.ho@ch.abb.com

S. Pettersson
e-mail: sami.pettersson@ch.abb.com

F. Vasca, L. Iannelli (eds.), *Dynamics and Control of Switched Electronic Systems*, Advances in Industrial Control,
DOI 10.1007/978-1-4471-2885-4_4, © Springer-Verlag London Limited 2012

The most widely accepted solution to provide information about the fundamental component of the grid voltage, which is somehow insensitive to frequency variations, consists in the use of PLL schemes [6]. Most PLLs deliver the fundamental component of the utility voltage (or a scaled or normalised version of it) plus an estimation of the phase angle and/or the fundamental frequency. The original schemes of PLLs were designed very intuitively, and were composed mainly of a phase detector, a filter, and a voltage controlled oscillator (VCO). It is perhaps at the level of the phase detector that many modifications were historically proposed.

One of the most appealing PLL schemes used for frequency-insensitive (positive-sequence) fundamental component detection is the three-phase *phase-locked loop* (PLL) based on the *synchronous reference frame* (SRF–PLL) [6, 9]. Different schemes have been proposed based on this scheme [1, 4, 19]. Under ideal utility conditions, i.e. without harmonic distortion or unbalance, a relatively high bandwidth feedback loop of the SRF–PLL yields a fast and precise detection of the phase angle and amplitude of the utility voltage vector. However, most of the schemes based on the SRF–PLL approach are very sensitive to harmonic distortion [8]. In case that the utility voltage is distorted with low-order harmonics, i.e. harmonics close to the fundamental frequency, the bandwidth of the SRF–PLL feedback loop can be reduced to reject the effect of these harmonics on the output. However, the PLL bandwidth reduction is not an acceptable solution as its speed of response is considerably reduced as well. It is worth noticing that the problem of estimating the fundamental component gets even more challenging in the case of an unbalanced utility voltage [18].

It is important to highlight the fact that the SRF–PLL and most modifications are based on the simple assumption that $\sin(x) \approx x$, which holds for x sufficiently small only. That is, they are based on a linearization argument, and thus the results can only be of local nature, which somehow restricts their range of operation. Moreover, these schemes are based on the detection of the phase angle, which gives a good intuitive understanding of the operation. However, as already noticed by several authors [9, 10], schemes following this philosophy present severe stability issues under utility fault conditions.

Although the SRF–PLL was originally intended for three-phase systems, several authors have tried to extend the idea of the SRF–PLL to the single-phase case [1, 15, 16]. For this purpose, it is usually proposed to emulate the single-phase system as a balanced three-phase system where the α-component is represented by the single-phase grid voltage, and the β-component, i.e. the quadrature companion signal,[1] is synthesised using many different approaches [2, 8]. Roughly speaking, these two signals form a virtual three-phase utility voltage vector in the fixed-frame representation to which the same ideas as in the SRF–PLL could be directly applied [2]. Different schemes have been proposed based on this simple scheme, where the contributions are mainly in the design of such a quadrature-signals generator (QSG) as

[1] Also referred as square phase signal, that is, a signal with a phase shift of 90 degrees with respect to the reference signal.

pointed out before. Conventional schemes based on the previously described concepts have been deeply studied in [9, 10], namely, inverse Park's transformation PLL [2, 9], enhanced PLL [1, 4], power-based PLL [19], Hilbert transformer-based PLL [8] and transport delay-based PLL [9]. However, it has been observed that most of these solutions are very sensitive to harmonic distortion [9].

Designs of PLL algorithms following alternative approaches have been presented recently in [17] and [12], which are based on the frequency detection instead of the phase angle detection. These methods have been referred to as frequency-locked loop (FLL) as, in these cases, the detected frequency is used to reconstruct the quadrature internally generated signal. Other methods based on the frequency detection approach have been reported in [11] and [20], where the authors propose the use of adaptive notch filters (ANF) for the reconstruction of a pure sinusoidal signal and online estimation of the fundamental frequency. This approach is aimed to alleviate the effect of harmonic distortion present in the external reference signal. These first works were aimed for single-phase systems, and then in [20] the extension for the three-phase case is presented. In this case, the authors propose the use of an ANF for each phase, where the frequency estimation is performed using the output information of all three ANF.

This chapter presents algorithms for both the three-phase and single-phase cases. They are based on the fundamental frequency detection [11, 13, 20], and thus, they belong to the family of FLL schemes. These algorithms provide an estimation of the angular frequency and the fundamental component of the grid voltage. In particular, the three-phase scheme estimates both positive and negative sequences of the fundamental component of an unbalanced three-phase signal. The design of the proposed FLL schemes is based on a complete dynamical model of the generator of the grid voltage signal, which considers operation under harmonic distortion, and in the three-phase case, unbalanced operation as well. The overall design of the proposed schemes does not involve at any moment linearization arguments as in most conventional PLLs. Instead, the design and analysis follow the Lyapunov's approach.

The proposed three-phase scheme is based on a description of the grid voltage in terms of the fixed-frame coordinates. Therefore, the three-phase scheme is referred to as fixed reference frame PLL with harmonics compensation (FRF–H–PLL). This scheme represents an extension to the scheme FRF–PLL in [3], which does not include an explicit harmonics compensation mechanism to cope with harmonic distortion present in the grid voltage signal. For the sake of completeness, a brief description of the scheme FRF–PLL is presented here.

The proposed single-phase FLL scheme is based on an adaptive estimation of the quadrature companion of the single-phase signal, and thus, it is referred to as PLL based on an adaptive quadrature signals generator (AQSG–PLL). It is shown that such a quadrature companion signal generator arises in a natural way from the model description of the system. An extension to this scheme, referred to as AQSG–H–PLL, is also presented that includes an explicit harmonic compensation mechanism.

Finally, simulations and experimental results are presented to show that the proposed schemes perform properly, providing a fast and precise response in spite of frequency variations, harmonic distortion and, in the three-phase case, unbalanced conditions. Moreover, it is shown that the harmonic distortion rejection is reinforced by the selective nature of the schemes.

4.2 Models of Grid Voltage Signals

This section presents the dynamical models of the generators for both single- and three-phase grid voltage signals. The models are instrumental for the design of the proposed FLL schemes. These models consider also the effect of a periodic distorted signal conformed by harmonic components of the fundamental frequency. In the three-phase case, the model also considers the effect of unbalanced operation in the fundamental component and also considers unbalance in the harmonic components.

4.2.1 Model of the Single-Phase Grid Voltage Signal

The design of the PLL based on an adaptive quadrature signals generator (AQSG–PLL) is based on the model of the generator of a periodic signal v_α, which is derived in what follows. For ease of presentation, let us consider first a simple sinusoidal signal disregarding the harmonic distortion.

Recall that the generator of a sinusoidal signal v_α can be described in a quite general form using the following basic representation

$$v_\alpha = \rho^\top v_{dq}, \quad \rho = \begin{pmatrix} \cos\theta_0 \\ \sin\theta_0 \end{pmatrix}, \quad v_{dq} = \begin{pmatrix} v_d \\ v_q \end{pmatrix}, \tag{4.1}$$

where ρ is a vector rotating at the fundamental frequency $\omega_0 = \dot{\theta}_0$, with θ_0 the phase angle; and v_{dq} is the vector of harmonic coefficients at frequency ω_0, also referred to as the *phasor* at frequency ω_0.

The time derivative of (4.1) is given by

$$\dot{v}_\alpha = \omega_0 \rho^\top J^\top v_{dq}, \quad J = \begin{pmatrix} 0 & -1 \\ 1 & 0 \end{pmatrix}, \tag{4.2}$$

where J is a skew symmetric matrix, i.e. $J^2 = -I_2$, $J^{-1} = J^\top$, with I_2 the 2×2 identity matrix.

Now, define the following auxiliary variable

$$v_\beta = \rho^\top J^\top v_{dq}. \tag{4.3}$$

Notice that the auxiliary variable v_β represents a signal in quadrature (90 degrees ahead) to v_α and owning the same magnitude as v_α, i.e. $|v_\alpha| = |v_\beta|$.

The time derivative of (4.3) is given by

$$\dot{v}_\beta = -\omega_0 \rho^\top v_{dq}, \tag{4.4}$$

which leads to the complete description, that is, the generator of the sinusoidal signal v_α, given by

$$\dot{v}_\alpha = \omega_0 v_\beta, \qquad v_\alpha(0) = v_d, \tag{4.5}$$

$$\dot{v}_\beta = -\omega_0 v_\alpha, \qquad v_\beta(0) = v_q. \tag{4.6}$$

The model (4.5)–(4.6) represents a second order *harmonic oscillator*, that is, a simple resonator, oscillating at frequency ω_0. Notice that the corresponding amplitude and phase shift are fixed by the initial conditions $v_\alpha(0) = v_d$ and $v_\beta(0) = v_q$, which complete the model description.

A matrix representation of generator (4.5)–(4.6) is

$$\dot{v}_{\alpha\beta} = -\omega_0 J v_{\alpha\beta}, \tag{4.7}$$

where $v_{\alpha\beta} = (v_\alpha \; v_\beta)^\top$.

Remark 4.2.1 Notice that vector $v_{\alpha\beta}$ can also be described as follows

$$\begin{pmatrix} v_\alpha \\ v_\beta \end{pmatrix} = \begin{pmatrix} \cos\theta_0 & \sin\theta_0 \\ -\sin\theta_0 & \cos\theta_0 \end{pmatrix} v_{dq} \tag{4.8}$$

which coincides with the representation of a balanced three-phase signal in terms of the fixed-frame coordinates.

In case of highly distorted grid voltage signals, the previous model (4.7) can be extended to the following more general model of the generator of the grid voltage signal:

$$\dot{v}_{\alpha\beta,1} = -\omega_0 J v_{\alpha\beta,1} \quad \begin{cases} \dot{v}_{\alpha,1} = \omega_0 v_{\beta,1}, \\ \dot{v}_{\beta,1} = -\omega_0 v_{\alpha,1}, \end{cases} \tag{4.9}$$

$$\dot{v}_{\alpha\beta,3} = -3\omega_0 J v_{\alpha\beta,3} \quad \begin{cases} \dot{v}_{\alpha,3} = 3\omega_0 v_{\beta,3}, \\ \dot{v}_{\beta,3} = -3\omega_0 v_{\alpha,3}, \end{cases} \tag{4.10}$$

$$\vdots \quad \vdots$$

$$\dot{v}_{\alpha\beta,k} = -k\omega_0 J v_{\alpha\beta,k} \quad \begin{cases} \dot{v}_{\alpha,k} = k\omega_0 v_{\beta,k}, \\ \dot{v}_{\beta,k} = -k\omega_0 v_{\alpha,k}, \end{cases} \quad \forall k \in \mathscr{H}, \tag{4.11}$$

where $\mathscr{H} = \{1, 3, 5, \ldots\}$ is the set of harmonics under consideration.
The overall single-phase grid voltage signal is given by the following sum of harmonic components

$$v_\alpha = \sum_{k\in\mathscr{H}} v_{\alpha,k}. \tag{4.12}$$

4.2.2 Model of the Three-Phase Grid Voltage Signal

In what follows, it is considered that the grid voltage is represented by an *unbalanced three-phase signal*, originally in three-phase coordinates $v_{123} = \begin{pmatrix} v_1 & v_2 & v_3 \end{pmatrix}^\top$. To facilitate the developments the grid voltage signal is transformed to (fixed-frame) $\alpha\beta$-coordinates using, for instance, the following Clarke's transformation or Park's transformation

$$v_{\alpha\beta} = \begin{pmatrix} v_\alpha \\ v_\beta \end{pmatrix} = \sqrt{\frac{2}{3}} \begin{pmatrix} 1 & -\frac{1}{2} & -\frac{1}{2} \\ 0 & \frac{\sqrt{3}}{2} & -\frac{\sqrt{3}}{2} \end{pmatrix} \begin{pmatrix} v_1 \\ v_2 \\ v_3 \end{pmatrix}.$$

Consider that the grid voltage signal comprises only a fundamental component, then a quite general description of such a three-phase unbalanced signal can be obtained, if both *positive symmetric sequence* and *negative symmetric sequence* are considered, as follows:

$$v_{\alpha\beta} = v_{\alpha\beta}^p + v_{\alpha\beta}^n = \mathrm{e}^{J\theta_0} v_{dq}^p + \mathrm{e}^{-J\theta_0} v_{dq}^n, \tag{4.13}$$

$$\mathrm{e}^{J\theta_0} = \begin{pmatrix} \cos\theta_0 & \sin\theta_0 \\ -\sin\theta_0 & \cos\theta_0 \end{pmatrix}, \qquad \mathrm{e}^{-J\theta_0} = \left(\mathrm{e}^{J\theta_0}\right)^\top, \quad J = \begin{pmatrix} 0 & -1 \\ 1 & 0 \end{pmatrix}, \tag{4.14}$$

where $v_{\alpha\beta}^p$ and $v_{\alpha\beta}^n$ represent the positive and negative symmetric components of $v_{\alpha\beta}$, respectively; $v_{dq}^p = \begin{pmatrix} v_d^p & v_q^p \end{pmatrix}^\top$ and $v_{dq}^n = \begin{pmatrix} v_d^n & v_q^n \end{pmatrix}^\top$ are the corresponding harmonic coefficient vectors, also referred to as *phasors*; and θ_0 represents the phase angle. Notice that matrix $\mathrm{e}^{J\theta_0}$ operates on vector v_{dq}^p by rotating it a phase angle θ_0 (counterclockwise), while matrix $\mathrm{e}^{-J\theta_0}$ rotates the vector v_{dq}^n by a phase angle $-\theta_0$ (clockwise).

In what follows, a dynamical model of the generator of this signal is obtained. For this purpose, the time derivative of the grid voltage signal (4.13) is calculated, which yields the following relationship

$$\dot{v}_{\alpha\beta} = \omega_0 J \left(v_{\alpha\beta}^p - v_{\alpha\beta}^n\right), \tag{4.15}$$

where we used the fact that $\dot{\theta}_0 = \omega_0$, with ω_0 representing the *fundamental frequency* of the grid voltage.

To complete the model description it is convenient to make the following definition

$$\varphi_{\alpha\beta} \triangleq v_{\alpha\beta}^p - v_{\alpha\beta}^n \tag{4.16}$$

which yields $\dot{v}_{\alpha\beta} = \omega_0 J \varphi_{\alpha\beta}$. Taking now the time derivative of the auxiliary variable $\varphi_{\alpha\beta}$ and using (4.13), we obtain

$$\dot{\varphi}_{\alpha\beta} = \omega_0 J \left(v_{\alpha\beta}^p + v_{\alpha\beta}^n\right). \tag{4.17}$$

Therefore, the model[2] that completely describes the generator of the unbalanced signal $v_{\alpha\beta}$ is given by

$$\dot{v}_{\alpha\beta} = \omega_0 J \varphi_{\alpha\beta}, \tag{4.18}$$

$$\dot{\varphi}_{\alpha\beta} = \omega_0 J v_{\alpha\beta}. \tag{4.19}$$

Based on definitions of $v_{\alpha\beta}$ in (4.13) and of $\varphi_{\alpha\beta}$ in (4.16), it is possible to establish the following relationship

$$\begin{pmatrix} v_{\alpha\beta} \\ \varphi_{\alpha\beta} \end{pmatrix} = \begin{pmatrix} I_2 & I_2 \\ I_2 & -I_2 \end{pmatrix} \begin{pmatrix} v_{\alpha\beta}^p \\ v_{\alpha\beta}^n \end{pmatrix}, \tag{4.20}$$

where I_2 is the 2×2 identity matrix.

In the case of presence of harmonic distortion in the grid voltage, the previous description (4.13) can be extended as follows:

$$v_{\alpha\beta} = \sum_{k \in \mathscr{H}} \left(v_{\alpha\beta,k}^p + v_{\alpha\beta,k}^n\right) = \sum_{k \in \mathscr{H}} \left(e^{Jk\theta_0} v_{dq,k}^p + e^{-Jk\theta_0} v_{dq,k}^n\right), \tag{4.21}$$

$$e^{Jk\theta} = \begin{pmatrix} \cos(k\theta_0) & \sin(k\theta_0) \\ -\sin(k\theta_0) & \cos(k\theta_0) \end{pmatrix}, \quad \forall k \in \mathscr{H}. \tag{4.22}$$

Following the same procedure as before, the model (4.18)–(4.19) can be extended to

$$\dot{v}_{\alpha\beta,k} = k\omega_0 J \varphi_{\alpha\beta,k}, \quad \forall k \in \{\mathscr{H}\}, \tag{4.23}$$

$$\dot{\varphi}_{\alpha\beta,k} = k\omega_0 J v_{\alpha\beta,k}, \tag{4.24}$$

where $\mathscr{H} = \{1, 3, 5, \dots\}$ is the set of harmonics under consideration. The voltage signal can now be reconstructed as

$$v_{\alpha\beta} = \sum_{k \in \mathscr{H}} v_{\alpha\beta,k}. \tag{4.25}$$

As before, it is possible to establish the following relationship[3] for the fundamental component

$$\begin{pmatrix} v_{\alpha\beta,1} \\ \varphi_{\alpha\beta,1} \end{pmatrix} = \begin{pmatrix} I_2 & I_2 \\ I_2 & -I_2 \end{pmatrix} \begin{pmatrix} v_{\alpha\beta}^{p_1} \\ v_{\alpha\beta}^{n_1} \end{pmatrix}. \tag{4.26}$$

---

[2]Notice that auxiliary variable $\varphi_{\alpha\beta}$ is strictly necessary to allow a complete model description in the unbalanced case.

[3]This expression is used later to recuperate the positive and negative sequences.

4.3 Single-Phase PLL Based on an Adaptive Quadrature Signals Generator (AQSG–PLL)

For ease of presentation, consider that the grid voltage comprises mainly a fundamental component. Thus, the dynamical model that generates such a fundamental component is given by (4.5)–(4.6).

Notice that the fundamental frequency parameter ω_0 appears in both rows of model (4.5)–(4.6), which may complicate the design of the adaptive algorithm to estimate ω_0. To overcome this issue it is proposed to re-parameterize the model using the following transformation

$$\psi = \varpi_0 \frac{v_\beta}{\omega_0}, \tag{4.27}$$

where ϖ_0 represents a nominal value of the fundamental frequency, i.e. a positive constant parameter, which can be a simple approximation of the *fundamental frequency* ω_0.

Out of this, the system (4.5)–(4.6) is transformed to

$$\dot{v}_\alpha = \Omega_0 \psi, \tag{4.28}$$

$$\dot{\psi} = -\varpi_0 v_\alpha, \tag{4.29}$$

where we have defined the following parameter

$$\Omega_0 = \frac{\omega_0^2}{\varpi_0}. \tag{4.30}$$

Notice that, after this simple transformation, the fundamental frequency parameter has been concentrated in one single row as a single unknown parameter Ω_0. This new parameter represents the scaled square of the fundamental frequency, and has also units of frequency, i.e. rad/s. Notice that the new state variable ψ is still a signal in quadrature to v_α, and thus, system (4.28)–(4.29) is still a second-order harmonic oscillator.

The objective of the single-phase AQSG–PLL algorithm consists in first estimating the fundamental component of the state variable v_α with zero phase shift, and second, in providing an estimate of the fundamental frequency ω_0 by including an adaptation scheme.

The design of such an adaptive estimator, which is referred to as adaptive quadrature signals generator (AQSG), is based on the structure of the above model description (4.28)–(4.29). The proposed scheme consists of a copy of system (4.28)–(4.29) to which a damping term is added as follows:

$$\dot{\hat{v}}_{\alpha,1} = \hat{\Omega}_0 \hat{\psi} + \gamma_1 \tilde{v}_\alpha, \tag{4.31}$$

$$\dot{\hat{\psi}}_1 = -\varpi \hat{v}_{\alpha,1}, \tag{4.32}$$

where $\tilde{\upsilon}_\alpha \triangleq \upsilon_\alpha - \hat{\upsilon}_{\alpha,1}$; $\hat{\upsilon}_{\alpha,1}$ and $\hat{\psi}_1$ represent the estimates of the fundamental components of υ_α and ψ, respectively; $\hat{\Omega}_0$ is the estimate of the parameter Ω_0; and $\gamma_1 > 0$ is a design parameter used here to introduce the required damping.

The design of the adaptive law to reconstruct $\hat{\Omega}_0$ involved in (4.31)–(4.32) follows the *Lyapunov approach*. For this, it is proposed to compute first the dynamics of the estimation error, i.e. the closed loop dynamics between system (4.28)–(4.29) and the AQSG (4.31)–(4.32), this yields

$$\dot{\tilde{\upsilon}}_\alpha = \Omega_0 \tilde{\psi}_1 - \tilde{\Omega}_0 \hat{\psi} + \gamma_1 \tilde{\upsilon}_\alpha, \tag{4.33}$$

$$\dot{\tilde{\psi}} = -\varpi \tilde{\upsilon}_\alpha, \tag{4.34}$$

where $\tilde{\psi} \triangleq \psi - \hat{\psi}_1$ and $\tilde{\Omega}_0 \triangleq \hat{\Omega}_0 - \Omega_0$.

Then, the following quadratic storage function is proposed

$$H = \frac{\tilde{\upsilon}_\alpha^2}{2} + \frac{\tilde{\psi}^2}{2} + \frac{\tilde{\Omega}_0^2}{2\lambda}, \tag{4.35}$$

where λ is a design parameter referred to as the adaptation gain. Its time derivative along the trajectories of the estimation error model (4.33)–(4.34) is made negative-semidefinite[4] by proposing the following adaptive law, which is referred to as the fundamental frequency estimator (FFE)

$$\dot{\hat{\Omega}}_0 = \lambda \hat{\psi}_1 \tilde{\upsilon}_\alpha. \tag{4.36}$$

After application of the FFE, the time derivative of the energy storage function yields $\dot{H} = -\gamma_1 \tilde{\upsilon}_\alpha^2$.

An estimate of the fundamental frequency[5] can now be obtained by using transformation (4.30) as follows

$$\hat{\omega}_0 = \sqrt{\varpi \hat{\Omega}_0}. \tag{4.37}$$

An estimate of the fundamental component of the quadrature companion signal υ_β can be reconstructed according to (4.27) as follows

$$\hat{\upsilon}_{\beta,1} = \hat{\omega}_0 \frac{\hat{\psi}_1}{\varpi}, \tag{4.38}$$

which is referred to as the quadrature companion generator (QCG).

Summarising, the proposed AQSG–PLL consists of the AQSG (4.31)–(4.32), the FFE (4.36) and the QCG (4.38). A block diagram of the overall proposed AQSG–PLL algorithm is depicted in Fig. 4.1.

---

[4] The stability analysis can be completed using properties of the involved signals. It is omitted here for the sake of space limitations.

[5] Notice that the estimate $\hat{\omega}_0$ is obtained for display purposes only, and is not used as a feedback variable.

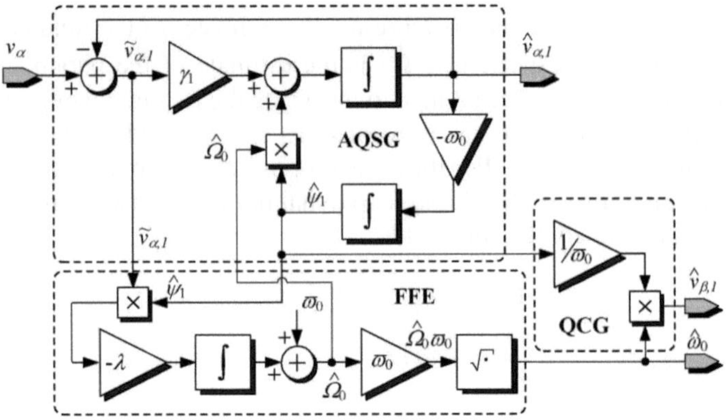

Fig. 4.1 Block diagram of the AQSG–PLL algorithm including the AQSG, FFE and QCG blocks

4.3.1 Single-Phase PLL Based on an Adaptive Quadrature Signals Generator with Harmonic Compensation (AQSG–H–PLL)

The same ideas from the previous scheme can be extended to the case where the grid voltage signal includes *harmonic distortion*. For this, consider the following transformation

$$\psi_k = \varpi_0 \frac{v_{\beta,k}}{\omega_0}, \quad \forall k \in \mathscr{H}, \tag{4.39}$$

out of which the model (4.9)–(4.11) can be rewritten as

$$\begin{cases} \dot{v}_{\alpha,1} = \Omega_0 \psi_1, \\ \dot{\psi}_1 = -\varpi_0 v_{\alpha,1}, \end{cases} \tag{4.40}$$

$$\begin{cases} \dot{v}_{\alpha,3} = 3\Omega_0 \psi_3, \\ \dot{\psi}_3 = -3\varpi_0 v_{\alpha,3}, \end{cases} \tag{4.41}$$

$$\vdots$$

$$\begin{cases} \dot{v}_{\alpha,k} = k\Omega_0 \psi_k, \quad \forall k \in \mathscr{H}. \\ \dot{\psi}_k = -k\varpi_0 v_{\alpha,k}, \end{cases} \tag{4.42}$$

Then the voltage signal can be represented by the following sum of harmonic components

$$v_\alpha = \sum_{k \in \mathscr{H}} v_{\alpha,k}. \tag{4.43}$$

Based on the structure of the above generic model (4.40)–(4.43), an estimator for the harmonic components of the grid voltage v_α can be proposed as follows:

$$\begin{cases} \dot{\hat{v}}_{\alpha,1} = \hat{\Omega}_0 \hat{\psi}_1 + \gamma_1 \tilde{v}_\alpha, \\ \dot{\hat{\psi}}_1 = -\varpi_0 \hat{v}_{\alpha,1}, \end{cases} \tag{4.44}$$

$$\begin{cases} \dot{\hat{v}}_{\alpha,3} = 3\hat{\Omega}_0 \hat{\psi}_3 + \gamma_3 \tilde{v}_\alpha, \\ \dot{\hat{\psi}}_3 = -3\varpi_0 \hat{v}_{\alpha,3}, \end{cases} \tag{4.45}$$

$$\vdots$$

$$\begin{cases} \dot{\hat{v}}_{\alpha,k} = k\hat{\Omega}_0 \hat{\psi}_k + \gamma_k \tilde{v}_\alpha, \quad \forall k \in \mathscr{H}, \\ \dot{\hat{\psi}}_k = -k\varpi_0 \hat{v}_{\alpha,k}, \end{cases} \tag{4.46}$$

where $\tilde{v}_\alpha \triangleq v_\alpha - \hat{v}_\alpha$ and

$$\hat{v}_\alpha = \sum_{k \in \mathscr{H}} \hat{v}_{\alpha,k}. \tag{4.47}$$

Notice that the estimated voltage signal $\hat{v}_\alpha$ can be decomposed as follows:

$$\hat{v}_\alpha = \hat{v}_{\alpha,1} + \hat{v}_{\alpha,h}, \tag{4.48}$$

where $\hat{v}_{\alpha,1}$ is computed as in (4.44). Notice that the estimator (4.44) coincides with the previous AQSG (4.31)–(4.32), except for the new definition of $\tilde{v}_\alpha \triangleq v_\alpha - \hat{v}_\alpha$, which includes the higher order harmonics, as shown in (4.48). Therefore, this estimator (4.44) is referred to as AQSG–H.
The higher order harmonics component $\hat{v}_{\alpha,h}$ is computed in a block referred to as harmonic compensation mechanism (HCM) as follows. First, the harmonic components are computed according to (4.46) for $k \in \{3, 5, \dots\}$, that is, all harmonics in $\mathscr{H}$ except for the fundamental. Second, these higher order harmonics are accumulated in the signal $\hat{v}_{\alpha,h}$, that is,

$$\hat{v}_{\alpha,h} = \sum_{k \in \{3,5,\dots\}} \hat{v}_{\alpha,k}. \tag{4.49}$$

This extended AQSG–H can be seen as the composition of the basic AQSG to which a HCM has been added as observed in the block diagram presented in Fig. 4.2. Therefore, the HCM can be seen as a plug-in block that can be easily added to the basic scheme AQSG–PLL. A detailed block diagram of the HCM is shown in Fig. 4.3.

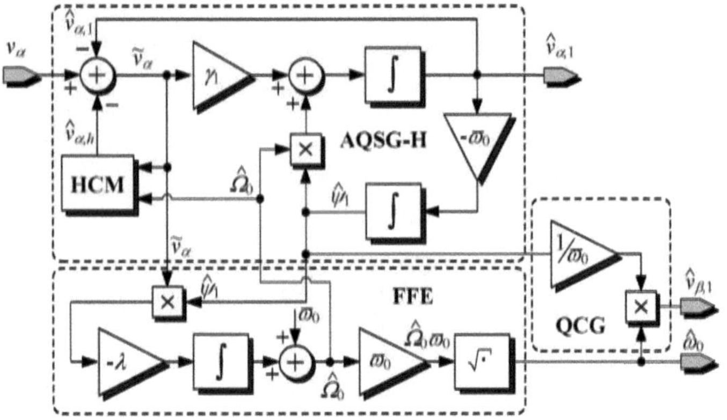

Fig. 4.2 Block diagram of the proposed AQSG–H–PLL algorithm including the AQSG–H (with the HCM), FFE and QCG blocks

Fig. 4.3 Block diagram of the proposed HCM

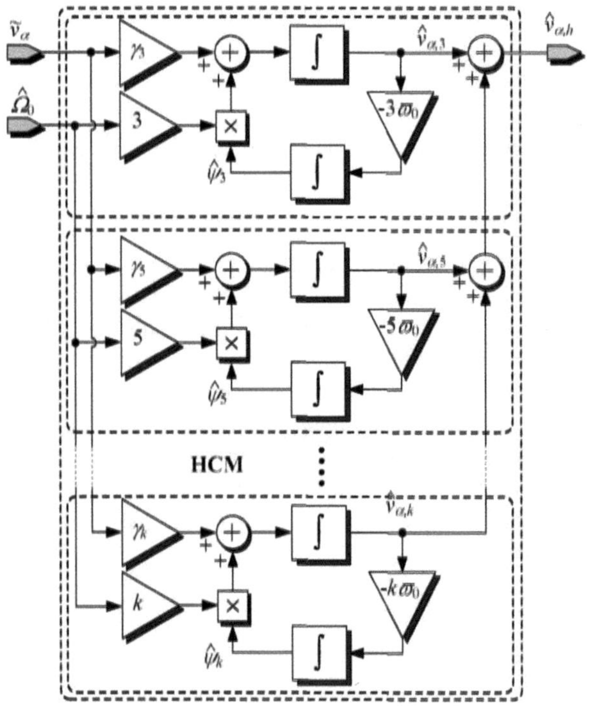

4.3.2 Numerical and Experimental Tests—Single-Phase Case

This scheme has been tested numerically under conditions of a highly distorted signal and under frequency variations. The grid voltage signal is composed of a fundamental component which is distorted by the third, fifth and seventh harmonics whose amplitudes and phase shift values are given in Table 4.1.

Table 4.1 Amplitudes and phases of the harmonic components of the grid voltage

No. Harmonic	Amplitude (RMS) (V)	Phase Shift (°)
1	230	0
3	50	14.3
5	25	8.6
7	15	5.7

Figure 4.4 shows a first test, in which the original AQSG–PLL operates from $t = 0$ s up to $t = 2$ s. It can be observed that, due to the highly distorted grid voltage v_α (in dashed line), the estimated fundamental component $\hat{v}_{\alpha,1}$ (in solid line) does not reach a perfect sinusoidal shape, but a slightly deformed signal containing still some remains of the harmonic distortion. Then, after $t = 2$ s the overall AQSG–H–PLL is enabled, i.e. the HCM is plugged in for the harmonics compensation. It can be observed in this same figure that, after a relatively short transient (about one and a half cycle), the estimated fundamental component $\hat{v}_{\alpha,1}$ reaches a pure sinusoidal shape, with the same amplitude and frequency as the actual fundamental component $v_{\alpha,1}$ (in dotted line), despite the relatively high distortion in the grid voltage signal v_α.

Then, at $t = 3$ s the fundamental frequency ω_0 of the grid voltage is step changed from 50 Hz to 42.5 Hz. Figure 4.5 shows the transient during such a disturbance. It can be observed that after a relatively short transient, of about two cycles, the estimated fundamental component reaches a pure sinusoidal shape, with the same amplitude and phase shift as the actual fundamental component of the grid voltage $v_{\alpha,1}$ (in dotted line).

Figure 4.6 (top plot) shows the transients of the estimated fundamental frequency $\hat{\omega}_0$ when the scheme is changed from AQSG–PLL to the AQSG–H–PLL at $t = 2$ s. It can be observed that in cases of relatively high harmonic distortion, the estimate of the fundamental frequency may present a considerable ripple, and also a considerable steady-state error if the scheme AQSG–PLL without harmonic compensation

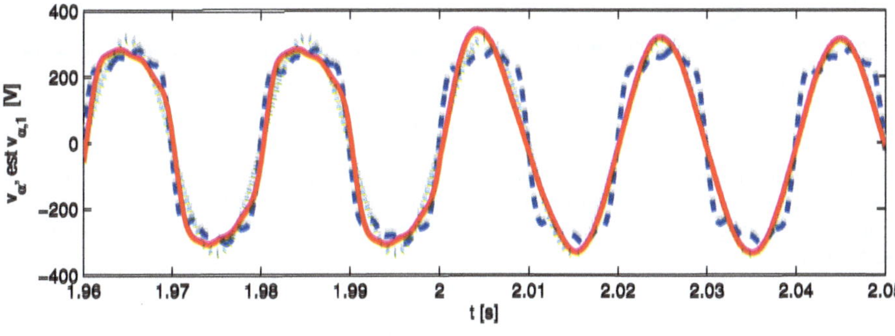

Fig. 4.4 Transient response of (*dashed*) grid voltage v_α, (*solid*) estimated fundamental component $\hat{v}_{\alpha,1}$, and (*dotted*) real fundamental component of the grid voltage $v_{\alpha,1}$. The AQSG–PLL is working from $t = 0$ s, then after $t = 2$ s the AQSG–H–PLL is enabled

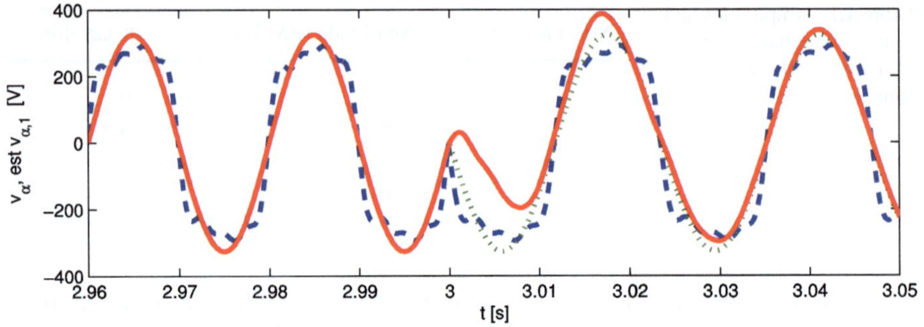

Fig. 4.5 Transient response of the proposed AQSG–H–PLL during a frequency step change from 314.16 rad/s (50 Hz) to 267.03 rad/s (42.5 Hz) at $t = 3$ s: (*dashed*) grid voltage v_α, (*solid*) estimated fundamental component $\hat{v}_{\alpha,1}$, and (*dotted*) real fundamental component of the grid voltage $v_{\alpha,1}$

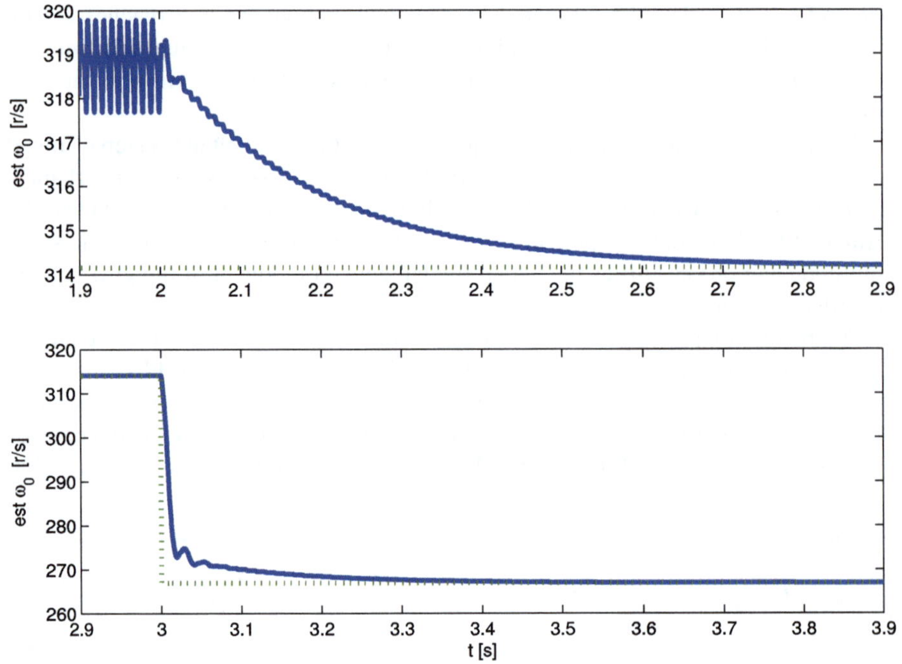

Fig. 4.6 Transients of (*solid*) the estimated fundamental frequency and (*dotted*) its reference, when (*top*) the scheme is changed from AQSG–PLL to the AQSG–H–PLL at $t = 2$ s; and (*bottom*) during a step change in the fundamental frequency from 314.16 rad/s (50 Hz) to 267.03 rad/s (42.5 Hz) at $t = 3$ s

is used. In contrast, it can be observed that after enabling the scheme AQSG–H–PLL at $t = 2$ s, the estimate of the fundamental frequency not only gets cleaner, but also reaches the expected constant value. In the same figure (bottom plot), the tran-

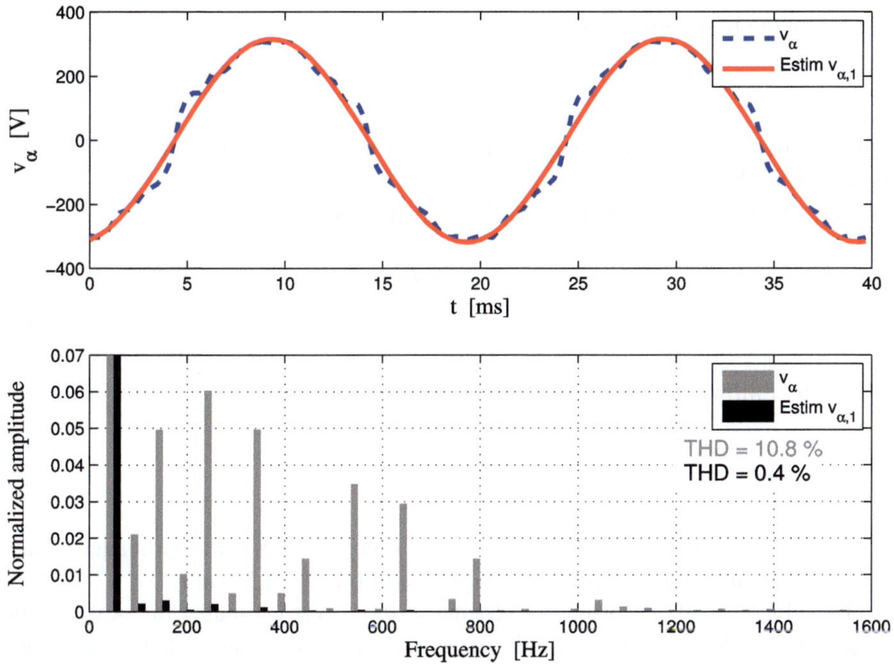

Fig. 4.7 (*Top plot*) Time response of: (*dotted*) distorted grid voltage signal v_α with harmonics according to IEC 61000-2-2, and (*solid*) estimated fundamental component $\hat{v}_{\alpha,1}$. (*Bottom plot*) Amplitude spectra of both signals

sient due to a step change in the fundamental frequency from 314.16 rad/s (50 Hz) to 267.03 rad/s (42.5 Hz) at $t = 3$ s is presented. Notice that after a relatively short transient the estimated frequency reaches the expected constant value as well.

The AGSG-H-PLL has also been implemented and experimentally tested in a DSP TMS320F28355 with a sampling time of 32 kHz. A distorted grid voltage signal with a total harmonic distortion (THD) of 10% was created according to standard IEC 61000-2-2. Figure 4.7 shows that the proposed AQSG–H–PLL delivers a quite clean sinusoidal signal estimate in phase with the grid voltage with a THD $= 0.4\%$ despite the existence of a high distortion in the reference signal.

4.4 Three-Phase PLL Based on the Fixed Reference Frame Representation

As observed in (4.18)–(4.19), parameter ω_0 appears in both differential equations of this system. This situation may complicate the estimation of parameter ω_0. Similar to the single-phase case, to facilitate the design and analysis of such an estimator, it is proposed to concentrate this parameter in the upper differential equation using the following transformation

$$\psi_{\alpha\beta} = \frac{\varpi_0}{\omega_0}\varphi_{\alpha\beta}, \tag{4.50}$$

where as before ϖ_0 is only an approximation of the nominal value of the fundamental frequency.[6]

The model (4.18)–(4.19) can thus be rewritten as

$$\dot{v}_{\alpha\beta} = \Omega_0 J \psi_{\alpha\beta}, \tag{4.51}$$

$$\dot{\psi}_{\alpha\beta} = \varpi_0 J v_{\alpha\beta}, \tag{4.52}$$

where the following definition has been used:[7]

$$\Omega_0 \triangleq \frac{\omega_0^2}{\varpi_0}. \tag{4.53}$$

Notice that the new parameter Ω_0 concentrates the unknown parameter ω_0 appearing now in the upper differential equation of (4.51)–(4.52) only.

The objective of the proposed scheme is to deliver estimates for positive and negative sequences of the grid voltage, as well as an estimate of the fundamental frequency ω_0. For this, an adaptive estimator for state variables $v_{\alpha\beta}$ and $\psi_{\alpha\beta}$ is designed based on model (4.51)–(4.52). This adaptive estimator will involve the unknown parameter Ω_0, and thus, it is necessary to incorporate an adaptive law to reconstruct Ω_0 (and thus ω_0). The adaptive estimator will be referred to as the three-phase adaptive quadrature signals generator (3ph-AQSG) as it generates two pairs of quadrature signals, while the adaptive law will be referred to as the three-phase fundamental frequency estimator (3ph-FFE).

The proposed estimator of the fundamental component of $v_{\alpha\beta}$ and $\psi_{\alpha\beta}$, referred to as 3ph-AQSG, consists of a copy of the system model (4.51)–(4.52) to which a damping term is added, that is,

$$\dot{\hat{v}}_{\alpha\beta,1} = \hat{\Omega}_0 J \hat{\psi}_{\alpha\beta,1} + \gamma_1 \tilde{v}_{\alpha\beta}, \tag{4.54}$$

$$\dot{\hat{\psi}}_{\alpha\beta,1} = \varpi_0 J \hat{v}_{\alpha\beta,1}, \tag{4.55}$$

where $\tilde{v}_{\alpha\beta} \triangleq v_{\alpha\beta} - \hat{v}_{\alpha\beta,1}$; $\hat{v}_{\alpha\beta,1}$ and $\hat{\psi}_{\alpha\beta,1}$ represent the estimates of the fundamental components of $v_{\alpha\beta}$ and $\psi_{\alpha\beta}$, respectively; $\hat{\Omega}_0$ is the estimate of parameter Ω_0; and γ_1 is a positive design parameter used to introduce the required damping.

The design of the adaptive law (3ph-FEE) to reconstruct $\hat{\Omega}_0$ (involved in (4.54)–(4.55)) follows the Lyapunov's approach. For this purpose, the dynamics describing

[6] ϖ_0 is used for scaling purposes only. This avoids numerical errors, and facilitates the tuning process.

[7] This coincides with definition (4.30) in the single-phase case.

the error due to the closed loop between (4.51)–(4.52) and (4.54)–(4.55) is computed as follows:

$$\dot{\tilde{v}}_{\alpha\beta} = \Omega_0 J \tilde{\psi}_{\alpha\beta} - \tilde{\Omega}_0 J \hat{\psi}_{\alpha\beta,1} - \gamma_1 \tilde{v}_{\alpha\beta}, \tag{4.56}$$

$$\dot{\tilde{\psi}}_{\alpha\beta} = \varpi_0 J \tilde{v}_{\alpha\beta}, \tag{4.57}$$

where $\tilde{\psi}_{\alpha\beta} \triangleq \psi_{\alpha\beta} - \hat{\psi}_{\alpha\beta}$, $\tilde{\Omega}_0 \triangleq \Omega_0 - \hat{\Omega}_0$.

For the design, the following quadratic storage function is used

$$W = \frac{\tilde{v}_{\alpha\beta}^{\top}\tilde{v}_{\alpha\beta}}{2} + \frac{\Omega_0 \tilde{\psi}_{\alpha\beta}^{\top}\tilde{\psi}_{\alpha\beta}}{2\varpi_0} + \frac{\tilde{\Omega}_0^2}{2\lambda}, \tag{4.58}$$

where $\lambda > 0$ is the adaptation gain.

The time derivative of W, along the trajectories of the error model (4.56)–(4.57), is made negative semidefinite[8] by proposing the adaptive law

$$\dot{\hat{\Omega}}_0 = \lambda \tilde{v}_{\alpha\beta}^{\top} J \hat{\psi}_{\alpha\beta,1} \tag{4.59}$$

which yields $\dot{W} = -\gamma_1 \tilde{v}_{\alpha\beta}^{\top}\tilde{v}_{\alpha\beta}$.

Based on definition (4.53), the fundamental frequency[9] can be reconstructed according to

$$\hat{\omega}_0 = \sqrt{\varpi_0 \hat{\Omega}_0}. \tag{4.60}$$

Based on estimates $\hat{v}_{\alpha\beta}$ and $\hat{\psi}_{\alpha\beta}$ (from estimator (4.54)–(4.55)), and $\hat{\omega}_0$ (from adaptation (4.59)), and based on definition (4.26), (4.50) and (4.60), the positive and negative sequences of the grid voltage can now be reconstructed as follows:

$$\begin{aligned} \hat{v}_{\alpha\beta,1}^{p} &= \frac{1}{2}\left(\hat{v}_{\alpha\beta,1} + \frac{\hat{\omega}_0}{\varpi_0}\hat{\psi}_{\alpha\beta,1}\right), \\ \hat{v}_{\alpha\beta,1}^{n} &= \frac{1}{2}\left(\hat{v}_{\alpha\beta,1} - \frac{\hat{\omega}_0}{\varpi_0}\hat{\psi}_{\alpha\beta,1}\right), \end{aligned} \tag{4.61}$$

which is referred to as the positive and negative sequences generator (PNSG).

Summarising, the 3ph-AQSG (4.54)–(4.55), the 3ph-FEE (4.59) and the PNSG (4.61) conform to a PLL referred to as FRF–PLL [3]. A block diagram of the FRF–PLL scheme is depicted in Fig. 4.8. Notice that in the 3ph-FFE a feedforward term has been included that prevents high transients during the startup operation. In this diagram, all thick lines represent vector variables while normal lines represent scalar variables.

[8] The stability and convergence analysis of the proposed scheme can be completed using the signals properties. It has been omitted here for the sake of space limitations.

[9] Notice that the estimate $\hat{\omega}_0$ is calculated for display purposes only, and is not used as a feedback variable.

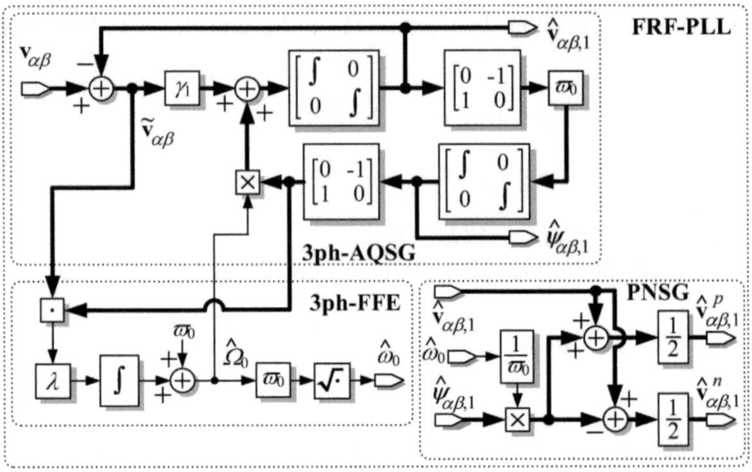

Fig. 4.8 Block diagram of the FRF–PLL algorithm

4.4.1 Three-Phase PLL Based on Fixed Reference Frame Representation with Harmonic Compensation

In what follows, the previous scheme FRF–PLL [3] is extended to consider harmonic distortion present in the grid voltage. For this purpose, it is proposed to introduce an unbalance harmonic compensation mechanism (UHCM) as shown in Fig. 4.9. The modified scheme is referred to as FRF–H–PLL as it considers harmonic distortion compensation in addition to unbalanced operation. Previous algorithms in [3] and [20] did not include any explicit mechanism for harmonic distortion cancelation. And thus, slight ripple was still present in the responses. This effect could be alleviated by limiting the bandwidth of the overall scheme, however, as a consequence, the speed of response is considerably reduced. Hence, a tradeoff between the speed of response and the harmonic compensation properties has to be established. In the proposed FRF–H–PLL scheme, this tradeoff is relaxed by the introduction of the UHCM, which allows faster and cleaner responses.

The idea behind the UHCM consists in designing an estimator to reconstruct the harmonic distortion of the grid voltage, which is later subtracted from the original signal as shown in the scheme of Fig. 4.9. Notice that the difference with respect to the diagram of Fig. 4.8 is simply the introduction of the feedback block UHCM. Therefore, the UHCM can be seen as a plug-in block that can be easily added to the basic scheme FRF–PLL. This scheme represents an alternative to the harmonic compensation scheme reported in [14].

The design of the UHCM considers the following transformation

$$\psi_{\alpha\beta,k} = \frac{\varpi_0}{\omega_0}\varphi_{\alpha\beta_k} \tag{4.62}$$

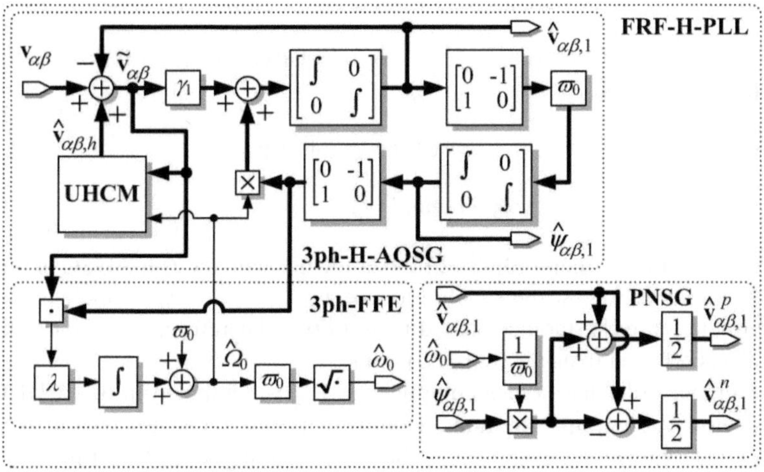

Fig. 4.9 Block diagram of the FRF–H–PLL scheme including the harmonic compensation mechanism UHCM for an unbalanced and distorted reference signal $v_{\alpha\beta}$

which is applied to (4.23)–(4.24) to get the following modified model:

$$\dot{v}_{\alpha\beta,k} = k\Omega_0 J \psi_{\alpha\beta,k}, \quad \forall k \in \mathscr{H}, \tag{4.63}$$

$$\dot{\psi}_{\alpha\beta,k} = k\varpi_0 J v_{\alpha\beta,k}, \tag{4.64}$$

where $\mathscr{H} = \{1, 5, 7, \ldots\}$ is the set of harmonics under consideration.

Based on the model (4.63)–(4.64), the estimator for the kth harmonic component ($k \in \mathscr{H}$) can be proposed as follows:

$$\dot{\hat{v}}_{\alpha\beta,k} = k\hat{\Omega}_0 J \hat{\psi}_{\alpha\beta,k} + \gamma_k \tilde{v}_{\alpha\beta}, \tag{4.65}$$

$$\dot{\hat{\psi}}_{\alpha\beta,k} = k\varpi_0 J \hat{v}_{\alpha\beta,k}, \tag{4.66}$$

where γ_k ($k \in \mathscr{H}$) is a positive design parameter used to introduce the required damping; $\hat{\Omega}_0$ is the estimate of parameter Ω_0; and $\tilde{v}_{\alpha\beta} \triangleq v_{\alpha\beta} - \hat{v}_{\alpha\beta}$, with $\hat{v}_{\alpha\beta}$ representing the estimated voltage.
In fact, the estimated voltage signal $\hat{v}_{\alpha\beta}$ can be decomposed as follows:

$$\hat{v}_{\alpha\beta} = \hat{v}_{\alpha\beta,1} + \hat{v}_{\alpha\beta,h}, \tag{4.67}$$

where $\hat{v}_{\alpha\beta,1}$ represents the estimate of the fundamental component, and $\hat{v}_{\alpha\beta,h}$ represents the estimate of the harmonic distortion components of the grid voltage.

The fundamental component $\hat{v}_{\alpha\beta,1}$ is reconstructed, according to (4.65)–(4.66)) as follows:

$$\dot{\hat{v}}_{\alpha\beta,1} = \hat{\Omega}_0 J \hat{\psi}_{\alpha\beta,1} + \gamma_1 \tilde{v}_{\alpha\beta}, \tag{4.68}$$

$$\dot{\hat{\psi}}_{\alpha\beta,1} = \varpi_0 J \hat{v}_{\alpha\beta,1}. \tag{4.69}$$

Notice that the estimator (4.68)–(4.69) coincides with the 3ph-AQSG in (4.54)–(4.55), except for the new definition of $\tilde{v}_{\alpha\beta} = v_{\alpha\beta} - \hat{v}_{\alpha\beta}$ which includes the higher order harmonics according to (4.67). Therefore, estimator (4.68)–(4.69) is referred to as the 3ph-H–AQSG.

The higher order harmonics component $\tilde{v}_{\alpha\beta,h}$ is computed in block UHCM as follows. First, the harmonic components are reconstructed according to (4.65)–(4.66) for $k \in \{5, 7, \dots\}$. Second, these higher order harmonics are accumulated in the signal $\hat{v}_{\alpha\beta,h}$, that is,

$$\hat{v}_{\alpha\beta,h} = \sum_{k \in \{5,7,\dots\}} \hat{v}_{\alpha\beta,k}. \tag{4.70}$$

A block diagram of the UHCM is presented in Fig. 4.10.

4.4.2 Numerical and Experimental Results—Three-Phase Case

For the numerical results, the following parameters have been selected $\gamma_1 = 300$ and $\lambda = 2.22$, which correspond to a settling time of $T_s = 0.03$ s, considering a nominal frequency of $\varpi_0 = 314.16$ rad/s (50 Hz), and a voltage amplitude of $|v_{\alpha\beta}| = 100$ V. All simulations are made with Simulink™, considering a continuous time implementation of the schemes with a fixed step of 0.000 25 s. The following test conditions have been considered for the utility voltage:

Balanced condition. The voltage source is formed only by a positive sequence of 100 V of amplitude, and fundamental frequency of 314.16 rad/s (50 Hz), with a zero phase shift.

Unbalanced condition. The voltage source includes both a positive and a negative sequence components. The positive sequence has 100 V of amplitude at 314.16 rad/s (50 Hz) and zero phase shift. For the negative sequence, an amplitude of 30 V and zero phase shift are considered.

Harmonic distortion. The fifth and seventh harmonics are added to the previous unbalanced signal to create a periodic distortion. Both harmonics have also a negative sequence component to allow unbalance in harmonics as well. Both positive and negative sequences of these harmonics have 10 V of amplitude and zero phase shift.

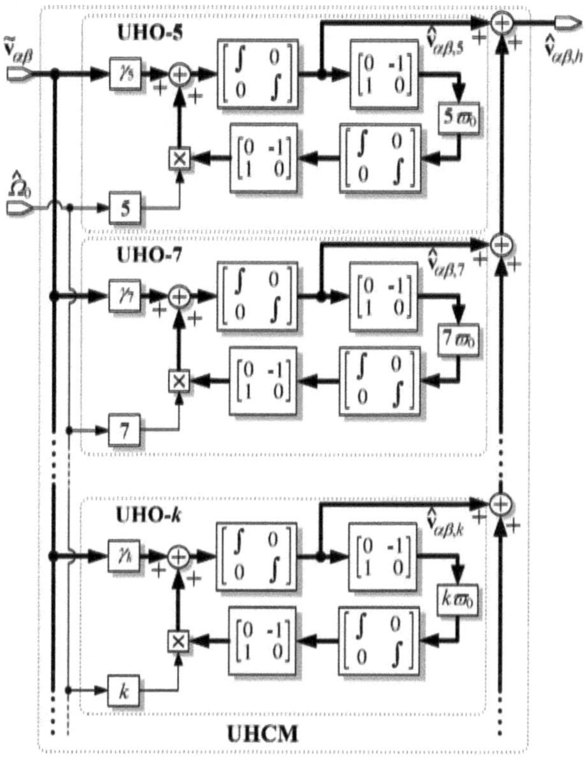

Fig. 4.10 Block diagram of the unbalanced harmonic compensation mechanism UHCM including unbalanced harmonic oscillators (UHO) tuned at the fifth, seventh and kth harmonics

Utility frequency changes. A step change is introduced in the fundamental frequency of the grid voltage, going from 314.16 rad/s (50 Hz) to 219.9 rad/s (35 Hz).

All the following plots show (*from top to bottom*):

- The utility voltage signal in the original three-phase coordinates v_{123};
- The estimated phase angle $\hat{\theta}_0$ (*solid line*), and the actual phase angle θ_0 (*dotted line*);
- The estimated fundamental frequency $\hat{\omega}_0$;
- The estimated positive-sequence voltages in three-phase coordinates $\hat{v}^p_{123}$.

Figure 4.11 shows the transient response obtained with the FRF–PLL scheme when the utility voltage goes from a balanced to an unbalanced operation condition at time $t = 1$ s. Notice that after a relatively short transient all signals return to their desired values. For instance, it is observed that the estimated phase angle (*solid line*) follows perfectly the true phase angle (*dashed line*) after an almost imperceptible transient. The estimated fundamental frequency $\hat{\omega}_0$ (*solid line*) is also maintained in its reference fixed to 314.16 rad/s (*dotted line*) after a small transient. Moreover, the estimated positive-sequence voltages $\hat{v}^p_{123}$ reach their final amplitude of 100 V after an almost imperceptible transient.

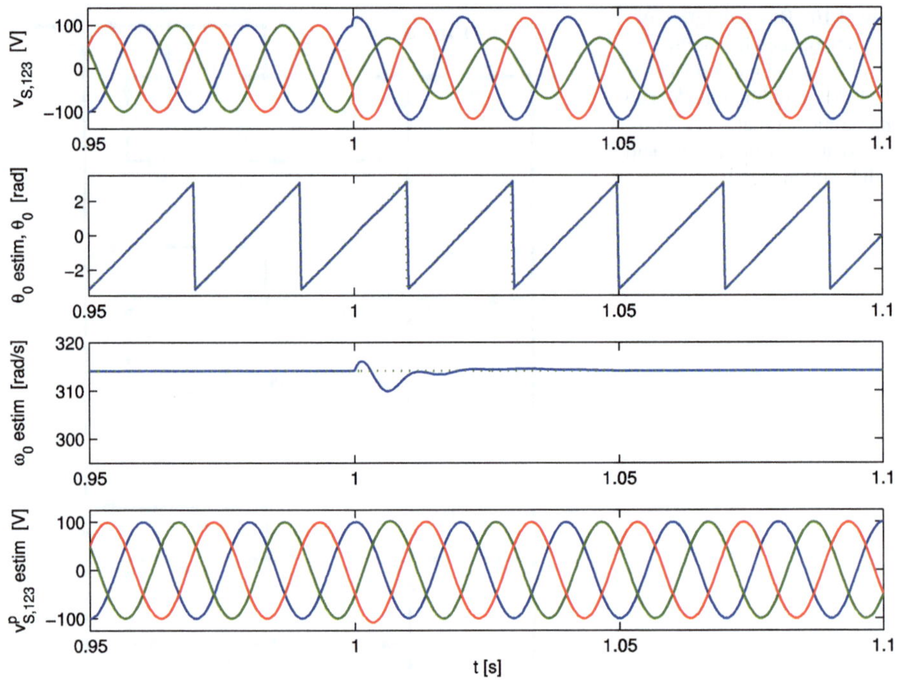

Fig. 4.11 Transient response of the FRF–PLL when the utility voltage goes from a balanced to an unbalanced condition at $t = 1$ s. (*from top to bottom*) Utility voltage v_{123}, estimated phase angle $\hat{\theta}$, estimated angular frequency $\hat{\omega}_0$, and estimated positive-sequence voltage in three-phase coordinates $\hat{v}^p_{123}$

Figure 4.12 presents the transient response of the FRF–PLL scheme when harmonic distortion is added to the already unbalanced utility voltage at $t = 2$ s. Notice that, due to this harmonic distortion, there is a persistent fluctuation in the fundamental frequency estimate $\hat{\omega}_0$ (*third plot*) which, however, keeps an average value of around 314.16 rad/s. This disturbance is also propagated to all other signals, causing mainly a slight deformation of (*bottom plot*) the estimated positive sequence voltages $\hat{v}^p_{123}$. It is worth mentioning that the response of FRF–PLL could be improved by reducing its bandwidth, with the consequence of deteriorating its dynamical response.

Figure 4.13 shows the transient response of the proposed FRF–H–PLL to a step change in the angular frequency of the source voltage going from 314.16 rad/s (50 Hz) to 219.9 rad/s (35 Hz). Notice that in this case the utility voltage includes both disturbances: unbalance and harmonic distortion. It is shown that after a short transient the estimated phase angle follows perfectly the true phase angle. It is shown that the estimated fundamental frequency $\hat{\omega}_0$, starting at a reference of 314.16 rad/s (50 Hz), reaches its new reference fixed to 219.9 rad/s (35 Hz) without persistent fluctuations, and in a relatively short time. The bottom plot shows that the estimated

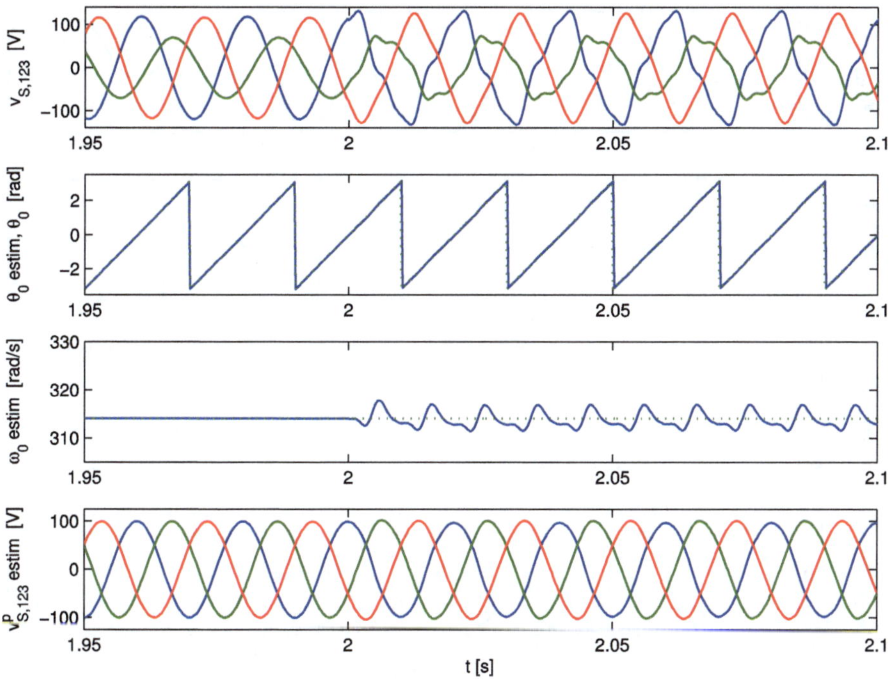

Fig. 4.12 Transient response of the FRF–PLL when harmonic distortion is added at $t = 2$ s to the already unbalanced utility voltage. (*from top to bottom*) Utility voltage v_{123}, estimated phase angle $\hat{\theta}$, estimated angular frequency $\hat{\omega}_0$, and estimated positive sequence voltage in three-phase coordinates $\hat{v}^p_{123}$

positive sequence voltages $\hat{v}^p_{123}$, after a relatively short transient, maintain their amplitude of 100 V without further distortion.

For the experimental results, the proposed FRF–H–PLL algorithm was programmed in a fixed-point digital signal processor (DSP) DSPIC33FJ128GP802 running at 46 MHz, where the sampling frequency has been fixed to 10 kHz. This DSP has been programmed in assembler to optimise the processing time.

The following parameters have been selected for the control: $\gamma_1 = 300$ and $\lambda = 2.22$, which correspond to a settling time of $T_s = 0.03$ s, and considering a nominal frequency $\varpi_0 = 314.16$ rad/s. The square root required in the algorithm is performed with a conventional successive approximation algorithm, which seems to be more effective, and requires less computational effort, than the Newton's iteration algorithm. More details on these algorithms can be seen in [7].

The same test conditions as in the numerical results section have been considered. The voltage signals are constructed using the same parameters as for the numerical results.

Figure 4.14 shows the response of the FRF–PLL when the utility voltage goes from a balanced to an unbalanced condition. It can be observed that after a relatively short transient the estimated fundamental frequency $\hat{\omega}_0$ maintains the correct value.

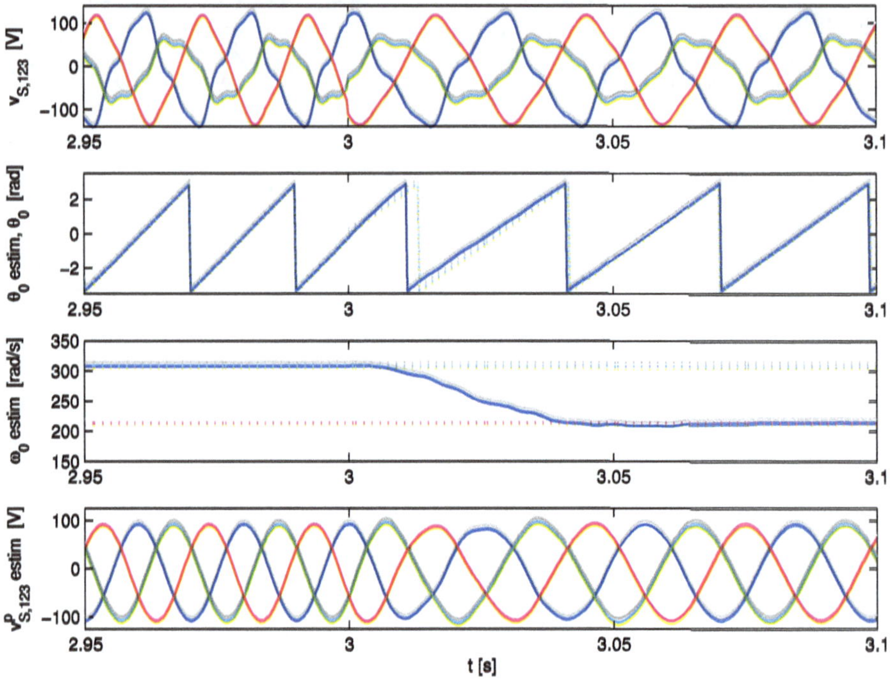

Fig. 4.13 Transient response of the FRF–PLL to an utility frequency step change from 50 Hz to 35 Hz at $t = 3$ s, where the utility voltage is unbalanced and includes harmonic distortion. (*from top to bottom*) Utility voltage v_{123}, estimated phase angle $\hat{\theta}_0$, estimated angular frequency $\hat{\omega}_0$, and estimated positive sequence voltage in three-phase coordinates $\hat{v}^p_{123}$

Notice that the estimated phase angle $\hat{\theta}_0$ has an almost imperceptible transient. Notice also that the estimated positive sequence voltages $\hat{v}^p_{123}$ are almost pure sinusoids with equal amplitude and phase shift of 120° between them.

Figure 4.15 presents the transient response of the FRF–PLL to a step change in the angular frequency of the source voltage going from 35 Hz to 50 Hz. Notice that the estimated angular frequency $\hat{\omega}_0$ reaches its desired value after a relatively short transient. Notice also that the estimated positive sequence voltages $\hat{v}^p_{123}$ maintain their amplitude and the appropriate phase after an almost imperceptible transient.

4.5 Summary

This chapter presented synchronisation methods for single and three-phase connections based on the detection of the fundamental frequency approach. They comprised an estimator for the fundamental component and an adaptive law to reconstruct the fundamental frequency. In addition, the three-phase scheme included estimators for the positive sequence component. The design was based on a complete

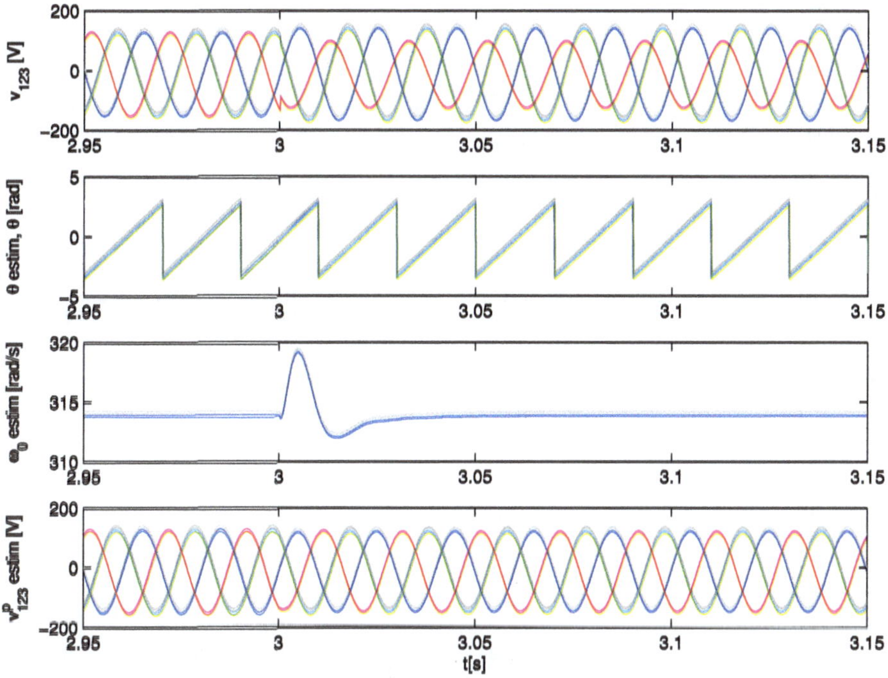

Fig. 4.14 Experimental transient response of the FRF–PLL when the utility voltage goes from a balanced to an unbalanced condition at $t = 3$ s. (*from top to bottom*) Utility voltage v_{123}, estimated phase angle $\hat{\theta}_0$, estimated angular frequency $\hat{\omega}_0$, and estimated positive sequence voltage in three-phase coordinates $\hat{v}^P_{123}$

model description in terms of fixed-frame coordinates of signals. The model description considered the effect of harmonic distortion, and in the three-phase case the effect of unbalanced operation as well. The methods presented were referred to as AQSG–PLL and AQSG–H–PLL for the single-phase case, and FRF–PLL and FRF–H–PLL for the three-phase case. In these schemes, the H stands for the harmonic compensation capability. It was observed that in spite of distorted reference signals both methods, AQSG–H–PLL and FRF–H–PLL, were able to deliver an almost pure sinusoidal signals vector representing fundamental component plus a ripple-free fundamental frequency estimated signal. In particular, the three-phase FRF–H–PLL was able to handle unbalanced operation, delivering a clean positive sequence component of the reference signal. It was also observed that both methods were able to compensate step variations on the frequency thanks to the adaptation mechanism. As the proposed methods includes an explicit mechanism to compensate harmonics, the effect of this type of disturbances is considerably reduced without compromising the speed of response of the overall scheme. The harmonic compensation mechanisms, either HCM or UHCM, appeared as plug-in blocks that could be used in cases of highly distorted reference signals. In cases of low distortion, these mechanisms can be easily disconnected and the harmonic compensation

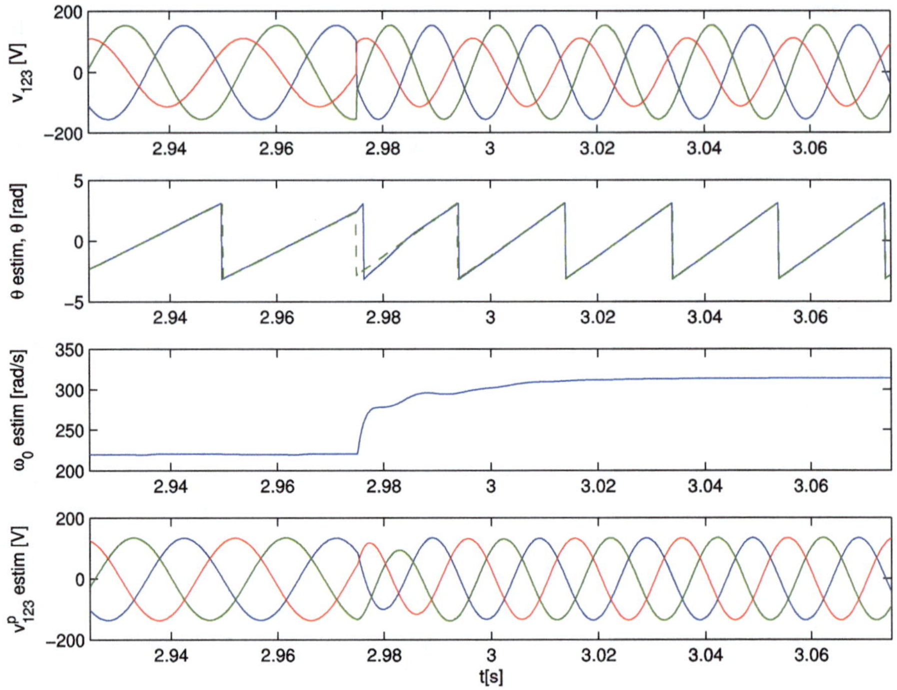

Fig. 4.15 Experimental transient response of the FRF–PLL to a utility frequency change from 35 Hz to 50 Hz. (*from top to bottom*) Utility voltage v_{123}, estimated phase angle $\hat{\theta}_0$, estimated angular frequency $\hat{\omega}_0$, and estimated positive sequence voltage in three-phase coordinates $\hat{v}^p_{123}$

can rely on the selective nature of a basic method, which can be enhanced at the expenses of reducing the speed of response of the overall scheme.

References

1. Arruda, L.N., Silva, S.M., Cardoso, B.J.: PLL structures for utility connected systems. In: Proc. of the IEEE Industry Applications Society Annual Meeting, Chicago, Illinois, USA, pp. 2655–2660 (2001)
2. Cichowlas, M., Malinowski, M., Kazmierkowski, M.P., Sobczuk, D.L., Rodriguez, P., Pou, J.: Active, filtering function of three-phase PWM boost rectifier under different line voltage conditions. IEEE Trans. Ind. Electron. **52**(2), 410–419 (2005)
3. Escobar, G., Martinez-Montejano, M.F., Valdez, A.A., Martinez, P.R., Hernandez-Gomez, M.: Fixed-reference-frame phase-locked loop for grid synchronization under unbalanced operation. IEEE Trans. Ind. Electron. **58**(5), 1943–1951 (2011)
4. Ghosh, A., Joshi, A.: A new algorithm for the generation of reference voltages of a DVR using the method of instantaneous symmetrical components. IEEE Power Eng. Rev. **22**(1), 63–65 (2002)
5. Hingorani, N.G., Gyugyi, L.: Understanding FACTS: Concepts and Technology of Flexible AC Transmission Systems, 1st edn. Wiley–IEEE Press, Piscataway (1999)

6. Hsieh, G.C., Hung, J.C.: Phase-locked loop techniques—a survey. IEEE Trans. Ind. Electron. **43**(6), 609–615 (1996)
7. Hwang, K.: Computer Arithmetic: Principles, Architecture, and Design. Wiley, New York (1979)
8. Karimi-Ghartemani, M., Iravani, M.R.: A method for synchronization of power electronic converters in polluted and variable-frequency environments. IEEE Trans. Power Syst. **19**(3), 1263–1270 (2004)
9. Kaura, V., Blasco, V.: Operation of a phase locked loop system under distorted utility conditions. IEEE Trans. Ind. Appl. **33**(1), 58–63 (1997)
10. Mattavelli, P.: A closed-loop selective harmonic compensation for active filters. IEEE Trans. Ind. Appl. **37**(1), 81–89 (2001)
11. Mojiri, M., Bakhshai, A.R.: An adaptive notch filter for frequency estimation of a periodic signal. IEEE Trans. Autom. Control **49**(2), 314–318 (2004)
12. Perez, M.A., Espinoza, J.R., Moran, L.A., Torres, M.A., Araya, E.A.: A robust phase-locked loop algorithm to synchronize static-power converters with polluted AC systems. IEEE Trans. Ind. Electron. **55**(5), 2185–2192 (2008)
13. Rodriguez, P., Luna, P., Ciobotaru, M., Teodorescu, R., Blaabjerg, F.: Advanced grid synchronization system for power converters under unbalance and distorted operation conditions. In: Proc. of the IEEE Industrial Electronics Conference, Paris, France, pp. 5173–5178 (2006)
14. Rodriguez, P., Luna, A., Candela, I., Teodorescu, R., Blaabjerg, F.: Grid synchronization of power converters using multiple second order generalized integrators. In: Proc. of the IEEE Industrial Electronics Conference, Orlando, Florida, USA, pp. 755–760 (2008)
15. Saitou, N., Matsui, M., Shimizu, T.: A control strategy of single-phase active filter using a novel d-q transformation. In: Proc. of the IEEE Industry Applications Society Annual Meeting, Shelby Township, Michigan, USA, pp. 1222–1227 (2003)
16. Santos, R.M., Seixas, P.F., Cortizo, P.C., Torres, L.A.B., Souza, A.F.: Comparison of three single-phase PLL algorithms for UPS applications. IEEE Trans. Ind. Electron. **55**(8), 2923–2931 (2008)
17. Shinnaka, S.: A robust single-phase pll system with stable and fast tracking. IEEE Trans. Ind. Appl. **44**(2), 624–633 (2008)
18. Song, H., Nam, K.: Instantaneous phase-angle estimation algorithm under unbalanced voltage-sag conditions. IEE Proc. **147**(6), 409–415 (2000)
19. Svensson, J.: Synchronisation methods for grid connected voltage source converters. IEE Proc. **148**(1), 229–235 (2001)
20. Yazdani, D., Mojiri, M., Bakhshai, A., Joos, G.: A fast and accurate synchronization technique for extraction of symmetrical components. IEEE Trans. Power Electron. **24**(3), 674–684 (2009)

Part II
Mathematical Modelling Perspectives

Chapter 5
Averaging for Power Converters

Carmen Pedicini, Luigi Iannelli, Francesco Vasca, and Ulf Jönsson

5.1 Introduction

Most power electronics systems have switching devices that more or less instantaneously change the dynamics of the system. Ideal models of diodes and controlled switches give rise to a class of switched systems which consists of differential equations with discontinuous nonlinearities and external excitation signals. Switched systems are appropriate models for many physical systems encountered in practice which exhibit a switching behaviour depending on various environmental factors. A switched system is a hybrid dynamical system consisting of a family of dynamical subsystems and a policy that selects the active subsystem at each time instant. This class of systems finds extensive application in various areas of engineering practice such as power electronics, network communication, hybrid control, traffic flow, biosystem modelling, etc. (see, e.g. [14, 24] and the references therein). So, switched systems have been studied for several decades in the systems and control literature and in the last few years every major control conference had several regular and invited session on switching systems and control.

Detailed analysis of switched models of power converters is difficult due to potentially complex interaction between the electrical dynamics and the high frequency modulation which determines the commutations of the switches. A possible approach to circumvent some of these difficulties is to average the hybrid dynamics

C. Pedicini (✉) · L. Iannelli · F. Vasca
Department of Engineering, University of Sannio, Piazza Roma 21, 82100 Benevento, Italy
e-mail: carmen.pedicini@unisannio.it

L. Iannelli
e-mail: luigi.iannelli@unisannio.it

F. Vasca
e-mail: francesco.vasca@unisannio.it

U. Jönsson
Department of Mathematics, Royal Institute of Technology (KTH), Stockholm, Sweden

F. Vasca, L. Iannelli (eds.), *Dynamics and Control of Switched Electronic Systems*, Advances in Industrial Control,
DOI 10.1007/978-1-4471-2885-4_5,

over the switching period. Analysis and control design can then be carried out on the simpler averaged system.

On the practical perspective, the averaging approach is widely used in the power electronics community [16, 22]. By assuming some typical operating conditions, such as the repetitive sequence of modes, the average model is obtained as a convex combination of the modes exhibited by the converter [20].

From the theoretical point of view, averaging is a classical mathematical tool for analysing time-scale separation in dynamical systems and it is based on two observations [4, 11]:

- Through an appropriate coordinate transformation, a rapidly time-varying system can be viewed as a small perturbation of a simplified, time-invariant, averaged system, and
- The qualitative behaviour of the solutions of a "classical" dynamical system is robust to small perturbations.

For a general class of Lipschitz-continuous systems, the averaging approach was theoretically justified already in the 1970s [26, 27], but only recently rigorous averaging analysis for nonsmooth and switched systems has been considered [9, 23, 25].

In this chapter, recent results on averaging for generic switched systems are recast in a framework suitable for power converters analysis. Initially, by using several electronic systems topologies, it is shown how power converters can be modelled as switched systems. Then the practical interest of some averaging theorems for nonsmooth systems with a quite arbitrary periodic modulation is presented by considering three different approaches: modes repetition, dithering and dynamic phasor. The theoretical findings are illustrated through several examples and simulated models of power converters. The chapter is concluded by presenting some open problems and directions for future research.

5.2 Switched Models for Power Converters

In this section, through practical examples we show why switched systems can be considered as a suitable class for modelling power converters.

5.2.1 Switched Linear Systems

A switched system can be described by a differential equation of the form [14]:

$$\dot{x}(t) = f_{\sigma(t)}\big(x(t), u(t)\big) \tag{5.1}$$

where the state $x \in \mathbb{R}^n$, the input $u \in \mathbb{R}^p$ and $\sigma : [0, \infty) \to \Sigma$ is a piecewise constant, right continuous function which selects at each time instant t the index of the active system from the index set Σ. The functions f_σ are assumed to be sufficiently regular at least locally Lipschitz, i.e. for each $\sigma \in \Sigma$ and for every pair (t_0, x_0) in $[t_0, \infty) \times \mathbb{R}^n$ there exists a constant $L > 0$ such that the inequality $|f_\sigma(t, x_1) - f_\sigma(t, x_2)| \leq L|x_1 - x_2|$ holds for all (t, x_1) and (t, x_2) in some neighbourhood of (t_0, x_0).[1] In the sequel, $\sigma(t)$ will be also called the *mode-selector function.*

By a solution of the switched system (5.1) we mean a pair $(x(t), \sigma(t))$ that satisfies (5.1) for almost all $t \in [0, \infty)$. Indeed, at the isolated switching time instants, i.e. the time instants at witch the mode-selector function has discontinuities (it is right-continuous), the state time derivative is not well defined in the classical sense. To simplify the notation, in some case we will omit the time arguments.

Switched systems with controlled time-dependent switchings can be described in a language that is more standard in control theory. Assume that Σ is a finite set, $\Sigma = \{1, 2, \dots, m\}$, and consider an invertible map that let correspond each element $\overline{\sigma} \in \Sigma$ to a vector $\overline{q} = (\overline{q}_1, \dots, \overline{q}_m)^\top$ with $\overline{q}_i = 1$ for $i = \overline{\sigma}$ and $\overline{q}_i = 0$ for $i \neq \overline{\sigma}$, with $i = 1, \dots, m$.

Then correspondingly to the mode-selector function $\sigma(t)$ one can define the *mode-switching functions*, or more simply indicated as *switching functions*:

$$q_i(t) : [0, \infty) \to \{0, 1\} \tag{5.2}$$

with

$$\sum_{i=1}^{m} q_i(t) = 1, \quad \forall t \in [0, \infty). \tag{5.3}$$

Since each index $\sigma(t) = i$ defines a different dynamic behaviour of the system, it is often referred to it as *mode* i. Figure 5.1 shows an example in the case $m = 4$.

Then the switching system (5.1) can be represented by the following equations:

$$\dot{x}(t) = \sum_{i=1}^{m} q_i(t) f_i\big(x(t), u(t)\big). \tag{5.4}$$

[1]Here and below, we denote by $|\cdot|$ the standard Euclidean norm on $\mathbb{R}^n$.

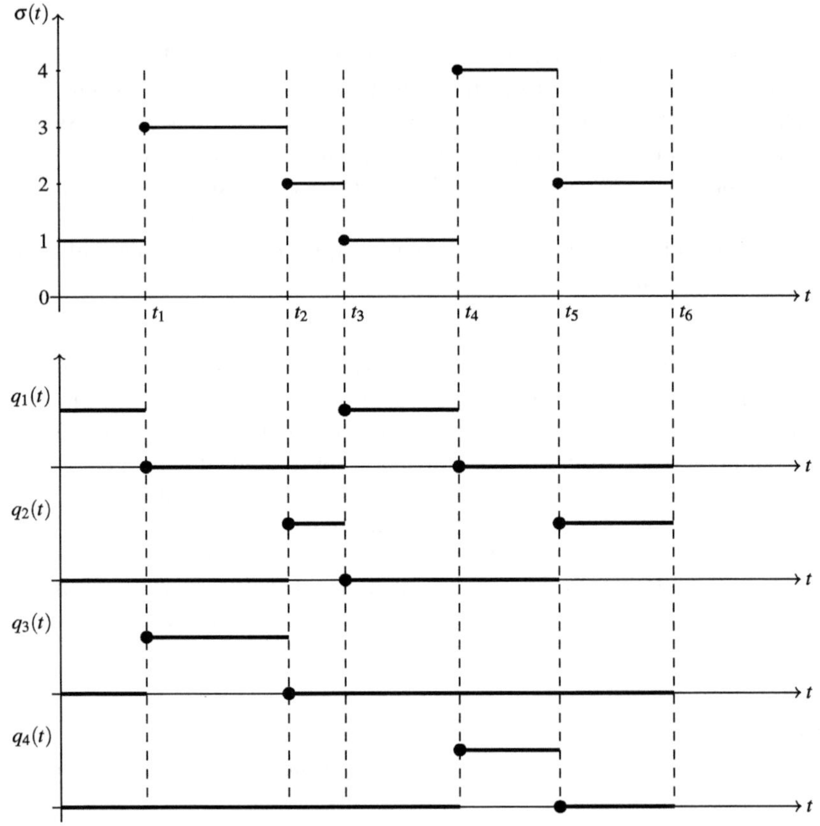

Fig. 5.1 Switching signals in the case of four modes

When all modes are described by linear differential equations, a switched linear system is obtained:

$$\dot{x}(t) = A_{\sigma(t)}x(t) + B_{\sigma(t)}u(t), \tag{5.5}$$

which by using (5.4) can be also represented as:

$$\dot{x}(t) = \sum_{i=1}^{m} q_i(t)\big(A_i x(t) + B_i u(t)\big), \tag{5.6}$$

where $A_i \in \mathbb{R}^{n\times n}$ and $B_i \in \mathbb{R}^{n\times p}$ are constant matrices.

This class of systems is one of the most commonly treated in the literature on switched systems and in such form dynamic models of power converters can be rep-

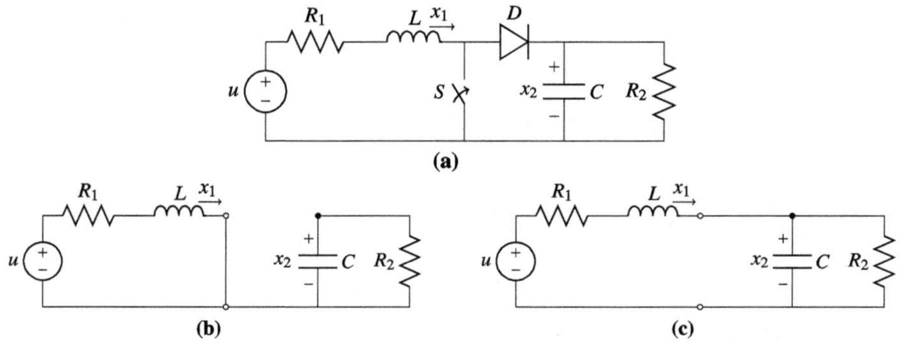

Fig. 5.2 (**a**) DC–DC boost converter, and the corresponding equivalent circuits for the mode with the switch (**b**) ON and (**c**) OFF

resented. In this section, problems will be formulated and discussed in the context of linear switched systems.

5.2.2 *Modes, Minimality and Jumps*

In order to obtain a model in the form (5.6), it is necessary to analyse the modes of the system according to the switching signals, and to obtain the matrices A_i and B_i for each mode.

Example 5.2.1 (DC–DC boost converter) Consider the DC–DC boost converter shown in Fig. 5.2 and assume it operates in continuous conduction mode (CCM), i.e. the current through the inductor never goes to zero. The converter switches between two different configurations according to the position of the switch S, see Figs. 5.2(b) and (c). There are two different modes, $\Sigma = \{1, 2\}$ and the mode-selector function is: $\sigma(t) = 1$ if the switch is ON and $\sigma(t) = 2$ if the switch is OFF. By choosing as state variables the inductor current x_1 and the capacitor voltage x_2, the matrices of the switched model (5.6) can be written as follows:

$$A_1 = \begin{pmatrix} -\frac{R_1}{L} & 0 \\ 0 & -\frac{1}{R_2 C} \end{pmatrix}, \qquad A_2 = \begin{pmatrix} -\frac{R_1}{L} & -\frac{1}{L} \\ \frac{1}{C} & -\frac{1}{R_2 C} \end{pmatrix},$$

$$B_1 = \begin{pmatrix} \frac{1}{L} \\ 0 \end{pmatrix}, \qquad B_2 = \begin{pmatrix} \frac{1}{L} \\ 0 \end{pmatrix}.$$

The model (5.6) is valid under the constraint that the inductor current x_1 is positive. This implicit assumption depends on the operating condition of the CCM for which it is assumed that the current through the diode is always positive.

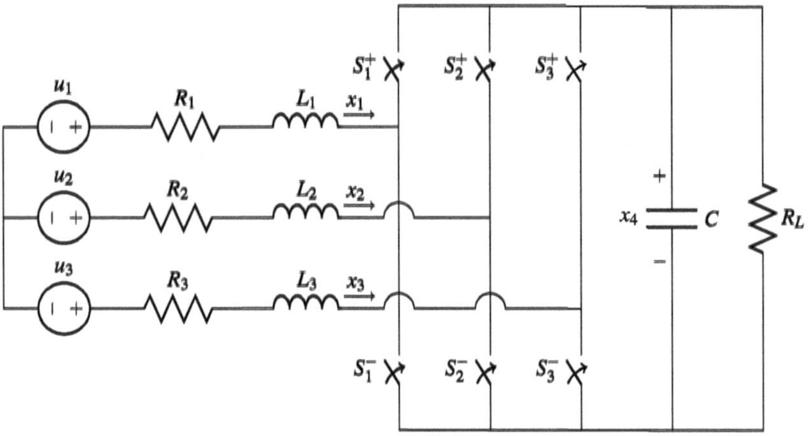

Fig. 5.3 Three-phase converter

Table 5.1 Modes and switching functions for three-phase converter; for each leg the lower switch is complementary controlled with respect to the corresponding upper switch, e.g. S_1^+ being ON implies that S_1^- is OFF

i (mode)	S_1^+	S_2^+	S_3^+	ℓ_1	ℓ_2	ℓ_3	q_i
1	ON	OFF	OFF	1	0	0	$\ell_1(1-\ell_2)(1-\ell_3)$
2	ON	ON	OFF	1	1	0	$\ell_1\ell_2(1-\ell_3)$
3	ON	OFF	ON	1	0	1	$\ell_1(1-\ell_2)\ell_3$
4	ON	ON	ON	1	1	1	$\ell_1\ell_2\ell_3$
5	OFF	OFF	OFF	0	0	0	$(1-\ell_1)(1-\ell_2)(1-\ell_3)$
6	OFF	ON	OFF	0	1	0	$(1-\ell_1)\ell_2(1-\ell_3)$
7	OFF	OFF	ON	0	0	1	$(1-\ell_1)(1-\ell_2)\ell_3$
8	OFF	ON	ON	0	1	1	$(1-\ell_1)\ell_2\ell_3$

The following example shows that the converter topology and some assumptions on its behaviour might be useful for simplifying the construction of the switched model, avoiding the need for a-priori computation of all modes.

Example 5.2.2 (Three-phase converter) Consider the three-phase power converter represented in Fig. 5.3 and assume that the switches of the same leg are never simultaneously ON or OFF. For the ith leg we define $\ell_i = 1$ if S_i^+ is ON (and S_i^- is OFF) and $\ell_i = 0$ if S_i^+ is OFF (and S_i^- is ON), for $i = 1, 2, 3$. There are 2^3 conducting modes that correspond to the different combinations of switches position as reported in Table 5.1.

By applying the Kirchhoff's laws, one obtains:

$$L_1\dot{x}_1 = -R_1x_1 + u_1 - u_2 + R_2x_2 + L_2\dot{x}_2 - (\ell_1 - \ell_2)x_4, \tag{5.7a}$$

$$L_2\dot{x}_2 = -R_2x_2 + u_2 - u_3 + R_3x_3 + L_3\dot{x}_3 - (\ell_2 - \ell_3)x_4, \tag{5.7b}$$

$$L_3\dot{x}_3 = -R_3x_3 + u_3 - u_1 + R_1x_1 + L_1\dot{x}_1 - (\ell_3 - \ell_1)x_4, \tag{5.7c}$$

$$C\dot{x}_4 = \ell_1x_1 + \ell_2x_2 + \ell_3x_3 - \frac{1}{R_L}x_4. \tag{5.7d}$$

By differentiating the equilibrium condition for the currents $x_1 + x_2 + x_3 = 0$, one can substitute $\dot{x}_2 = -\dot{x}_1 - \dot{x}_3$ in (5.7a) and, by using (5.7c), one obtains:

$$\begin{aligned}\left(L_1 + L_2 + \frac{L_1L_2}{L_3}\right)\dot{x}_1 = &-\left(R_1 + R_2 + \frac{L_2}{L_3}R_1\right)x_1 + \left(\frac{L_2}{L_3}R_3 - R_2\right)x_3\\ &-\left[\ell_1 - \ell_2 - \frac{L_2}{L_3}(\ell_3 - \ell_1)\right]x_4 + \left(1 + \frac{L_2}{L_3}\right)u_1\\ &- u_2 - \frac{L_2}{L_3}u_3.\end{aligned} \tag{5.8a}$$

With similar arguments one can write

$$\begin{aligned}\left(L_2 + L_3 + \frac{L_2L_3}{L_1}\right)\dot{x}_2 = &-\left(R_2 + R_3 + \frac{L_3}{L_1}R_2\right)x_2 + \left(\frac{L_3}{L_1}R_1 - R_3\right)x_1\\ &-\left[\ell_2 - \ell_3 - \frac{L_3}{L_1}(\ell_1 - \ell_2)\right]x_4 + \left(1 + \frac{L_3}{L_1}\right)u_2\\ &- u_3 - \frac{L_3}{L_1}u_1,\end{aligned} \tag{5.8b}$$

$$\begin{aligned}\left(L_3 + L_1 + \frac{L_3L_1}{L_2}\right)\dot{x}_3 = &-\left(R_3 + R_1 + \frac{L_1}{L_2}R_3\right)x_3 + \left(\frac{L_1}{L_2}R_2 - R_1\right)x_2\\ &-\left[\ell_3 - \ell_1 - \frac{L_1}{L_2}(\ell_2 - \ell_3)\right]x_4 + \left(1 + \frac{L_1}{L_2}\right)u_3\\ &- u_1 - \frac{L_1}{L_2}u_2,\end{aligned} \tag{5.8c}$$

$$C\dot{x}_4 = \ell_1x_1 + \ell_2x_2 + \ell_3x_3 - \frac{1}{R_L}x_4. \tag{5.8d}$$

By specifying the values of the switching functions ℓ_i and by using Table 5.1, it is simple to obtain the matrices of the modes corresponding to the model (5.6) for the three-phase converter.

The switched model presented in Example 5.2.2 is clearly not minimal because of the zero sum constraint among the three currents x_1, x_2 and x_3. On the other hand, it is straightforward to obtain a minimal representation by substituting, for instance, $x_3 = -x_1 - x_2$ in (5.8a)–(5.8d).

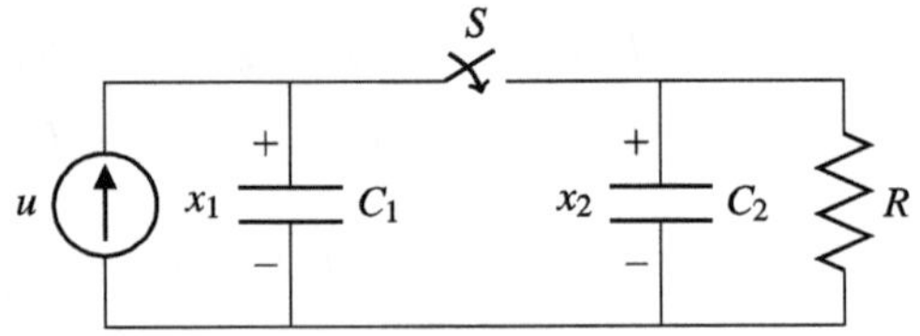

Fig. 5.4 Two switched parallel capacitors

There are other power converters for which only some of the modes (and not the entire model) can be represented with a reduced-order model. Also in these cases, the model (5.6) remains valid because one can use a redundant state for some modes.

This situation is described by the following example.

Example 5.2.3 (Two switched parallel capacitors) Consider the circuit in Fig. 5.4, which has two modes, corresponding to the switch being OFF ($\sigma = 1$) and ON ($\sigma = 2$). To characterise the circuit behaviour, we choose as state variables the voltages on the capacitors C_1 and C_2. In the OFF mode, the model is:

$$C_1\dot{x}_1 = u, \qquad C_2\dot{x}_2 = -\frac{1}{R}x_2. \tag{5.9}$$

When the switch is conducting, the model reduces to a single equation because the capacitors are in parallel:

$$\dot{x}_1 = -\frac{1}{R(C_1 + C_2)}x_1 + \frac{1}{(C_1 + C_2)}u, \tag{5.10}$$

with $x_1 \equiv x_2$. The matrices for the switched model (5.6) can be obtained by using twice Eq. (5.10) for the ON mode.

In general, it is possible to have state jumps (discontinuities) at the switching time instants. For instance, this occurs for the Example 5.2.3 if S is turned on when the two capacitor voltages are different. On the other hand, such jumps are not allowed in practical operating conditions of power converters and therefore will be excluded from our analysis.

Equation (5.6) is not always representative for the entire converter behaviour because some algebraic constraints resulting from the Kirchhoff's laws might have not been included. Another complication in the representation (5.6) arises in power converters with diodes, for which the mode-selector functions and the mode-switching functions become dependent also on the state $x(t)$. Such aspects will be considered in the following subsection.

5.2.3 Modulations, Sequence Repetitiveness and Periodicity

For the averaging analysis, it is of primary importance how the mode-selector function $\sigma(t)$ (or equivalently, the mode-switching function $q_i(t)$) is determined. In the previous discussion, the mode-selector functions and the mode-switching functions are assumed to be only functions of time t; however, in many situations the values of σ and q_i might depend also on the state $x(t)$. In these cases, Eq. (5.6) can be written as:

$$\dot{x}(t) = \sum_{i=1}^{m} q_i\big(t, x(t)\big)\big(A_i x(t) + B_i u(t)\big). \tag{5.11}$$

A typical example of this situation is the use of pulse-width modulation (PWM).

Example 5.2.4 (PWM with two modes) Consider a switched system with $\Sigma = \{1, 2\}$, and the step function $n : \mathbb{R} - \{0\} \to \{0, 1\}$ defined as

$$n(z) = \begin{cases} 1 & \text{if } z > 0, \\ 0 & \text{if } z < 0, \end{cases} \tag{5.12}$$

and not defined for $z = 0$. In order to control the operating mode of the system with a PWM, a typical choice is

$$z = k^\top x(t) + r - \delta(t) \tag{5.13}$$

where $k^\top$ and r are constant and $\delta(t) : [0, \infty) \to [\Delta_{\min}, \Delta_{\max}]$ is a periodic signal (typically a sawtooth signal).

The switching function $q_1(t, x(t))$ can be defined as equal to the nonlinearity n given by (5.12)–(5.13), almost everywhere. The other switching function is defined as

$$q_2\big(t, x(t)\big) = 1 - q_1\big(t, x(t)\big) = 1 - n\big(k^\top x(t) + r - \delta(t)\big). \tag{5.14}$$

The value of q_1 (and q_2) at the isolated switching instants is defined by the assumption that the mode-switching functions are right-continuous. The dynamic model of the switched system is still represented by the model (5.6).

The determination of the mode-selector function becomes more complicated when the number of modes increases, as shown by the following example.

Example 5.2.5 (PWM for the three-phase converter) Consider the converter presented in Example 5.2.2. A typical open-loop PWM approach for determining the

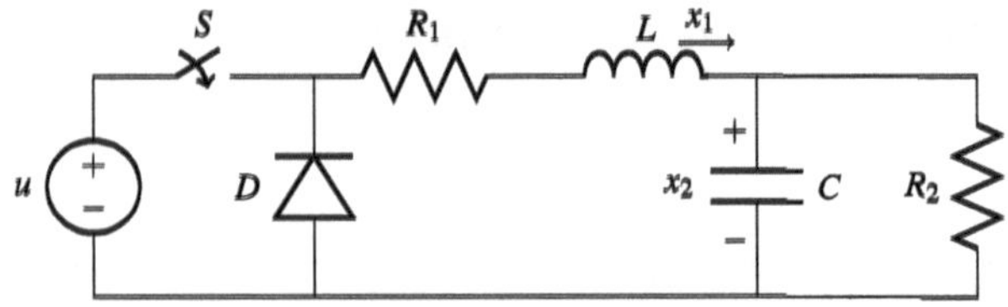

Fig. 5.5 DC–DC buck converter

modes corresponds to

$$\ell_i(t) = n\big(r_i(t) - \delta(t)\big) \tag{5.15}$$

almost everywhere with $r_i(t) = \sin(\omega t - 2\pi(i-1)/3)$ and $\omega = 2\pi/T_r$ being the frequency. Typically, $\delta(t)$ is a periodic signal (a triangular waveform) with period $T_\delta \ll T_r$. By using Table 5.1, we can obtain the mode-switching functions whose values at the switching time instants can be determined by the right-continuous assumption and (5.3).

In power converters that include also diodes, the *discontinuous conduction mode* (DCM) operations make the mode-selector function dependent on the state also in open-loop. In this case, it is not easy to define the mode-switching functions by using step nonlinearity but they can be simply defined by using an automaton, as shown in the following example.

Example 5.2.6 (DC–DC buck converter in DCM) Consider the DC–DC buck converter shown in Fig. 5.5. The DCM operation differs from the CCM operation by an additional third mode in which neither the switch nor the diode conducts current. Then the converter switches among three different modes: switch ON and the diode is blocking ($\sigma = 1$), switch OFF and the inductor current is positive determining the diode to be conducting ($\sigma = 2$), switch OFF and the current is zero with the diode being blocking ($\sigma = 3$). Since u is typically strictly positive, the mode with the switch ON and the diode conducting is not possible. By choosing as state variables the inductor current x_1 and the capacitor voltage x_2, the dynamic model of the converter in mode 1 is

$$L\dot{x}_1 = -R_1 x_1 - x_2 + u, \qquad C\dot{x}_2 = x_1 - \frac{1}{R_2}x_2; \tag{5.16}$$

in mode 2 it is

$$L\dot{x}_1 = -R_1 x_1 - x_2, \qquad C\dot{x}_2 = x_1 - \frac{1}{R_2}x_2, \tag{5.17}$$

and in mode 3 it is

$$\dot{x}_1 = 0, \qquad C\dot{x}_2 = -\frac{1}{R_2}x_2. \tag{5.18}$$

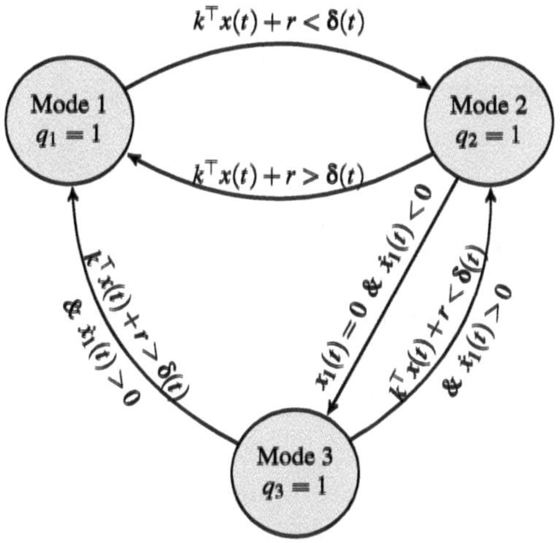

Fig. 5.6 Mode-selector function automaton for the switched model of buck converter in DCM

Note that in mode 3 the current x_1 is identically zero. From (5.16)–(5.17), it is straightforward to determine the matrices of the model (5.6). Assuming the switch ON and the switch OFF states are controlled by means of a PWM, the mode-switching functions can be determined by using the automaton shown in Fig. 5.6. The mode-switching functions at the switching time-instants are determined according to the conditions on the edges of the automaton, i.e. when the condition on the edge $i \to j$ holds we assume $q_i = 0$ and $q_j = 1$. In this case, it is assumed that for all x and t, the mode-switching function q_i is feasible. For instance, suppose that the condition on the edge $i \to j$ is verified but immediately also the condition on the edge $j \to k$ is verified so the system does not stop in mode j but goes directly into mode k. If it always happens and for all edges, the system never stops in one state. It is assumed that this situation does not occur, i.e. the system remains in one state for a finite time interval.

The examples above have shown that the PWM is characterised by high-frequency periodic signals $\delta(t)$ that influence the sequence of modes. Indeed, $\delta(t)$ is an external periodic carrier signal which establishes the switching frequency. The sequence of modes is not always repetitive in time. For instance, the buck converter might operate in DCM ($m = 3$ and $123123\ldots$ as a sequence of modes) for some time intervals during transients and in the CCM at the steady-state ($121212\ldots$ as a sequence of modes). The easiest way to tackle the problem is to first analyse the sequence of modes. To this aim, we can consider the definition of a cycle.

A *cycle* is any sequence of modes in which each mode appears at most once. Then any sequence of modes can be represented as a sequence of cycles. For instance, in the case of four modes, considering two different sequences of modes, we have the following partitioning into cycles: $1234123212314 = \{1234, 123, 21, 2314\}$, $3421232131412 = \{3421, 23, 213, 14, 12\}$. Consider a sequence of modes; without loss of generality, we denote t_k^{in} the time instant at which the kth cycle starts and t_{k+1}^{in} the time instant when the kth cycle ends and the $(k+1)$th cycle starts. The time interval $t^{\rm in}_{k+1} - t^{\rm in}_{k} = T_k$ is called *cycle-time* of the kth cycle. The duration of a mode i in the kth cycle is denoted by $\tau_{i,k}$. The fraction of cycle-time T_k that the system spends in that mode is called *duty ratio* $d_{i,k}$:

$$d_{i,k} = \frac{\tau_{i,k}}{T_k}. \tag{5.19}$$

From the above definitions, it is clear that the duty ratios satisfy: $d_{i,k} \in [0, 1]$ and $\sum_{i=1}^{m} d_{i,k} = 1$; $d_{i,k} = 0$ with $i = M$ indicates that the mode $\sigma = M$ is not visited in the kth cycle time. In the case where in each cycle the sequence of modes, the cycle-times T_k, the duty ratios $d_{i,k}$, and the state at switching instants are identical, the system is called periodic.

For the analysis of power converters, people are interested in the periodic case only at steady-state operating conditions. Since $d_{i,k}$ might be zero, it is possible to intuitively accept that the cycle time is chosen (constant) by some 'external' forcing signal. For instance, in the case of Examples 5.2.4 and 5.2.5, a possible choice could be $T_k = T = T_\delta\ \forall k$. Indeed, for a wide class of pulse-modulated power converter we have $T_k = T$ for any k. In what follows, we assume that the cycle time is constant; on the other hand, some of the averaging results can be extended to the case of different cycle times, for instance, by choosing $T = \max_k\{T_k\}$. Note that the assumption of a constant cycle time does not imply that the system is periodic or that all modes are visited in each cycle time (the switching functions do not need to be periodic). Moreover, the definition of a cycle implies that each mode is visited at most once in each cycle time from kT and $(k+1)T$ with k any integer. If a common period T can be chosen, the duty ratio of the ith mode can be written as:

$$d_i(t) = \frac{1}{T}\int_{t-T}^{t} q_i(s)\,\mathrm{d}s, \tag{5.20}$$

with $q_i(t) = 0$ for $t < 0$. Note that if $q_i(t)$ is periodic of period T then $d_i(t)$ is constant and is equal to (5.19). The duty ratio represents the moving average over T of the mode-switching function [20].

5.3 Averaged Systems

Averaging theory is a powerful tool to analyse nonlinear systems. This method has a rich history and was proposed initially by Bogoliuboff and Mitropolsky [4] and developed by several authors as Sethna [21] and Hale [8]. The basic idea exploited is the time-scale separation between the time variations of the state of a dynamical system and the time variations of the derivative of the state; from this point of view, the dynamics of a (slowly) time-varying system are approximated by the dynamics of a time invariant system, that is called averaged system. The theory justifies the closeness of the trajectories of the averaged system to the unaveraged system and the relationship between the stability properties of the two systems [11]. To the best of our knowledge, the first averaging theorems for discontinuous time systems were presented in [3, 12]; such theorems were based over an infinite time. Hereafter the averaging techniques have been used extensively in the analysis of switched systems: in [7, 18] the averaging for the switched linear system and the corresponding error bound have been discussed, and in [9, 13] results on averaging for dithered systems have been obtained. Recently, averaging theories have been extended to PWM systems and switched systems with disturbances [23, 25].

An explicit formulation of averaging for switching systems is possible when a two time scale property is exhibited in the system, i.e. the switching signal is faster than the continuous state space variables that can be considered the slow variables. Then the system can be approximated by a model consisting only of the slow continuous states, that is, the averaged model. Making an analogy with the averaging for continuous case, the approximation properties depend on a small parameter ε which is related to the frequencies of the switching signal, i.e. the approximation is a function of the switching time intervals. A typical assumption for the switched systems in the averaging theory is that there do not exist Zeno solutions, i.e. infinitely many jumps in finite time. Furthermore, in the following it will be assumed that the switched system (5.4) has at least one absolutely continuous solution $x(t, x_0)$ (in the sense of Carathéodory) and that the solutions have no sliding modes. Thus, similarly to standard averaging approaches, it can be expected that the behaviour of the slow states can be described by an averaged model [25].

Consider the switched system (5.6), assume that for small cycle time T the continuous state $x(t)$ varies slower than the mode-switching functions $q_i(t)$. In this case, an averaged model can be defined as [11, 20]:

$$\dot{w}(t) = \frac{1}{T}\int_{t-T}^{t}\sum_{i=1}^{m} q_i(s)\big(A_i w(t) + B_i u(t)\big)\,\mathrm{d}s = \sum_{i=1}^{m} d_i(t)\big(A_i w(t) + B_i u(t)\big), \tag{5.21}$$

with $d_i(t)$ expressed by (5.20).

The following example shows that the trajectories of the switched system may remain close to those of the averaged system.

Example 5.3.1 (DC–DC buck converter in CCM) Consider DC–DC buck converter operating in CCM. The switching matrices of the system are:

$$A_1 = \begin{pmatrix} -\frac{R_1}{L} & -\frac{1}{L} \\ \frac{1}{C} & -\frac{1}{R_2 C} \end{pmatrix}, \qquad A_2 = A_1, \qquad B_1 = \begin{pmatrix} \frac{1}{L} \\ 0 \end{pmatrix}, \qquad B_2 = \begin{pmatrix} 0 \\ 0 \end{pmatrix}.$$

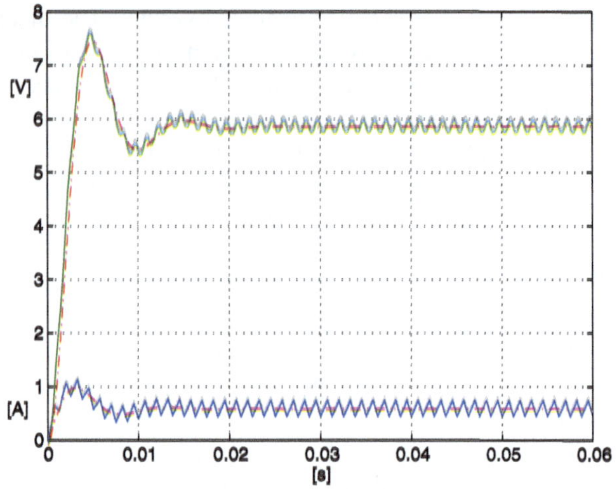

Fig. 5.7 Capacitor voltage and inductor current of a DC–DC buck converter

Consider an open-loop PWM, with the step function expressed by (5.12), $z = r - \delta(t)$, $r = 2.5V$ and $\delta(t) : [0, \infty) \to [0, \ \Delta]$, with $\Delta = 5$. Let $T_k = T_\delta \ \forall k$ the constant length of a switching cycle, and $d(t)$ the duty ratio. The dynamic model of the converter is represented by Eq. (5.6), and, according to (5.21), the corresponding averaged system is

$$\dot{w}(t) = A_2 w(t) + B_2 u(t) + \big[(A_1 - A_2)w(t) + (B_1 - B_2)u(t)\big]d(t) \qquad (5.22)$$

with

$$d(t) = D = \begin{cases} 1 & \text{if } r > \Delta, \\ \frac{r}{\Delta} & \text{if } 0 \leq r \leq \Delta, \\ 0 & \text{if } r < 0. \end{cases} \qquad (5.23)$$

Consider the following converter parameters: $R = 8.9\,\Omega$, $L = 10\,\text{mH}$, $C = 220\,\mu\text{F}$, $u(t) = E = 10\,\text{V}$, $T = 1.25 \times 10^{-3}\,\text{s}$. Figure 5.7 shows the simulation results of the switched system and those corresponding to the averaged system. The frequency of the switching function is larger than the dynamics of the converter, so starting with the same initial conditions the solutions of the averaged system are close to those of the switched system.

5.4 Dithering and Averaging

The basic principle of dithering is that if we add a suitable high frequency signal to the input of a nonlinearity, the averaged input–output relation is smoothed. Of course, this effect depends on the features of the dither signal. In the case when the original nonlinearity is Lipschitz continuous, Zames and Shneydor [26] showed that the effect of the dither on the behaviour of nonlinear systems depends on its amplitude distribution function (ADF). Stability of the dithered system is related to that of its corresponding averaged system (defined by using an averaging operation on the original dithered nonlinearity). Mossaheb [17] proved that the dither with a sufficiently high frequency may result in getting the smoothed system's output and dithered system's output as close as desired. The Lipschitz continuity assumption on the nonlinearity of the dithered system is often violated in practice, it is the case of several power converters controlled using PWM. In [13], an averaging theorem for PWM systems with triangular dither is proved, and in [23] similar results are obtained for PWM systems with measurable or slowly varying inputs and in this case also stability conditions are given. An extension of the previous results is given in [9] where the case of dither applied to discontinuous nonlinearities is studied. In this section, we briefly present some averaging results for a general class of switched systems with a quite arbitrary periodic dither; see [9, 10] for details. The results state that the averaged and the dithered systems have qualitatively the same behaviour when the nonlinearities have bounded variation and the dither has an absolutely continuous amplitude distribution function with bounded derivative and a sufficiently high frequency. Furthermore, the approximation error is of the order of the dither period.

5.4.1 Dithered System

A dither is a high-frequency signal injected into a system with the objective of modifying its nonlinear characteristics, and it is a frequently used technique by looking at modulation schemes in power electronic converters as shown in Examples 5.2.4 and 5.2.5. Indeed, the dithered model for a wide class of power converter controlled using PWM can be written in the following form:

$$\dot{x}(t) = \sum_{l=1}^{m} \big(A_l x(t) + B_l u(t)\big) n\big(k_l^\top x(t) + r_l(t) - \delta(t)\big), \tag{5.24a}$$

$$x(0) = x_0, \tag{5.24b}$$

where A_l, b_l and $k_l^\top$ are constant matrices of appropriate dimensions, δ is the high-frequency carrier signal of period T_δ and n is the step function defined as in (5.12).

5.4.2 Amplitude Distribution Function

In order to consider the averaged model of the dithered system (5.24a), (5.24b), let us introduce the following ADF definition.

Definition 5.4.1 The amplitude distribution function $F_\delta : \mathbb{R} \to [0, 1]$ of a T_δ-periodic dither signal $\delta : [0, \infty) \to \mathbb{R}$ is defined as:

$$F_\delta(\xi) = \frac{1}{T_\delta}\mu\big(\{t \in [0, T_\delta) : \delta(t) \le \xi\}\big), \tag{5.25}$$

where μ denotes the Lebesgue measure.
The ADF simply says how large a fraction of the dither signal lies below the level ξ in each period. When the amplitude distribution function is absolutely continuous (with respect to Lebesgue measure), the amplitude density function $f_\delta(\xi)$ is defined as:

$$f_\delta(\xi) = \frac{dF_\delta}{d\xi}(\xi). \tag{5.26}$$

The amplitude density and amplitude distribution functions play in a deterministic framework the same role as probability density and cumulative distribution functions play in a stochastic framework. In particular, the amplitude distribution function is bounded, monotonously increasing, continuous from the right, and, if it is differentiable, its derivative obviously corresponds to the amplitude density function.

5.4.3 Averaged System

If the dither frequency is sufficiently high, the behaviour of the dithered system (5.24a), (5.24b) will be qualitatively the same as that of an *averaged system*. Analysis and control design can then be carried out on the averaged system, which in most cases is simpler to analyse.

The averaged system corresponding to the dithered system (5.24a), (5.24b) is defined as

$$\dot{w}(t) = \sum_{i=1}^{m}\big(A_i w(t) + B_i u(t)\big)N\big(k_i^\top w(t) + r_i(t)\big), \tag{5.27}$$

where $w(0) = x_0$ and $N : \mathbb{R} \to \mathbb{R}$ is the averaged nonlinearity, which is the time average of the original step nonlinearity n by assuming g constant:

$$N(g) = \frac{1}{T_\delta}\int_{[0,T_\delta)} n\big(g - \delta(s)\big)\,ds. \tag{5.28}$$

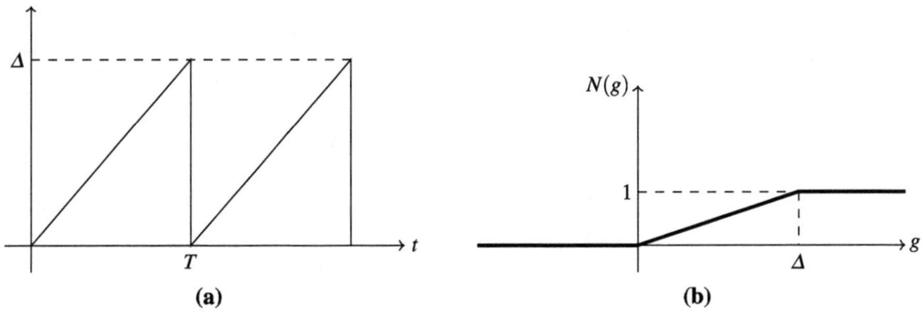

Fig. 5.8 (**a**) Sawtooth dither signal, and (**b**) the corresponding averaged nonlinearity

The *averaged nonlinearity* corresponding to the step function is equal to the ADF of the dither:

$$\begin{aligned} N(g) &= \frac{1}{T_\delta}\int_{[0,T_\delta)} n\big(g-\delta(s)\big)\,\mathrm{d}s = \int_{\mathbb{R}} n(g+\xi)\,\mathrm{d}F_\delta(\xi) \\ &= \int_{-\infty}^{g} \mathrm{d}F_\delta(\xi) = F_\delta(g), \end{aligned} \tag{5.29}$$

where the second integral is a Lebesgue–Stieltjes integral. Therefore, the averaged nonlinearity is determined by the shape of the dither as shown in the following examples.

Example 5.4.2 (Sawtooth dither signal) A typical dither signal is a sawtooth waveform of amplitude $\Delta > 0$ and period $T_\delta > 0$:

$$\delta(t) = \frac{\Delta}{T_\delta}t, \quad t \in [0, T_\delta). \tag{5.30}$$

By applying the definition (5.29), it is easy to show that

$$N(g) = \begin{cases} 1 & \text{if } g > \Delta, \\ \frac{g}{\Delta} & \text{if } 0 \le g \le \Delta, \\ 0 & \text{if } g < 0. \end{cases} \tag{5.31}$$

In Fig. 5.8, the sawtooth dither signal and the corresponding averaged nonlinearity are shown.

Another typical dither signal used for the PWM is a triangular signal.

Example 5.4.3 (Triangular dither signal) Consider triangular dither signal $\delta(t) : [0, \infty) \to [-\Delta, \Delta]$ of period T_δ:

$$\delta(t) = \begin{cases} \frac{4\Delta}{T_\delta} t & \text{if } t \in [0, \frac{T_\delta}{4}), \\ 2\Delta - \frac{4\Delta}{T_\delta} t & \text{if } t \in [\frac{T_\delta}{4}, 3\frac{T_\delta}{4}), \\ -4\Delta + \frac{4\Delta}{T_\delta} t & \text{if } t \in [3\frac{T_\delta}{4}, T_\delta). \end{cases} \tag{5.32}$$

By applying the definition (5.29), in this case the averaged nonlinearity is

$$N(g) = \begin{cases} 1 & \text{if } g > \Delta, \\ \frac{1}{2}(1 + \frac{g}{\Delta}) & \text{if } -\Delta \le g \le \Delta, \\ 0 & \text{if } g < -\Delta. \end{cases} \tag{5.33}$$

The dynamics of the dithered system can be approximated with those of the averaged system. Indeed, by assuming that the averaged nonlinearity and the external signals $r(t)$ and $u(t)$ are Lipschitz continuous, the averaging theorem in [9] allows us to conclude that the averaged system (5.27) will have a unique absolutely continuous solution. Moreover, if the dither frequency is sufficiently high, the behaviour of the dithered system (5.24a), (5.24b) will be qualitatively the same as that of the averaged system over any finite time horizon, provided that the two systems have the same initial condition. The approximation error between the averaged system and the dithered system is of the order of the dither period. Note that the theorem in [9] can be also applied to the more general class of nonsmooth systems in the form (5.24a), (5.24b) with n being a nonsmooth nonlinearity of bounded variation, its argument being a Lipschitz continuous function, and with the terms $A_l x + B_l u$ in (5.24a), (5.24b) being generalised as Lipschitz continuous, eventually time-dependent functions.

Dealing with stability issues, in [23] by neglecting the exogenous input u and by generalising the terms $A_l x$ in (5.24a), (5.24b) with continuous functions of the state, it is proved that if the averaged system is asymptotically stable then the switched system is practically stable for any dither period lower than a suitable value. In [10], a detailed discussion on the implications of the dither shape on a similar type of stability inference is presented.

Example 5.4.4 (Controlled DC–DC buck converter in CCM) Consider the DC–DC buck converter of Example 5.3.1 controlled in voltage-mode with proportional control $g(x) = k_p(x_2 - x_2^{\text{ref}})$, and δ a sawtooth signal. The averaged system is:

$$\dot{w}(t) = A_2 w(t) + B_2 u(t) + \big[(A_1 - A_2)w(t) + (B_1 - B_2)u(t)\big] N\big(g\big(w(t)\big)\big), \tag{5.34}$$

with averaged nonlinearity $N(g)$ expressed by (5.31). Consider $k_p = 0.5$ and $x_2^{\text{ref}} = 5$ V. Figure 5.9 shows the simulation results of the switched system and that

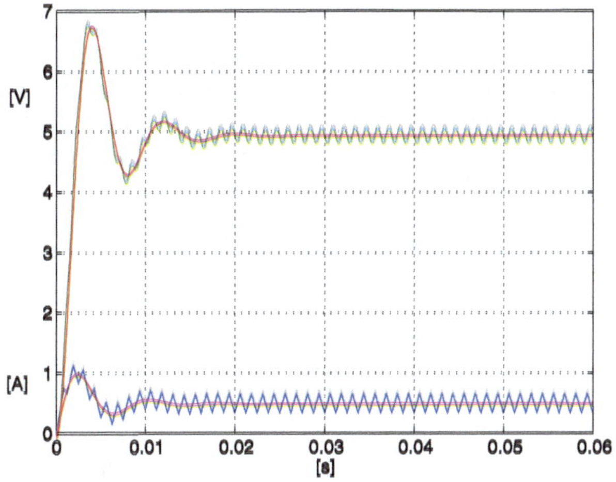

Fig. 5.9 Capacitor voltage and inductor current of a closed-loop controlled DC–DC buck converter

of the averaged system. Clearly, taking the same initial condition, the solutions of the two systems remain close to each other.

5.5 Dynamic Phasor Modelling

A useful tool which can be interpreted as a generalisation of classical averaged systems is represented by the so-called dynamic phasor model [1, 2]. The dynamic phasor model is obtained from a Fourier series expansion of the system state over a moving time window. It can be interpreted as a generalisation of the averaged model which includes an arbitrary number of harmonics (whereas the averaged model only represents the fundamental), and it has proved to be an efficient tool for modelling PWM systems.

Consider a PWM system in the form

$$\dot{\zeta}(t) = \sum_{l=1}^{m} q_l(t)\big(A_l \zeta(t) + B_l\big), \tag{5.35}$$

where $\zeta(t) \in \mathbb{R}^n$ and

$$q_l(t) = \begin{cases} 1 & \text{if } t \in [kT + t_{l,k}, kT + t_{l,k} + \tau_{l,k}], \\ 0 & \text{if otherwise.} \end{cases} \tag{5.36}$$

Here $T > 0$ is the period time and $t_{i,k}$ is the time instant when the switching signal becomes equal to 1. Ideally, the system attains a periodic solution $\zeta^0(t) = \zeta^0(t + T)$ when the duty ratios are constant, say equal to d_i^0 $\forall k$. Denoting the deviation from ζ^0 as $x = \zeta - \zeta^0$, it is possible to consider the error dynamics

$$\dot{x}(t) = A(t, x)x(t) + B(t, x), \tag{5.37}$$

where

$$A(t, x) = \sum_{i=1}^{m} q_i(t)A_i, \tag{5.38}$$

$$B(t, x) = \sum_{i=1}^{m} \left(q_i(t) - q_i^0\right)\left(A_i\zeta^0(t) + B_i\right), \tag{5.39}$$

where q_i^0 are the mode-switching functions with constant duty cycles fixed to d_i^0.

5.5.1 The Dynamic Phasor Model

The dynamic phasor model provides a frequency domain representation of (5.35) which allows for the cyclically switched dynamics to be represented by continuous equations.

The lth phasor (Fourier coefficient) of the state x is defined as

$$\langle x\rangle_l = \frac{1}{T}\int_{t-T}^{t} x(s)\mathrm{e}^{-\mathrm{j}l\omega s}\,\mathrm{d}s \tag{5.40}$$

with $\omega = 2\pi/T$. It should be noted that the phasors are defined over a moving time window. They are thus time-dependent and describe how the harmonics of the solution develop over time.

The time domain signal x is reconstructed on the interval $[t - T, t]$ according to

$$\mathrm{x}(t, s) = \sum_{l=-\infty}^{\infty} \langle x\rangle_l \mathrm{e}^{\mathrm{j}l\omega(s+t)}, \quad s \in [-T, 0]. \tag{5.41}$$

Note that $x(t) \neq \mathrm{x}(t, s)$ but the equality $x(t + s) = \mathrm{x}(t, s)$ holds for almost all $s \in [-T, 0]$. The approximation of the variables x with a Fourier series can be seen as a generalised averaged model [5, 19]. The essence in phasor dynamic modelling is to retain only the relatively large Fourier coefficients to capture the interesting behaviour of the system. If only the index-zero coefficient is retained for a fast switching system, only the low frequency behaviour is captured and the result would be the averaged model (5.21). However, when there is a ripple in the state variables or when the switching frequency is not high enough, we need to add other Fourier coefficients in order to capture, through the dynamic phasor model, the system behaviour. Considering Eq. (5.40) and using the partial integration, we obtain:

$$\frac{\mathrm{d}}{\mathrm{d}t}\langle x\rangle_l = \left\langle \frac{\mathrm{d}}{\mathrm{d}t}x\right\rangle_l - \mathrm{j}n\omega\langle x\rangle_l. \tag{5.42}$$

Using the fact that the state x is bounded continuous and the mode-switching functions q_i are bounded piecewise continuous functions over any finite time interval, the phasor coefficients of the product $q_i x$ satisfy

$$\langle q_i x\rangle_l = \sum_{h=-\infty}^{\infty} \langle q_i\rangle_{l-h}\langle x\rangle_h, \tag{5.43}$$

where

$$\langle q_i\rangle_l = \frac{1}{T}\int_{t-T}^{t} q_i(s)\mathrm{e}^{-\mathrm{j}l\omega s}\,\mathrm{d}s. \tag{5.44}$$

Note that by comparing (5.20) and (5.44) it simply follows that

$$d_i(t) = \langle q_i\rangle_0. \tag{5.45}$$

The switched system dynamic model (5.42) can then be rewritten in terms of the phasor dynamics:

$$\frac{\mathrm{d}}{\mathrm{d}t}\langle x\rangle_l = -\mathrm{j}l\omega\langle x\rangle_l + \sum_{h=-\infty}^{\infty}\sum_{i=1}^{m} A_i\langle q_i\rangle_{l-h}\langle x\rangle_h + \sum_{i=1}^{m}\left(A_i\zeta^0(t)+B_i\right)\left\langle q_i - q_i^0\right\rangle_l. \tag{5.46}$$

5.5.2 *State Space Representation*

The phasor dynamics (5.46) can be written in compact form:

$$\frac{\mathrm{d}}{\mathrm{d}t}\widehat{x} = (-\mathrm{j}\omega\mathcal{N} + \widehat{A})\widehat{x} + \widehat{B}, \tag{5.47}$$

where the state vector is the infinite dimensional phasor vector

$$\widehat{x} = \left(\dots \langle x\rangle_1^\top \; \langle x\rangle_0^\top \; \langle x\rangle_{-1}^\top \dots\right)^\top, \tag{5.48}$$

where $\mathscr{N} = \mathrm{diag}(\dots, 2I_n, I_n, 0, -I_n, -2I_n, \dots)$ is a doubly infinite dimensional block diagonal matrix and

$$\widehat{A} = \sum_{l=1}^{m} (I \otimes A_l)\mathscr{T}[\widehat{q}_l], \tag{5.49}$$

$$\widehat{B} = \left(\sum_{l=1}^{m} (I \otimes B_l) + \sum_{l=1}^{m} (I \otimes A_l)\mathscr{T}\left[\widehat{\zeta}^0\right]\right)(\widehat{q}_l - \widehat{q}_l^0). \tag{5.50}$$

In (5.50), we used

$$\widehat{q}_l = \left(\dots \langle q_l\rangle_1^\top \; \langle q_l\rangle_0^\top \; \langle q_l\rangle_{-1}^\top \dots\right)^\top, \tag{5.51}$$

$$\widehat{q}_l^0 = \left(\dots \langle q_l^0\rangle_1^\top \; \langle q_l^0\rangle_0^\top \; \langle q_l^0\rangle_{-1}^\top \dots\right)^\top. \tag{5.52}$$

The transformation $\mathscr{T}$ maps an infinite complex vector $\xi = \left(\dots \xi_1^\top \; \xi_0^\top \; \xi_{-1}^\top \dots\right)$ where $\xi_k \in \mathbb{C}$ to a doubly infinite dimensional block Toeplitz matrix according to

$$\mathscr{T}(\xi) = \begin{pmatrix} \ddots & & \vdots & & \ddots \\ & \xi_0 I_n & \xi_1 I_n & \xi_2 I_n & \\ \cdots & \xi_{-1} I_n & \xi_0 I_n & \xi_1 I_n & \cdots \\ & \xi_{-2} I_n & \xi_{-1} I_n & \xi_0 I_n & \\ \ddots & & \vdots & & \ddots \end{pmatrix}. \tag{5.53}$$

By comparing (5.47) with (5.21), we can say that (5.21) can be obtained from (5.47) when the definition (5.40) is taken by considering only the zero-index component, i.e. only for $l = 0$:

$$w(t) = \langle x\rangle_0. \tag{5.54}$$

In that sense, (5.47) can be interpreted as a generalisation of (5.21).

In the open loop case (where the duty cycle is constant and equal to d_l^0), $\mathscr{T}$ is constant and the affine term $\widehat{B}$ disappears. In this case, the time periodic switched system (5.37) is represented by a linear time invariant system in the frequency domain. However, when q is determined by the feedback, the model (5.46) is an infinite dimensional, nonlinear system and the phasor coefficients (5.44) depend on the sampled state with a delay. In this case, in order to obtain a tractable model, it is possible to introduce an approximation of the model (5.46). The approximation is obtained in two steps. In the first step, the phasor coefficients (5.44) can be expressed with a nonlinear averaged approximation [2]:

$$q_{l,av,l} = \begin{cases} d_l & \text{if } l = 0, \\ \frac{\mathrm{j}}{l2\pi}(\mathrm{e}^{-\mathrm{j}l2\pi d_l} - 1) & \text{if } l \neq 0. \end{cases} \tag{5.55}$$

This implies that if the duty cycle varies slowly (compared to modulation period T), then (5.55) is a good approximation of (5.44). In the second step, the model (5.46) is truncated by redefining all (5.40)–(5.53), with the phasors (5.44) approximated by (5.55), for $l \in \{0, \dots, \Lambda\}$ and $h \in \{-\Lambda, \dots, 0, \dots, \Lambda\}$ with a finite Λ.

Some results on the particular case of *state-dependent* switching functions have been obtained in [2]. In that paper, by considering a two-modes switched system, the authors assumed state-dependent duty ratio functions which are piecewise constant in each modulation period. They analysed the error between the phasor dynamic model (5.46) and the truncated system. In particular, it was proved that if the origin of the autonomous truncated system is locally exponentially stable, then for each $\varepsilon > 0$ there exists a modulation period T_0 such that for any $T \leq T_0$ the norm of the error between the phasor dynamic model and the truncated system will not exceed ε.

The phasor dynamics can be also used for an efficient simulation of a switched power converter. Due to the switchings, in classical simulation tools for power converters it is necessary to keep the step length of the differential equation solver very short, which results in long simulation times. An alternative is to simulate a truncated phasor model and then reconstruct the time-domain signal.

This section is concluded by showing an example on how the dynamic phasor technique can be used to represent the behaviour of a power converter.

Example 5.5.1 (Buck converter in CCM) Consider the DC–DC buck converter of Example 5.4.4. By choosing as state variables the inductor current ζ_1 and the capacitor voltage ζ_2, the switching system can be written as:

$$\dot{\zeta}(t) = A\zeta(t) + Bq(t), \tag{5.56}$$

with $A = A_1$ and $B = B_1 E$. Let d^0 be the nominal duty cycle and let $\zeta^0(t) = \zeta^0(t + T_s)$ be the corresponding periodic solution attained when the duty cycle is constant ($d_k = d^0 \; \forall k$) and the dynamics of the error $x = \zeta - \zeta^0$ are:

$$\dot{x} = Ax + B\big(q - q^0\big). \tag{5.57}$$

To obtain a tractable model, we truncate the infinite state vector to obtain a finite dimensional system which approximates the low order phasor coefficients:

$$\frac{\mathrm{d}}{\mathrm{d}t}\langle x\rangle_0 = A\langle x\rangle_0 + B\langle q - q^0\rangle_0, \tag{5.58a}$$

$$\frac{\mathrm{d}}{\mathrm{d}t}\langle x\rangle_1 = -\mathrm{j}\omega_s\langle x\rangle_1 + A\langle x\rangle_1 + B\langle q - q^0\rangle_1, \tag{5.58b}$$

$$\frac{\mathrm{d}}{\mathrm{d}t}\langle x\rangle_{-1} = -\mathrm{j}\omega_s\langle x\rangle_{-1} + A\langle x\rangle_{-1} + B\langle q - q^0\rangle_{-1}. \tag{5.58c}$$

In a compact form:

$$\frac{\mathrm{d}}{\mathrm{d}t}\widehat{x} = (-\mathrm{j}\omega_s\mathcal{N} + \widehat{A})\widehat{x} + \widehat{B}, \tag{5.59}$$

where the state vector is a two-dimensional phasor vector

$$\widehat{x} = \left(\langle x\rangle_1^\top \quad \langle x\rangle_0^\top \quad \langle x\rangle_{-1}^\top\right)^\top \tag{5.60}$$

and where $\mathcal{N} = \mathrm{diag}(1, 0, 1)$ and

$$\widehat{A} = \mathrm{diag}(A, A, A), \tag{5.61}$$

$$\widehat{B} = \mathrm{diag}(B, B, B)\left(\widehat{q} - \widehat{q}^0\right). \tag{5.62}$$

In the last equation, we used

$$\widehat{q} = \left(\langle q\rangle_1^\top \quad \langle q\rangle_0^\top \quad \langle q\rangle_{-1}^\top\right)^\top, \tag{5.63}$$

$$\widehat{q}^0 = \left(\langle q^0\rangle_1^\top \quad \langle q^0\rangle_0^\top \quad \langle q^0\rangle_{-1}^\top\right)^\top. \tag{5.64}$$

5.6 Summary

In this chapter, it has been shown that a wide class of power converters can be recast in the framework of switched systems. By exploiting the theory of averaging for switched systems, the behaviour of power converters has been studied though examples and simulations, showing also subtleties under some typical assumption made for modelling power converters. The results presented propose several interesting research problems for further investigation. For instance, it could be of interest to prove closeness of solutions between the switched system and the phasor dynamic model. One should expect that the approximation between the switched and the averaged system reduces when a larger number of harmonics is considered. That opens another interesting line of research. Indeed, many practical examples show that the bound on closeness of switched and averaged solutions is much smaller than its theoretical prediction: it might be interesting to investigate techniques able to obtain more realistic bounds. Also, the possible usefulness of an averaged system for the stability analysis of the switched system is an interesting research field which still needs more efforts. Other directions of future research could be to investigate averaging for switched models with modes appearing more than once within the modulation period, with multiple modulation periods, with non-periodic state-dependent switching functions. A different tome should be also dedicated to the averaging analysis for stochastic systems [15] and differential inclusions [6].

References

1. Almér, S., Jönsson, U.: Dynamic phasor analysis of pulse modulated systems. In: Proc. of the IEEE Conference on Decision and Control, New Orleans, Louisiana, USA, pp. 3252–3259 (2007)
2. Almér, S., Jönsson, U.: Harmonic analysis of pulse-width modulated systems. Automatica **45**, 851–862 (2009)
3. Bainov, D.D., Milusheva, S.D.: Justification of the averaging method for a system of functional-differential equation with variable structure and impulses. Appl. Math. Optim. **16**(1), 19–36 (1987)
4. Bogoliuboff, N.N., Mitropolskii, Y.A.: Asymptotic Methods in the Theory of Nonlinear Oscillators. Gordon & Breach, New York (1961)
5. Caliskan, V.A., Verghese, G.C., Stanković, A.M.: Multifrequency averaging of DC/DC converters. IEEE Trans. Power Electron. **14**(1), 124–132 (1999)
6. Donchev, T., Slalov, I.: On the limit of the solution set of singularly perturbed control differential inclusions. In: Proc. of the IEEE Conference on Decision and Control, Phoenix, Arizona, USA, pp. 3340–3345 (1999)
7. Ezzine, J., Haddad, A.H.: Error bounds in the averaging of hybrid systems. IEEE Trans. Autom. Control **34**(11), 1188–1192 (1989)
8. Hale, J.K.: Ordinary Differential Equations. Krieger, Huntington (1980)
9. Iannelli, L., Johansson, K.H., Jönsson, U., Vasca, F.: Averaging of nonsmooth systems using dither. Automatica **42**(4), 669–676 (2006)
10. Iannelli, L., Johansson, K.H., Jönsson, U., Vasca, F.: Subtleties in the averaging of a class of hybrid systems with applications to power converters. Control Eng. Pract. **18**, 961–975 (2008)
11. Khalil, H.K.: Nonlinear Systems, 3rd edn. Prentice Hall, Reading (2002)
12. Khapaev, M.M.: On the method of averaging and on certain problems connected with averaging. Differ. Equ. **2**(5), 310–314 (1966)
13. Lehman, B., Bass, R.: Extensions of averaging theory for power electronics systems. IEEE Trans. Power Electron. **11**(4), 542–553 (1996)
14. Liberzon, D.: Switching in System and Control, 1st edn. Birkhäuser, Boston (2003)
15. Liu, S.J., Krstic, M.: Continuous-time stochastic averaging on the infinite interval for locally Lipschitz systems. SIAM J. Control Optim. **48**(5), 3589–3622 (2010)
16. Mohan, N., Undeland, T.M., Robbins, W.P.: Power Electronics: Converters, Applications and Design, 3rd edn. Wiley, New York (2002)
17. Mossaheb, S.: Application of a method of averaging to the study of dither in non-linear systems. Int. J. Control **38**(3), 557–576 (1983)
18. Porfiri, M., Roberson, D.G., Stilwell, D.J.: Fast switching analysis of linear switched systems using exponential splitting. SIAM J. Control Optim. **47**(5), 2582–2597 (2008)
19. Sanders, S.R., Noworolski, J.M., Liu, X.Z., Verghese, G.C.: Generalized averaging method for power conversion circuit. IEEE Trans. Power Electron. **6**(2), 251–259 (1991)
20. Sedghi, B., Srinivasav, B., Longchamp, R.: Control of hybrid systems via dehybridization. In: Proc. of the American Control Conference, Anchorage, Alaska, USA, pp. 692–697 (2002)

21. Sethna, P.R.: Method of averaging for systems bounded for positive time. J. Math. Anal. Appl. **41**, 621–631 (1973)
22. Sun, J., Michell, D.M., Greuel, M.F., Krein, P.T., Bass, R.M.: Averaged modeling of PWM converters operating in discontinuous conduction mode. IEEE Trans. Power Electron. **16**(4), 482–492 (2001)
23. Teel, A., Moreau, L., Nešić, D.: Input-to-state set stability of pulse width modulated systems with disturbances. Syst. Control Lett. **51**(1), 23–32 (2004)
24. van der Schaft, A., Schumacher, H.: An Introduction to Hybrid Dynamical Systems. Lecture Notes in Control and Information Science, vol. 251. Springer, London (2000)
25. Wang, W., Nešić, D.: Input-to-state set stability and averaging of linear fast switching systems. IEEE Trans. Autom. Control **55**(5), 1274–1279 (2010)
26. Zames, G., Shneydor, N.A.: Dither in non-linear systems. IEEE Trans. Autom. Control **21**(5), 660–667 (1976)
27. Zames, G., Shneydor, N.A.: Structural stabilization and quenching by dither in non-linear systems. IEEE Trans. Autom. Control **22**(3), 352–361 (1977)

Chapter 6
Switched Differential Algebraic Equations

Stephan Trenn

6.1 Introduction

In this chapter, an electrical circuit with switches is modelled as a *switched differential algebraic equation* (switched DAE; DAEs are also known as descriptor systems or singular systems):

$$E_{\sigma(t)}\dot{x}(t) = A_{\sigma(t)}x(t) + B_{\sigma(t)}u(t), \quad \text{or in short form} \quad E_\sigma \dot{x} = A_\sigma x + B_\sigma u, \tag{6.1}$$

where $\sigma : \mathbb{R} \to \{1, \ldots, p\}$ is the switching signal, $E_p, A_p \in \mathbb{R}^{n\times n}$, $B_p \in \mathbb{R}^{n\times m}$ for $p \in \{1, \ldots, \bar{p}\}$, $\bar{p} \in \mathbb{N}$ is the number of different subsystems, $x : \mathbb{R} \to \mathbb{R}^n$ is the state variable and $u : \mathbb{R} \to \mathbb{R}^m$ is the input. In other words, the system is modelled as a time-varying linear differential-algebraic equation whose coefficient matrices are piecewise-constant. The time-variance follows from the action of the switches present in the circuit, but can also be induced by faults occurring in the circuit. Hence the proposed framework can be used to study the behaviour of the circuit for nominal switching as well as in the case of sudden component faults.

For a motivation and an illustration of the notation, consider the following example.

Example 6.1.1 (ODE vs. DAE description) Consider the two simple circuits as shown in Fig. 6.1.

Standard circuit analysis of these two circuits yields the following two ordinary differential equations (ODEs)

$$\frac{\mathrm{d}}{\mathrm{d}t} i_L = \frac{1}{L} u$$

S. Trenn (✉)
Department for Mathematics, University of Kaiserslautern, Erwin-Schrödinger-Str. Geb 48, Kaiserslautern, Germany
e-mail: trenn@mathematik.uni-kl.de

F. Vasca, L. Iannelli (eds.), *Dynamics and Control of Switched Electronic Systems*, Advances in Industrial Control,
DOI 10.1007/978-1-4471-2885-4_6,

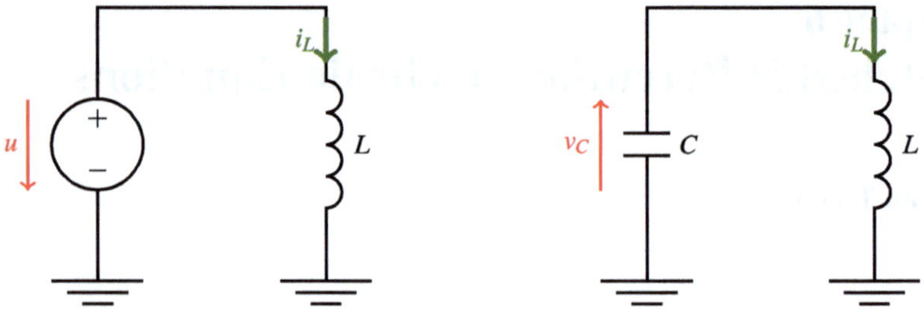

Fig. 6.1 Two simple electrical circuits

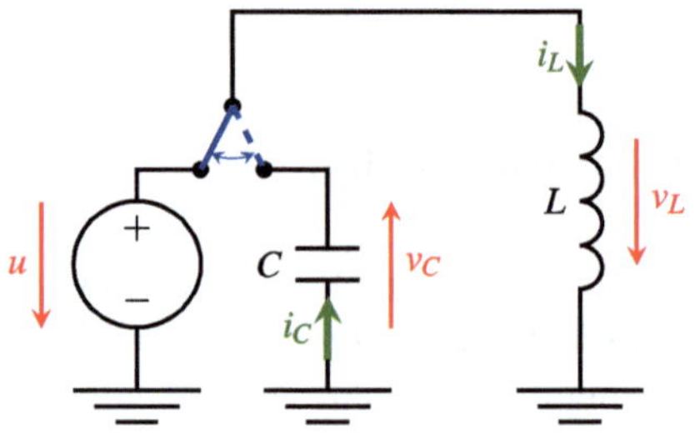

Fig. 6.2 A circuit with a switch whose two modes can be described by the two circuits from Fig. 6.1

and

$$\frac{\mathrm{d}}{\mathrm{d}t} i_L = -\frac{1}{L} v_C,$$
$$\frac{\mathrm{d}}{\mathrm{d}t} v_C = \frac{1}{C} i_L,$$

i.e. each circuit is modelled as an ODE of the form $\dot{x} = Ax + Bu$ (possibly with $B = 0$), where $x = i_L$ or $x = (i_L, v_C)^\top$.

Now assume that the two circuits from Fig. 6.1 originate from the analysis of the two modes of the circuit with a switch as shown in Fig. 6.2. Obviously, it is not possible to directly model the overall switched circuit as a switched ODE of the form

$$\dot{x}(t) = A_{\sigma(t)} x(t) + B_{\sigma(t)} u(t)$$

because the two modes are modelled with different state variables, in particular, the variable v_C does not exist in the first mode's description at all. Furthermore, it is not clear from the ODE description alone how v_C is initialised after a switch. The underlying problem is that standard circuit analysis eliminates algebraic constraints (like the Kirchhoff's law) to obtain an ODE description. However, in the presence of switches this elimination might be different for the different modes, so that the resulting ODEs are not compatible anymore. If, on the other hand, the algebraic constraints are not eliminated then this problem disappears. This approach leads to a DAE description of the form $E\dot{x} = Ax + Bu$ for each mode; for the two modes of the switched circuit from Fig. 6.2, these are given by, with $x = (i_L, v_L, i_C, v_C)^\top$,

$$\text{Mode 1:} \quad \begin{bmatrix} L & 0 & 0 & 0 \\ 0 & 0 & 0 & C \\ 0 & 0 & 0 & 0 \\ 0 & 0 & 0 & 0 \end{bmatrix} \dot{x} = \begin{bmatrix} 0 & 1 & 0 & 0 \\ 0 & 0 & 1 & 0 \\ 0 & 1 & 0 & 0 \\ 0 & 0 & 1 & 0 \end{bmatrix} x + \begin{bmatrix} 0 \\ 0 \\ -1 \\ 0 \end{bmatrix} u$$

and

$$\text{Mode 2:} \quad \begin{bmatrix} L & 0 & 0 & 0 \\ 0 & 0 & 0 & C \\ 0 & 0 & 0 & 0 \\ 0 & 0 & 0 & 0 \end{bmatrix} \dot{x} = \begin{bmatrix} 0 & 1 & 0 & 0 \\ 0 & 0 & 1 & 0 \\ 0 & 1 & 0 & 1 \\ -1 & 0 & 1 & 0 \end{bmatrix} x + \begin{bmatrix} 0 \\ 0 \\ 0 \\ 0 \end{bmatrix} u.$$

Now the behaviour of the circuit can directly be modelled as a switched DAE of the form (6.1). In addition, if the switch does not change the modes instantaneously but spends some time between the two contacts then this can easily be modelled with a third mode given by

$$\text{Mode 3:} \quad \begin{bmatrix} L & 0 & 0 & 0 \\ 0 & 0 & 0 & C \\ 0 & 0 & 0 & 0 \\ 0 & 0 & 0 & 0 \end{bmatrix} \dot{x} = \begin{bmatrix} 0 & 1 & 0 & 0 \\ 0 & 0 & 1 & 0 \\ 1 & 0 & 0 & 0 \\ 0 & 0 & 1 & 0 \end{bmatrix} x + \begin{bmatrix} 0 \\ 0 \\ 0 \\ 0 \end{bmatrix} u.$$

In general, switches or component faults induce jumps in certain state variables, and it is common to define additional jump-maps based on physical arguments [9]. However, it turns out that the appropriate formulation as a switched DAE already implicitly defines these jumps, no additional jump-map must be given. In fact, an easy way to calculate these jumps will be presented in terms of the *consistency projectors*.

In order to allow for jumps in the solution, the problem is embedded into a distributional solution framework. It turns out that general switched DAEs can have not only jumps in the solutions but also Dirac impulses and/or their derivatives. Actually, this is in agreement with the observation of a small experiment (see also the last part of Example 6.1.1): When connecting a coil via a switch to a constant voltage source, one can observe a spark when opening the switch, which can be explained by a voltage peak induced by the rapid drop of the current in the coil. For ideal elements this peak is, in fact, the Dirac impulse because it is the (distributional) derivative of a jump. Unfortunately, it is not possible to just take the classical distribution space as formally introduced by Schwartz in the 1950s [31] as a solution space for switched DAEs. Roughly speaking, the reason is that this space is too large. For example, it is not possible to define a restriction of a general distribution to some interval or, equivalently, to multiply a distribution with a piecewise-constant function, but these operations are needed to formulate the problem as a switched DAEs. To overcome this problem, the smaller space of *piecewise-smooth distributions* [33, 34] is considered as an underlying solution space.

With the right underlying solution space it is possible to study existence and uniqueness of solutions of switched DAEs. The latter is strongly related to the so-called *regularity* of the coefficient matrices and it will turn out that regularity of the matrix pairs is necessary and sufficient for the existence of unique solutions. A remarkable consequence of this general existence and uniqueness result of solutions of switched DAEs is the above mentioned property that the jumps induced by switches (or faults) are already determined uniquely by the switched DAE description.

With the help of the aforementioned consistency projectors it is easy to formulate conditions whether a switch (or fault) can induce jumps or even Dirac impulses in the solutions. This has important application for designing, e.g. fault tolerant systems, because if a component fault can induce Dirac impulses in the solution, this peak of voltages or currents of the circuit might destroy other components, possibly leading to a cascading total destruction of the circuit.

In the context of control theory, asymptotic stability of a switched DAE is an important property. It is well known for switched systems (not in DAE form) that switching between stable subsystems can yield an unstable overall system. This is not the case when one is able to find a common Lyapunov function. It will be shown that this result can be generalised to switched DAEs. The consistency projectors play a prominent role again.

The proposed framework also has limitations. The major drawback is that the distributional solution framework cannot be used in a general nonlinear context. In particular, state dependent switching is not covered by the presented theory. Since diodes yield state dependent switching and play an important role for power converters, extensions of the presented theory to encompass at least diodes is a topic of ongoing research.

6.2 Mathematical Preliminaries: Distribution Theory

Before introducing piecewise-smooth distributions in the second part of this section, classical distributions as formalised by Schwartz [31] are summarised and important properties are highlighted. A (real-valued) distribution is a continuous linear map from the space of test functions $\mathscr{C}_0^\infty$ into the real numbers $\mathbb{R}$, where $\mathscr{C}_0^\infty$ is the set of all functions $\varphi : \mathbb{R} \to \mathbb{R}$ which are smooth (i.e. arbitrarily often differentiable) and are zero outside some compact set. Continuity of a distribution is defined in terms of a certain topology on the space of test functions $\mathscr{C}_0^\infty$, however, this topology is rarely used, instead one works with the following characterisation of continuity.

Lemma 6.2.1 *A linear map* $D : \mathscr{C}_0^\infty \to \mathbb{R}$ *is continuous if and only if* $\lim_{n\to\infty} D(\varphi_n) = 0$ *for all sequences* $(\varphi_n) \in (\mathscr{C}_0^\infty)^{\mathbb{N}}$ *with the following properties:*

(C_1) $\exists$ *compact* $K \subseteq \mathbb{R}\ \forall n \in \mathbb{N}\ \forall t \notin K : \varphi_n(t) = 0$ *and*
(C_2) $\forall i \in \mathbb{N} : \lim_{n\to\infty} \|\varphi_n^{(i)}\|_\infty = 0$,

where $\|\cdot\|_\infty$ *denotes the supremum norm of a (bounded) function.*

The space of all distributions is denoted by $\mathbb{D}$, i.e.

$$\mathbb{D} := \{D : \mathscr{C}_0^\infty \to \mathbb{R} \mid D \text{ is linear and continuous}\}.$$

Distributions are also called generalised functions because of the following result which shows that the fairly large class of locally integrable functions $f : \mathbb{R} \to \mathbb{R}$, (i.e. $\int_K f < \infty$ for any compact $K \subseteq \mathbb{R}$) is a "subspace" of $\mathbb{D}$.

Lemma 6.2.2 *Each locally integrable $f : \mathbb{R} \to \mathbb{R}$ induces a distribution $f_\mathbb{D} \in \mathbb{D}$ given by*

$$f_\mathbb{D} : \mathscr{C}_0^\infty \to \mathbb{R}, \quad \varphi \mapsto \int_\mathbb{R} \varphi(t) f(t)\,\mathrm{d}t,$$

and the correspondence is one-to-one in the following sense:

$$f_\mathbb{D} = g_\mathbb{D} \quad \Leftrightarrow \quad f = g \text{ almost everywhere.}$$

Distributions induced by locally integrable functions are called regular distributions.

A very important and useful property of distributions is that all distributions have a derivative within $\mathbb{D}$.

Lemma 6.2.3 *The distributional derivative of $D \in \mathbb{D}$ is given by*

$$D'(\varphi) := -D(\varphi'), \quad D \in \mathbb{D},\ \varphi \in \mathscr{C}_0^\infty,$$

and is a distribution again. Furthermore, for differentiable $f : \mathbb{R} \to \mathbb{R}$ it holds that $(f')_\mathbb{D} = (f_\mathbb{D})'$.

Finally, distributions can be multiplied with smooth functions as follows:

$$\alpha D(\varphi) := D(\alpha\varphi), \quad D \in \mathbb{D},\ \alpha \in \mathscr{C}^\infty,\ \varphi \in \mathscr{C}_0^\infty,$$

and again this generalises the multiplication of functions: $(\alpha f)_\mathbb{D} = \alpha f_\mathbb{D}$. Furthermore, the product rule for derivatives holds, i.e.

$$(\alpha D)' = \alpha' D + \alpha D'.$$

The most famous distribution which is not induced by a function and which can be seen as the initiating object to study distributions in the first place is the *Dirac impulse* (or Dirac-Delta) given by

$$\delta(\varphi) := \varphi(0), \quad \varphi \in \mathscr{C}_0^\infty.$$

In general, the Dirac impulse at $t \in \mathbb{R}$ is given by $\delta_t(\varphi) := \varphi(t)$. The Dirac impulse is the (distributional) derivative of the Heaviside step function $\mathbb{1}_{[0,\infty)}$, i.e. $\delta = \mathbb{1}_{[0,\infty)}{}'$.

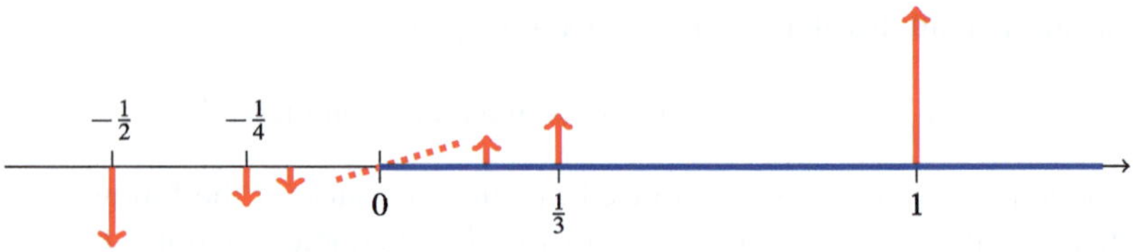

Fig. 6.3 A "bad" distribution for which a restriction to the interval $(0, \infty)$ is not possible

The main contribution of Schwartz was the embedding of the Dirac impulse into a general functional analytical framework (viewing it as a linear operator on the space of test functions). However, this approach yielded a very large space of distributions where many distributions do not have intuitive properties and cannot be handled easily. For example, in theory there exists a continuous function which is nowhere differentiable, but in the distributional framework this function has a derivative and there is no intuition as to how this distribution looks like. This existence of "weird" distributions makes it impossible to simply plug in a distribution x into the switched DAE (6.1) because it is not clear how the product of the piecewise-constant coefficient matrices E_σ and A_σ with $\dot{x}$ or x should be defined. If one rewrites (6.1) with the help of restrictions to intervals as

$$(E_{p_i}\dot{x})_{[t_i,t_{i+1})} = (A_{p_i}x + B_{p_i}u)_{[t_i,t_{i+1})}, \quad \forall i \in \mathbb{Z}, \tag{6.2}$$

where $\sigma(t) = p_i$ for $t \in [t_i, t_{i+1})$ and $i \in \mathbb{Z}$ then the following remark shows that there is no suitable definition for the terms in (6.1) in a general distributional framework.

Remark 6.2.4 (Restriction impossible, [34, Rem. 9], see also [33, Thm. 2.2.2]) Consider the following (well defined!) distribution, see also Fig. 6.3

$$D = \sum_{i\in\mathbb{N}} d_i \delta_{d_i}, \quad d_i := \frac{(-1)^i}{i+1}, \ i \in \mathbb{N}.$$

The restriction to the interval $(0, \infty)$ should then be

$$D_{(0,\infty)} = \sum_{k\in\mathbb{N}} \frac{1}{2k+1}\delta_{\frac{1}{2k+1}},$$

but for any test function $\varphi \in \mathscr{C}_0^\infty$ with $\varphi(0) \neq 0$ the infinite sum does not converge, hence the restriction is not well defined.

The problem in the above counter-example is the accumulation of Dirac impulses at zero. If, however, the Dirac impulses are isolated then it seems very trivial to define a restriction to intervals. For example, the restriction of the Dirac impulse δ to the closed interval $[0, \infty)$ should be the Dirac impulse itself again, while the restriction to the open interval $(0, \infty)$ should be the zero distribution. In order to be able to define a restriction for distributions, it therefore seems reasonable to consider first the space of *piecewise-regular distributions*

$$\mathbb{D}_{\text{pwreg}} := \left\{ D = f_{\mathbb{D}} + \sum_{t \in T} D_t \;\middle|\; \begin{array}{l} f \text{ is locally integrable, } T \subset \mathbb{R} \text{ locally finite,} \\ \forall t \in T : D_t \in \operatorname{span}\{\delta_t, \delta_t', \delta_t'', \ldots\} \end{array} \right\},$$

where $\operatorname{span}\{\delta_t, \delta_t', \delta_t'', \ldots\}$ denotes the set of all finite linear combinations of the Dirac impulse at t and its derivatives. For a piecewise-regular distribution $D = f_{\mathbb{D}} + \sum_{t \in T} D_t$, the restriction to some interval $M \subseteq \mathbb{R}$ is simply defined by

$$D_M := (f_M)_{\mathbb{D}} + \sum_{t \in T \cap M} D_t,$$

where $f_M : \mathbb{R} \to \mathbb{R}$ is the restriction of f given by $f_M(t) = f(t)$ if $t \in M$ and $f_M(t) = 0$ otherwise. Although this space is suitable for defining a restriction, it is still too big as a solution space for the switched DAE (6.1) because of the following two reasons:

1. $\mathbb{D}_{\text{pwreg}}$ is not closed under differentiation, i.e. $x \in \mathbb{D}_{\text{pwreg}}$ does not imply $\dot{x} \in \mathbb{D}_{\text{pwreg}}$, in particular the restriction of $\dot{x}$ is still not well defined, and
2. It is not possible to specify initial values since neither $x(t)$, $x(t-)$ nor $x(t+)$ are well defined quantities for $x \in \mathbb{D}_{\text{pwreg}}$.

Cobb [5] solved the second problem by considering the space of *piecewise-continuous distributions*, i.e. he considered piecewise-regular distributions $D = f_{\mathbb{D}} + \sum_{t \in T} D_t$ with piecewise-continuous f and defined $D(t-) := f(t-)$ and $D(t+) := f(t+)$. However, Cobb seemed to have overlooked the first problem. The following definition of the space of *piecewise-smooth distributions* resolves this problem.

Definition 6.2.5 (Piecewise-smooth distributions, [33, 34]) First, define the space of *piecewise-smooth functions* as

$$\mathscr{C}^\infty_{\text{pw}} := \left\{ \alpha = \sum_{i \in \mathbb{Z}} \mathbb{1}_{[t_i, t_{i+1})} \alpha_i \;\middle|\; \begin{array}{l} \{t_i \in \mathbb{R} \mid i \in \mathbb{Z}\} \text{ locally finite,} \\ t_i < t_{i+1},\ (\alpha_i)_{i \in \mathbb{Z}} \in \left(\mathscr{C}^\infty\right)^{\mathbb{Z}} \end{array} \right\}.$$

The space of *piecewise-smooth distributions* is defined as

$$\mathbb{D}_{\text{pw}\mathscr{C}^\infty} := \left\{ D = f_{\mathbb{D}} + \sum_{t \in T} D_t \;\middle|\; \begin{array}{l} f \in \mathscr{C}^\infty_{\text{pw}},\ T \subset \mathbb{R} \text{ locally finite,} \\ \forall t \in T : D_t \in \operatorname{span}\{\delta_t, \delta_t', \delta_t'', \ldots\} \end{array} \right\}.$$

The space of piecewise-smooth distributions combines the idea from Cobb's piecewise-continuous distributions (position of impulses not fixed a-priori) and from the space of impulsive-smooth distributions (closed under differentiation) introduced in [17] in the context of optimal control and later used for studying DAEs, see e.g. [13, 29]. The latter allows Dirac impulses and its derivatives only at zero but it was already suggested in [14] (without carrying out the details) to allow Dirac impulses everywhere (without accumulation points). Another viewpoint of piecewise-smooth distributions is based on the axiomatic definition of general distributions as locally finite derivatives of continuous functions (see e.g. [8]) because any piecewise-smooth distribution can be represented locally as a finite derivative of a piecewise-smooth function. One important feature of the piecewise-smooth distributions is the existence of the *Fuchssteiner multiplication* [10, 11, 33] which defines an associative (but not commutative) multiplication between two arbitrary piecewise-smooth distributions and which fulfils the Leibniz's product rule. Here this multiplication will not be used in its full generality because for studying (6.1) only the product of a piecewise-smooth function with a piecewise-smooth distribution is needed. The latter is given by

$$\alpha D := \sum_{i\in\mathbb{Z}} \alpha_i D_{[t_i,t_{i+1})}, \quad \alpha \in \mathscr{C}^\infty_{\mathrm{pw}},\ D \in \mathbb{D}_{\mathrm{pw}\mathscr{C}^\infty},$$

where $\alpha = \sum_{i\in\mathbb{Z}} (\alpha_i)_{[t_i,t_{i+1})}$ with $\alpha_i \in \mathscr{C}^\infty$. The matrix–vector product in (6.1) is defined in the classical way, i.e. for $M \in (\mathscr{C}^\infty_{\mathrm{pw}})^{m\times n}$ and $x \in (\mathbb{D}_{\mathrm{pw}\mathscr{C}^\infty})^n$ the product $Mx \in (\mathbb{D}_{\mathrm{pw}\mathscr{C}^\infty})^m$ is a vector of sums of scalar products.[1] Note that the Fuchssteiner multiplication makes the standard formulation (6.1) of a switched DAE equivalent to the restriction based formulation (6.2).

Definition 6.2.6 (Point-wise evaluation) Let $t \in \mathbb{R}$ and $D = f_{\mathbb{D}} + \sum_{\tau\in T} D_\tau$ then the left/right evaluation of D at t is given by

$$D(t-) := f(t-) = \lim_{\varepsilon \searrow 0} f(t-\varepsilon), \qquad D(t+) := f(t+) = f(t)$$

and the impulsive part of D at t is

$$D[t] := \begin{cases} D_t & \text{if } t \in T, \\ 0 & \text{if } t \notin T. \end{cases}$$

[1] This is actually a consequence from the standard identification of $x = (x_1,\ldots,x_n) \in (\mathbb{D}_{\mathrm{pw}\mathscr{C}^\infty})^n$ with the corresponding functional $x : \mathscr{C}_0^\infty \to \mathbb{R}^n$ given by $x(\varphi) = (x_1(\varphi),\ldots,x_n(\varphi))^\top$. Some authors [19, 28] identify $x = (x_1,\ldots,x_n) \in (\mathbb{D}_{\mathrm{pw}\mathscr{C}^\infty})^n$ with $x : (\mathscr{C}_0^\infty)^n \to \mathbb{R}$ and $x(\varphi) := \sum_{i=1}^n x_i(\varphi_i)$. This different viewpoint of the "vector" x makes it necessary to define the matrix–vector product differently; however, these authors do not give a motivation for this different viewpoint.

6.3 Regularity of Matrix Pairs

It is assumed that each mode of the switched DAE (6.1) is described by a *regular* matrix pair (E, A). In this section, some properties and consequences of regularity are collected.

Definition 6.3.1 (Regularity) A matrix pair $(E, A) \in \mathbb{R}^{m\times n} \times \mathbb{R}^{m\times n}$, $n, m \in \mathbb{N}$, is called regular if and only if $m = n$ and the polynomial $\det(sE - A)$ is not the zero polynomial.

In the following theorem, characterisations of regularity of (E, A) are given which highlight the importance of regularity of (E, A) with respect to solvability of the corresponding DAE $E\dot{x} = Ax + f$. Note that here "solution" refers to a classical solution, i.e. a differentiable function x fulfilling the DAE.

Theorem 6.3.2 (Characterisations of regularity) *The following statements are* equivalent *for matrix pairs (E, A) with square matrices $E, A \in \mathbb{R}^{n\times n}$:*

1. *The matrix pair (E, A) is regular.*
2. *There exist invertible matrices $S, T \in \mathbb{R}^{n\times n}$ such that (E, A) is transformed into a* quasi-Weierstrass form
$$(SET, SAT) = \left(\begin{bmatrix} I & 0 \\ 0 & N \end{bmatrix}, \begin{bmatrix} J & 0 \\ 0 & I \end{bmatrix}\right), \tag{6.3}$$
where J is some matrix and N is a nilpotent matrix.
3. *For all smooth $f : \mathbb{R} \to \mathbb{R}^n$ there exists a solution x of $E\dot{x} = Ax + f$ and x is uniquely given by the value $x(t_0)$ for any fixed $t_0 \in \mathbb{R}$.*
4. *The only solution x of $E\dot{x} = Ax$ with $x(0) = 0$ is the trivial solution.*

Proof 1 $\Leftrightarrow$ 2 This is a classical result going back to Weierstrass [36], for a proof see e.g. the textbook [19]. The prefix "quasi" in "quasi-Weierstrass form" is used because it is not assumed here that J and N are in Jordan canonical form [2].

2 $\Rightarrow$ 3 It suffices to show that $\dot{v} = Jv + f_1$ and $N\dot{w} = w + f_2$ are solvable for all smooth f_1, f_2 and the solutions are uniquely determined by the values $v(t_0)$ and $w(t_0)$. Classical ODE theory provides this already for v. Consider the operator $(N\frac{\mathrm{d}}{\mathrm{d}t} - I) : \mathcal{C}^\infty \to \mathcal{C}^\infty$, which is invertible with inverse given by

$$\left(N\frac{\mathrm{d}}{\mathrm{d}t} - I\right)^{-1} = -\sum_{i=0}^{\nu-1}\left(N\frac{\mathrm{d}}{\mathrm{d}t}\right)^i,$$

where $\nu \in \mathbb{N}$ is the nilpotency index of N, i.e. ν is the minimal value such that $N^\nu = 0$. Hence the unique solution of $N\dot{w} = w + f_2$ is given by

$$w = -\sum_{i=0}^{\nu}\left(N\frac{\mathrm{d}}{\mathrm{d}t}\right)^i (f_2) = -\sum_{i=0}^{\nu} N^i f_2^{(i)}.$$

3 $\Rightarrow$ 4 Choosing $f = 0$ in 3 implies 4.

4 $\Rightarrow$ 1 This is shown in [19, Thm. 2.14]. The basic idea is to choose $n+1$ different pairs $(\lambda_i, v_i) \in \mathbb{R} \times \mathbb{R}^n \setminus \{0\}$, $i = 1, \dots, n+1$ such that $(E\lambda_i - A)v_i = 0$ and a vanishing nontrivial linear combination $\sum_{i=1}^{n+1} \alpha_i v_i = 0$ for $\alpha_i \in \mathbb{R}$ not all zero. Then $x(t) = \sum_{i=1}^{n+1} \alpha_i v_i e^{\lambda_i t}$ is not identically zero but solves $E\dot{x} = Ax$, $x(0) = 0$. □

Since the quasi-Weierstrass form (6.3) will play an important role in the following, a convenient method to obtain the transformation matrices S, T is presented with the following theorem which utilises the *Wong sequences* (named after [37]).

Theorem 6.3.3 (Wong sequences and the quasi-Weierstrass form [1, 2]) *Let (E, A) be a regular matrix pair. Define the* Wong sequences *of subspaces by, for $i \in \mathbb{N}$,*

$$\mathscr{V}^0 := \mathbb{R}^n, \qquad \mathscr{V}^{i+1} := A^{-1}(E\mathscr{V}^i), \qquad \mathscr{W}^0 := \{0\}, \qquad \mathscr{W}^{i+1} := E^{-1}(A\mathscr{W}^i),$$

where $M^{-1}\mathscr{S} := \{x \in \mathbb{R}^n \mid Mx \in \mathscr{S}\}$ denotes the pre-image of the set $\mathscr{S} \subseteq \mathbb{R}^n$ under the matrix $M \in \mathbb{R}^{n\times n}$ and $M\mathscr{S} := \{y = Mx \mid x \in \mathscr{S}\}$ denotes the image of $\mathscr{S}$ under M. These sequences get stationary after finitely many steps; denote the limits with $\mathscr{V}^*$ and $\mathscr{W}^*$. Choose full rank matrices V, W such that $\operatorname{im} V = \mathscr{V}^*$ *and* $\operatorname{im} W = \mathscr{W}^*$ *then*

$$T := [V, W], \qquad S := [EV, AW]^{-1}$$

are invertible and put (E, A) into a quasi-Weierstrass form (6.3).

6.4 Explicit Solution Formula for Non-switched DAE

In this section, two explicit formulas for the solutions of the DAE

$$E\dot{x} = Ax + f \tag{6.4}$$

are presented. The first is based on certain projections defined with the help of the quasi-Weierstrass form (6.3), the second is based on the *Drazin inverses* of E and A. The Drazin inverse approach only works when E and A commute, i.e. $EA = AE$ must hold. This is in general not true, but by multiplying both matrices with $(E\lambda_0 - A)^{-1}$ for some λ_0 with $\det(E\lambda_0 - A) \neq 0$ this assumption always holds [3].

Definition 6.4.1 (Consistency, differential and impulse projector, [32]) Consider a regular matrix pair (E, A) and its quasi-Weierstrass form (6.3) obtained by corresponding transformation matrices $S, T \in \mathbb{R}^{n\times n}$. Let the block sizes in the quasi-Weierstrass form be $n_1 \times n_1$ and $n_2 \times n_2$. The *consistency projector* is given by

$$\Pi_{(E,A)} := T \begin{bmatrix} I & 0 \\ 0 & 0 \end{bmatrix} T^{-1},$$

the *differential projector* is given by

$$\Pi^{\text{diff}}_{(E,A)} := T \begin{bmatrix} I & 0 \\ 0 & 0 \end{bmatrix} S$$

and the *impulse projector* is given by

$$\Pi^{\text{imp}}_{(E,A)} := T \begin{bmatrix} 0 & 0 \\ 0 & I \end{bmatrix} S,$$

where I is the identity matrix of size $n_1 \times n_1$ for the consistency and differential projector and of size $n_2 \times n_2$ for the impulse projector.

Note that, in contrast to the consistency projector, the differential and impulse projectors are not projectors in the usual sense because they are in general not idempotent matrices. Furthermore, it is not difficult to see that the definition of the projectors does not depend on the specific choices of the transformations S and T.

Definition 6.4.2 (Drazin inverse) Let $M \in \mathbb{R}^{n\times n}$. Any matrix $M^D \in \mathbb{R}^{n\times n}$ is called *Drazin inverse* of M if

1. $MM^D = M^D M$,
2. $M^D M M^D = M^D$, and
3. $\exists \nu \in \mathbb{N}:\ M^D M^{\nu+1} = M^\nu$.

Lemma 6.4.3 ([3], see also [2]) *For all $M \in \mathbb{R}^{n\times n}$ there exists a unique Drazin inverse M^D. Furthermore, if $M = T^{-1}\left[\begin{smallmatrix} C & 0 \\ 0 & N \end{smallmatrix}\right]T$, where $T \in \mathbb{R}^{n\times n}$ is invertible, $C \in \mathbb{R}^{n_1\times n_1}$, $n_1 \in \mathbb{N}$, is an invertible matrix and $N \in \mathbb{R}^{n_2\times n_2}$, $n_2 = n - n_1$, is nilpotent, then*

$$M^D = T^{-1} \begin{bmatrix} C^{-1} & 0 \\ 0 & 0 \end{bmatrix} T.$$

The solution of the DAE (6.4) based on the above defined projectors as well as on the Drazin inverse is now given in the following theorem.

Theorem 6.4.4 (Explicit solution formula) *Let (E, A) be a regular matrix pair and let $\Pi_{(E,A)}$, $\Pi^{\text{diff}}_{(E,A)}$ and $\Pi^{\text{imp}}_{(E,A)}$ be the consistency, differential and impulse projector, resp., as in Definition* 6.4.1. *Furthermore, let*

$$A^{\text{diff}} := \Pi^{\text{diff}}_{(E,A)} A \quad \text{and} \quad E^{\text{imp}} := \Pi^{\text{imp}}_{(E,A)} E.$$

Then all solutions of (6.4) *are given by, for $c \in \mathbb{R}^n$,*

$$x(t) = e^{A^{\text{diff}} t} \Pi_{(E,A)} c + \int_0^t e^{A^{\text{diff}}(t-s)} \Pi^{\text{diff}}_{(E,A)} f(s)\,\mathrm{d}s - \sum_{i=0}^{n-1} (E^{\text{imp}})^i \Pi^{\text{imp}}_{(E,A)} f^{(i)}(t). \tag{6.5}$$

If, in addition, $EA = AE$ then also

$$\begin{aligned} x(t) = e^{E^D A t} E^D E c &+ \int_0^t e^{E^D A(t-s)} E^D f(s)\,\mathrm{d}s \\ &- (I - E^D E) \sum_{i=0}^{n-1} (E A^D)^i A^D f^{(i)}(t). \end{aligned}$$

Proof The quasi-Weierstrass form (6.3) with corresponding transformation matrices $S, T \in \mathbb{R}^{n\times n}$ yields that x solves $E\dot{x} = Ax + f$ if and only if $\binom{v}{w} := T^{-1}x$ solves $\dot{v} = Jv + [I\ 0]Sf$ and $N\dot{w} = w + [0\ I]Sf$, hence, together with the step 2 $\Rightarrow$ 3 in the proof of Theorem 6.3.2,

$$\begin{aligned} x(t) &= T\begin{pmatrix} v \\ w \end{pmatrix} = T\begin{pmatrix} v \\ 0 \end{pmatrix} + T\begin{pmatrix} 0 \\ w \end{pmatrix} \\ &= T\begin{pmatrix} e^{Jt}v_0 \\ 0 \end{pmatrix} + T\begin{pmatrix} \int_0^t e^{J(t-s)}[I\ 0]Sf(s)\,\mathrm{d}s \\ 0 \end{pmatrix} + T\begin{pmatrix} 0 \\ -\sum_{i=0}^{n-1} N^i [0\ I] S f^{(i)}(t) \end{pmatrix} \\ &= T e^{\left[\begin{smallmatrix} J & 0 \\ 0 & 0 \end{smallmatrix}\right] t} \begin{bmatrix} I & 0 \\ 0 & 0 \end{bmatrix} T^{-1} c + \int_0^t T e^{\left[\begin{smallmatrix} J & 0 \\ 0 & 0 \end{smallmatrix}\right](t-s)} \begin{bmatrix} I & 0 \\ 0 & 0 \end{bmatrix} Sf(s)\,\mathrm{d}s \\ &\quad - \sum_{i=0}^{n-1} T \begin{bmatrix} 0 & 0 \\ 0 & N^i \end{bmatrix} \begin{bmatrix} 0 & 0 \\ 0 & I \end{bmatrix} Sf^{(i)}(t) \\ &= e^{T\left[\begin{smallmatrix} J & 0 \\ 0 & 0 \end{smallmatrix}\right] T^{-1} t} T \begin{bmatrix} I & 0 \\ 0 & 0 \end{bmatrix} T^{-1} c + \int_0^t e^{T\left[\begin{smallmatrix} J & 0 \\ 0 & 0 \end{smallmatrix}\right] T^{-1}(t-s)} T \begin{bmatrix} I & 0 \\ 0 & 0 \end{bmatrix} Sf(s)\,\mathrm{d}s \\ &\quad - \sum_{i=0}^{n-1} \left(T \begin{bmatrix} 0 & 0 \\ 0 & N \end{bmatrix} T^{-1} \right)^i T \begin{bmatrix} 0 & 0 \\ 0 & I \end{bmatrix} Sf^{(i)}. \end{aligned}$$

Together with

$$T\begin{bmatrix} J & 0 \\ 0 & 0 \end{bmatrix} T^{-1} = T\begin{bmatrix} I & 0 \\ 0 & 0 \end{bmatrix} SS^{-1}\begin{bmatrix} J & 0 \\ 0 & I \end{bmatrix} T^{-1} = \Pi^{\text{diff}}_{(E,A)} A = A^{\text{diff}}$$

and

$$T\begin{bmatrix} 0 & 0 \\ 0 & N \end{bmatrix} T^{-1} = T\begin{bmatrix} 0 & 0 \\ 0 & I \end{bmatrix} SS^{-1}\begin{bmatrix} I & 0 \\ 0 & N \end{bmatrix} T^{-1} = \Pi^{\text{imp}}_{(E,A)} E = E^{\text{imp}},$$

the first proposed solution formula is obtained. The second solution formula is standard and can be found in [3, 19]. □

Remark 6.4.5 (Remarks on the solution formulas)

1. If $EA = AE$ then both solution formulas are almost identically because then it can be shown, see [2], that $E^D = \Pi^{\text{diff}}_{(E,A)}$ and $E^D E = \Pi_{(E,A)}$. However, it is in general *not true* that $A^D = \Pi^{\text{imp}}_{(E,A)}$ or $EA^D = E^{\text{imp}}$, therefore the second formula needs the "correction term" $I - E^D E$.
2. From the solution formula (6.5) it follows that

$$x(0) = \Pi_{(E,A)} c - \sum_{i=0}^{n-1} (E^{\text{imp}})^i \Pi^{\text{imp}}_{(E,A)} f^{(i)}(0),$$

   in particular, the initial value problem (6.4), $x(0) = x_0 \in \mathbb{R}^n$ has a solution if and only if

$$x_0 + \sum_{i=0}^{n-1} (E^{\text{imp}})^i \Pi^{\text{imp}}_{(E,A)} f^{(i)}(0) \in \operatorname{im} \Pi_{(E,A)} = \mathcal{V}^*,$$

   which characterises consistency of the initial value.
3. If x is allowed to have jumps or even Dirac impulses, i.e. $x \in (\mathbb{D}_{\text{pw}\mathcal{C}^\infty})^n$, then Theorem 6.4.4 is still true, i.e. considering distributional solutions doesn't add any new solutions to the problem (for which the DAE $E\dot{x} = Ax + f$ should hold globally). In particular, also all distributional solutions have consistent initial values. Furthermore, if f contains jumps or Dirac impulses, i.e. $f \in \mathbb{D}_{\text{pw}\mathcal{C}^\infty}$ then the solution formula also holds; however, it is necessary to define the notion of the antiderivative $H = \int_0 D$ of the distribution D which fulfils $H' = D$ and $H(0-) = 0$, see [34, Prop. 3].
4. The proof of Theorem 6.4.4 reveals that, instead of the consistency projector $\Pi_{(E,A)}$ in formula (6.5), any matrix M with $\operatorname{im} M = \operatorname{im} \Pi_{(E,A)} = \mathcal{V}^*$ could be used. However, the special choice of $\Pi_{(E,A)}$ yields that this formula holds also true when an inconsistent initial value is given, see the next section.

6.5 Inconsistent Initial Values

In the presence of switches, the initial conditions need not to be consistent so that no solution (classical or distributional) with this initial value exists. It is therefore necessary to make precise what a "solution" to an inconsistent initial value problem should be. The viewpoint here is the following.

An inconsistent initial value can only occur if the considered DAE was not active before the initial time (say $t_0 = 0$). This gives rise to the following *initial trajectory problem (ITP)*, where $x^0 : (-\infty, 0) \to \mathbb{R}^n$ is some initial trajectory:

$$\begin{aligned} x_{(-\infty,0)} &= x^0{}_{(-\infty,0)}, \\ (E\dot{x})_{[0,\infty)} &= (Ax)_{[0,\infty)} + f_{[0,\infty)}. \end{aligned} \tag{6.6}$$

If $x^0(0)$ is not consistent for $E\dot{x} = Ax + f$ then no classical solution exists; however, it will be shown in the following that a distributional solution exists. Therefore, (6.6) is considered as an equation of piecewise-smooth distributions, in particular, the inhomogeneity f and the initial trajectory are also piecewise-smooth distributions.

Theorem 6.5.1 (Solvability of the ITP) *Let (E, A) be a regular matrix pair. Then for every initial trajectory $x^0 \in (\mathbb{D}_{\mathrm{pw}\mathscr{C}^\infty})^n$ and any inhomogeneity $f \in (\mathbb{D}_{\mathrm{pw}\mathscr{C}^\infty})^n$ the ITP* (6.6) *has a unique solution $x \in (\mathbb{D}_{\mathrm{pw}\mathscr{C}^\infty})^n$. In particular, the jump from $x^0(0-)$ to $x(0+)$ and the impulsive part $x[0]$ is uniquely determined. In fact, if $f[0] = 0$ then*

$$x(0+) = \Pi x^0(0-) - \sum_{i=0}^{n-1} (E^{\mathrm{imp}})^i \Pi^{\mathrm{imp}} f^{(i)}(0+),$$

$$\begin{aligned} x[0] = &- \sum_{i=0}^{n-1} (E^{\mathrm{imp}})^{i+1} (I - \Pi) x^0(0-) \delta^{(i)} \\ &- \sum_{i=0}^{n-1} (E^{\mathrm{imp}})^{i+1} \sum_{j=0}^{i} \Pi^{\mathrm{imp}} f^{(i-j)}(0+) \delta^{(j)}, \end{aligned}$$

where Π is the consistency projector and $E^{\mathrm{imp}} = \Pi^{\mathrm{imp}} E$ with impulse projector Π^{imp} as in Definition 6.4.1. *In particular, if $f = 0$ then*

$$x(0+) = \Pi x^0(0-)$$

and, on the open interval $(0, \infty)$,

$$\dot{x} = A^{\text{diff}} x.$$

Furthermore, if f *is smooth then the solution* x *restricted to the open interval* $(0, \infty)$ *is induced by the smooth function given by* (6.5) *where* $c = x^0(0-)$.

Proof Choose $S, T \in \mathbb{R}^{n\times n}$ invertible such that (E, A) is put into a quasi-Weierstrass form (6.3) with block sizes $n_1 \times n_1$ and $n_2 \times n_2$, resp., and let $T^{-1}x = \binom{v}{w}$. In the new coordinates, the ITP (6.6) then decouples into

$$\begin{aligned} v_{(-\infty,0)} &= {v^0}_{(-\infty,0)}, & w_{(-\infty,0)} &= {w^0}_{(-\infty,0)}, \\ \dot{v}_{[0,\infty)} &= \big(Jv + [I \quad 0]Sf\big)_{[0,\infty)}, & (N\dot{w})_{[0,\infty)} &= \big(w + [0 \quad I]Sf\big)_{[0,\infty)}, \end{aligned}$$

where $\binom{v^0}{w^0} = T^{-1}x^0$. It can be shown (see e.g. the proof of [33, Thm. 3.3.8]) that the ITP for v just yields the same solutions on $[0, \infty)$ as the classical solution of the ODE initial value problem $\dot{v} = Jv + f$, $v(0) = v^0(0-)$. In particular, $v(0+) = v(0-)$ and $v[0] = 0$. As shown in [33, Thm. 3.1.7], the DAE for w can equivalently be written as a special "switched" DAE without initial trajectory

$$N_{\text{itp}}\dot{w} = w + f_{\text{itp}},$$

where $N_{\text{itp}} := N_{[0,\infty)}$ and $f_{\text{itp}} = -{w^0}_{(-\infty,0)} + \big[0 \; I\big] Sf_{[0,\infty)}$. Since the operator $N_{\text{itp}}\frac{\mathrm{d}}{\mathrm{d}t} : (\mathbb{D}_{\text{pw}\mathscr{C}^\infty})^{n_2} \to (\mathbb{D}_{\text{pw}\mathscr{C}^\infty})^{n_2}$ is still nilpotent, the solution formula derived in the proof of Theorem 6.3.2 still holds, i.e. w is uniquely given by

$$\begin{aligned} w &= -\sum_{i=0}^{n_2-1}\left(N_{\text{itp}}\frac{\mathrm{d}}{\mathrm{d}t}\right)^i (f_{\text{itp}}) \\ &= \sum_{i=0}^{n-1}\left(N_{[0,\infty)}\frac{\mathrm{d}}{\mathrm{d}t}\right)^i \big({w^0}_{(-\infty,0)}\big) - \sum_{i=0}^{n-1}\left(N_{[0,\infty)}\frac{\mathrm{d}}{\mathrm{d}t}\right)^i \big([0 \quad I]Sf_{[0,\infty)}\big). \end{aligned}$$

This shows existence and uniqueness of a solution x of the ITP (6.6) and $x = T\binom{v}{w}$. Some calculations within the piecewise-smooth distributional framework yield

$$\left(N_{[0,\infty)}\frac{\mathrm{d}}{\mathrm{d}t}\right)^i \big({w^0}_{(-\infty,0)}\big) = \begin{cases} w^0_{(-\infty,0)}, & \text{if } i = 0, \\ -N^i w^0(0-)\delta^{(i-1)}, & \text{if } i > 0, \end{cases}$$

and, with the abbreviation $\tilde{f} := [0 \; I]Sf$,

$$\left(N_{[0,\infty)}\frac{\mathrm{d}}{\mathrm{d}t}\right)^i (\tilde{f}_{[0,\infty)}) = \begin{cases} \tilde{f}_{(0,\infty)}, & \text{if } i = 0, \\ N^i \tilde{f}^{(i)}_{(0,\infty)} + N^i \sum_{j=0}^{i-1} \tilde{f}^{(i-1-j)}(0+)\delta^{(j)}, & \text{if } i > 0. \end{cases}$$

Hence,

$$w(0+) = -\sum_{i=0}^{n-1} N^i [0 \quad I] S f^{(i)}(0+)$$

and

$$w[0] = -\sum_{i=0}^{n-2} N^{i+1} w^0(0-)\delta^{(i)} - \sum_{i=0}^{n-2} N^{i+1} \sum_{j=0}^{i} [0 \quad I] S f^{(i-j)}(0+)\delta^{(j)}.$$

Together with analogous rearrangements of matrices as in the proof of Theorem 6.4.4, this yields the claimed expressions for $x(0+)$ and $x[0]$. □

Viewing now the switched DAE (6.1) as a repeated ITP (where the switching times are the initial times) one obtains the following result. Note that it has to be assumed that the switching times do not accumulate, otherwise this approach does not work.

Corollary 6.5.2 (Existence and uniqueness of solutions of a switched DAE) *Consider the switched DAE* (6.1) *with regular matrix pairs* (E_p, A_p), $p \in \{1, \dots, \bar{p}\}$ *and assume for the switching signal* σ

$$\sigma \in \Sigma_0 := \left\{ \sigma : \mathbb{R} \to \{1, \dots, \bar{p}\} \;\middle|\; \begin{array}{l} \sigma \text{ has locally finitely many switches,} \\ \sigma_{(-\infty,0)} \text{ is constant} \end{array} \right\}.$$

Then there exists a globally defined solution $x \in (\mathbb{D}_{\mathrm{pw}\mathscr{C}^\infty})^n$ *which is uniquely given by* $x(0-)$.

Remark 6.5.3 (On the assumption that $\sigma_{(-\infty,0)}$ is constant) The assumption in Corollary 6.5.2 that the switching signal σ in (6.1) is constant on $(-\infty, 0)$ is just a technicality to ensure uniqueness of the solution also backward in time. If one is interested only in the behaviour of the solution on $[0, \infty)$ then this assumption is not necessary to obtain the same result. More formally, one could also just study the corresponding ITP

$$x_{(-\infty,0)} = x^0{}_{(-\infty,0)},$$
$$(E_\sigma \dot{x})_{[0,\infty)} = (A_\sigma x + B_\sigma u)_{[0,\infty)}$$

with some initial trajectory $x^0 \in (\mathbb{D}_{\mathrm{pw}\mathscr{C}^\infty})^n$ then it doesn't matter how σ is defined on $(-\infty, 0)$.

Example 6.5.4 (Example 6.1.1 continued) Consider the switched circuit from Fig. 6.2 and the corresponding switched DAE (6.1) with matrices (E_1, A_1, B_1),

(E_2, A_2, B_2), (E_3, A_3, B_3) as given at the end of Example 6.1.1, i.e. (E_1, A_1, B_1) correspond to the switch in the left position, (E_2, A_2, B_2) correspond to the switch in the right position and (E_3, A_3, B_3) describes the system when the switch is in between. To check whether the switched DAE is uniquely solvable for all input signals u, regularity of the matrix pairs (E_1, A_1), (E_2, A_2), (E_3, A_3) must be checked:

$$\det(sE_1 - A_1) = CLs^2, \qquad \det(sE_2 - A_2) = CLs^2 + 1, \qquad \det(sE_3 - A_3) = Cs.$$

Hence, Corollary 6.5.2 implies that the switched DAE (6.1) has a solution for every switching signal $\sigma \in \Sigma_0$ and for every input $u \in \mathbb{D}_{\mathrm{pw}\mathscr{C}^\infty}$ and this solution is uniquely determined by $x(0-)$. To characterise the jumps at the switching instances the Wong sequences are calculated first. The corresponding limit spaces are given by

$$\mathscr{V}^*_{(E_1,A_1)} = \operatorname{im}\begin{bmatrix} 1 & 0 \\ 0 & 0 \\ 0 & 0 \\ 0 & 1 \end{bmatrix}, \qquad \mathscr{V}^*_{(E_2,A_2)} = \operatorname{im}\begin{bmatrix} 1 & 0 \\ 0 & 1 \\ 1 & 0 \\ 0 & -1 \end{bmatrix}, \qquad \mathscr{V}^*_{(E_3,A_3)} = \operatorname{im}\begin{bmatrix} 0 \\ 0 \\ 0 \\ 1 \end{bmatrix},$$

$$\mathscr{W}^*_{(E_1,A_1)} = \operatorname{im}\begin{bmatrix} 0 & 0 \\ 1 & 0 \\ 0 & 1 \\ 0 & 0 \end{bmatrix}, \qquad \mathscr{W}^*_{(E_2,A_2)} = \operatorname{im}\begin{bmatrix} 0 & 0 \\ 1 & 0 \\ 0 & 1 \\ 0 & 0 \end{bmatrix},$$

$$\mathscr{W}^*_{(E_3,A_3)} = \operatorname{im}\begin{bmatrix} 1 & 0 & 0 \\ 0 & 1 & 0 \\ 0 & 0 & 1 \\ 0 & 0 & 0 \end{bmatrix},$$

and the consistency, differential and impulse projectors are given by

$$\Pi_{(E_1,A_1)} = \begin{bmatrix} 1 & 0 & 0 & 0 \\ 0 & 0 & 0 & 0 \\ 0 & 0 & 0 & 0 \\ 0 & 0 & 0 & 1 \end{bmatrix}, \qquad \Pi_{(E_2,A_2)} = \begin{bmatrix} 1 & 0 & 0 & 0 \\ 0 & 0 & 0 & -1 \\ 1 & 0 & 0 & 0 \\ 0 & 0 & 0 & 1 \end{bmatrix},$$

$$\Pi_{(E_3,A_3)} = \begin{bmatrix} 0 & 0 & 0 & 0 \\ 0 & 0 & 0 & 0 \\ 0 & 0 & 0 & 0 \\ 0 & 0 & 0 & 1 \end{bmatrix},$$

$$\Pi^{\mathrm{diff}}_{(E_1,A_1)} = \begin{bmatrix} \frac{1}{L} & 0 & \frac{-1}{L} & 0 \\ 0 & 0 & 0 & 0 \\ 0 & 0 & 0 & 0 \\ 0 & \frac{1}{C} & 0 & \frac{-1}{C} \end{bmatrix}, \qquad \Pi^{\mathrm{diff}}_{(E_2,A_2)} = \begin{bmatrix} \frac{1}{L} & 0 & \frac{-1}{L} & 0 \\ 0 & \frac{-1}{C} & 0 & \frac{1}{C} \\ \frac{1}{L} & 0 & \frac{-1}{L} & 0 \\ 0 & \frac{1}{C} & 0 & \frac{-1}{C} \end{bmatrix},$$

$$\Pi^{\text{diff}}_{(E_3,A_3)} = \begin{bmatrix} 0 & 0 & 0 & 0 \\ 0 & 0 & 0 & 0 \\ 0 & 0 & 0 & 0 \\ 0 & \frac{1}{C} & 0 & \frac{-1}{C} \end{bmatrix},$$

$$\Pi^{\text{imp}}_{(E_1,A_1)} = \begin{bmatrix} 0 & 0 & 0 & 0 \\ 0 & 0 & 1 & 0 \\ 0 & 0 & 0 & 1 \\ 0 & 0 & 0 & 0 \end{bmatrix}, \qquad \Pi^{\text{imp}}_{(E_2,A_2)} = \begin{bmatrix} 0 & 0 & 0 & 0 \\ 0 & 0 & 1 & 0 \\ 0 & 0 & 0 & 1 \\ 0 & 0 & 0 & 0 \end{bmatrix},$$

$$\Pi^{\text{imp}}_{(E_3,A_3)} = \begin{bmatrix} 0 & 0 & 1 & 0 \\ 1 & 0 & 0 & 0 \\ 0 & 0 & 0 & 1 \\ 0 & 0 & 0 & 0 \end{bmatrix}.$$

With the help of these projectors the re-initialisation of $x(0+)$ for a given (possibly inconsistent) initial value $x(0-) =: (i_L^0, u_L^0, i_C^0, u_C^0)$ and the corresponding impulse in the solution can be calculated for each mode by the formula given in Theorem 6.5.1

$$\begin{aligned}
&\text{Mode 1:} \quad i_L(0+) = i_L^0, && v_L(0+) = u(0+), && i_C(0+) = 0, && v_c(0+) = v_C^0, \\
&\qquad\qquad i_L[0] = 0, && v_L[0] = 0, && i_C[0] = 0, && v_c[0] = 0, \\
&\text{Mode 2:} \quad i_L(0+) = i_L^0, && v_L(0+) = -v_C^0, && i_C(0+) = i_L^0, && v_c(0+) = u_C^0, \\
&\qquad\qquad i_L[0] = 0, && v_L[0] = 0, && i_C[0] = 0, && v_c[0] = 0, \\
&\text{Mode 3:} \quad i_L(0+) = 0, && v_L(0+) = 0, && i_C(0+) = 0, && v_c(0+) = v_C^0, \\
&\qquad\qquad i_L[0] = 0, && v_L[0] = -Li_L^0\delta, && i_C[0] = 0, && v_c[0] = 0.
\end{aligned}$$

6.6 Stability

In this section, only the homogeneous switched DAE (6.1), i.e.

$$E_\sigma \dot{x} = A_\sigma x \tag{6.7}$$

is considered, because stability analysis usually considers a closed loop where the input is already replaced by a feedback. Furthermore, certain simple inputs can be incorporated as new state variables, for example, a constant input signal u can be rewritten as the state equation $\dot{u} = 0$ and a sinusoidal input u can be rewritten as $\dot{u} = -\omega\overline{u}$, $\dot{\overline{u}} = \omega u$, for some $\omega \in \mathbb{R}$. Hence the switched DAE (6.1) with constant or sinusoidal inputs can be written as the DAE (6.7). Throughout this section, it is assumed that the switching signal σ is such that the switching times do not accumulate.

Definition 6.6.1 (Asymptotic stability) The switched DAE (6.7) is called asymptotically stable if and only if for all solutions $x \in (\mathbb{D}_{\mathrm{pw}\mathscr{C}^\infty})^n$ the following properties hold

(S) $\forall \varepsilon > 0\ \exists \delta > 0: \ \|x(0-)\| < \delta \Rightarrow \forall t > 0: \ \|x(t\pm)\| < \varepsilon$,
(A) $x(t\pm) \to 0$ as $t \to \infty$,
(I) $\forall t \geq 0: \ x[t] = 0$.

The assumptions (S) (stability) and (A) (attractivity) are standard for the definition of asymptotic stability. The assumption (I) (impulse-freeness) is motivated by the following observation: Assume (6.7) has a solution x with $x[t] \neq 0$ for some t. By linearity of (6.7), the scaling of x by some $\varepsilon > 0$ will yield also a solution, in particular, the initial value $\varepsilon x(0-)$ can be made arbitrarily small, but the corresponding impulse $\varepsilon x[t]$ does not vanish. Since an impulse can be interpreted as an infinite peak (or the limit of an unbounded sequence of functions) this shows that the classical stability assumption (S) cannot be fulfilled in the sense that small initial values yield a bounded and small solution. The problem whether the switched system (6.7) has impulses in the solution or not, i.e. whether (I) holds or not, is also an interesting question on its own. Consider, for example, the situation where the switched DAE (6.7) models a nominal system with additional sudden faults (like a shortcut in an circuit element). If the corresponding switch is able to produce a Dirac impulse in some of the variables (e.g. voltages and currents) then this impulse might destroy other components of the system, possibly leading to a cascading destruction of the system. Therefore, the topic of impulse-freeness is studied first.

6.6.1 Impulse Detection

Theorem 6.6.2 (Impulse-freeness) *Consider the switched DAE (6.7) with regular matrix pairs (E_p, A_p), $p \in \{1, \dots, \bar{p}\}$ and a switching from mode $p \in \{1, \dots, \bar{p}\}$ to mode $q \in \{1, \dots, \bar{p}\}$ at some switching time $t_s \in \mathbb{R}$, i.e. $\sigma(t_s-) = p$ and $\sigma(t_s+) = q$. Then this switch cannot produce an impulse, i.e. $x[t_s] = 0$ for all solutions $x \in (\mathbb{D}_{\mathrm{pw}\mathscr{C}^\infty})^n$ of (6.7) if the following impulse-freeness-condition holds:*

$$E_q(I - \Pi_q)\Pi_p = 0, \tag{6.8}$$

where $\Pi_p := \Pi_{(E_p, A_p)}$, $\Pi_q := \Pi_{(E_q, A_q)}$ are the consistency projectors as in Definition 6.4.1. In particular, if (6.8) holds for all $p, q \in \{1, \dots, \bar{p}\}$ then all solutions of (6.7) are impulse-free independently of the switching signal.

Proof Let $x \in (\mathbb{D}_{\mathrm{pw}\mathscr{C}^\infty})^n$ be a solution of (6.7) with a switching signal σ such that $\sigma(t_s-) = p$ and $\sigma(t_s+) = p$. Since $x(t_s-) \in \mathscr{V}_p^*$, where $\mathscr{V}_p^*$ is the limit of the first

Wong sequence corresponding to (E_p, A_p), it follows that $x(t_s-) = \Pi_p x(t_s-)$. The impulse formula from Theorem 6.5.1 now yields

$$\begin{aligned} x[t_s] &= -\sum_{i=0}^{n-2} (E_q^{\mathrm{imp}})^{i+1} (I - \Pi_q) x(t_s-) \\ &= -\sum_{i=0}^{n-2} (E_q^{\mathrm{imp}})^{i} \Pi_q^{\mathrm{imp}} \underbrace{E_q (I - \Pi_q) \Pi_p}_{=0} x(t_s-) = 0. \end{aligned}$$

□

Example 6.6.3 (Example 6.1.1 continued) Consider the circuit as shown in Fig. 6.2 and assume the input u is constant. As mentioned above, the input is reinterpreted as a state variable given by $\dot{u} = 0$, hence one obtains the following three DAEs with $x = (u, i_L, v_L, i_C, v_C)^\top$ describing the three different modes:

$$\text{Mode 1:} \quad \begin{bmatrix} 1 & 0 & 0 & 0 & 0 \\ 0 & L & 0 & 0 & 0 \\ 0 & 0 & 0 & 0 & C \\ 0 & 0 & 0 & 0 & 0 \\ 0 & 0 & 0 & 0 & 0 \end{bmatrix} \dot{x} = \begin{bmatrix} 0 & 0 & 0 & 0 & 0 \\ 0 & 0 & 1 & 0 & 0 \\ 0 & 0 & 0 & 1 & 0 \\ -1 & 0 & 1 & 0 & 0 \\ 0 & 0 & 0 & 1 & 0 \end{bmatrix} x,$$

$$\text{Mode 2:} \quad \begin{bmatrix} 1 & 0 & 0 & 0 & 0 \\ 0 & L & 0 & 0 & 0 \\ 0 & 0 & 0 & 0 & C \\ 0 & 0 & 0 & 0 & 0 \\ 0 & 0 & 0 & 0 & 0 \end{bmatrix} \dot{x} = \begin{bmatrix} 0 & 0 & 0 & 0 & 0 \\ 0 & 0 & 1 & 0 & 0 \\ 0 & 0 & 0 & 1 & 0 \\ 0 & 0 & 1 & 0 & 1 \\ 0 & -1 & 0 & 1 & 0 \end{bmatrix} x,$$

$$\text{Mode 3:} \quad \begin{bmatrix} 1 & 0 & 0 & 0 & 0 \\ 0 & L & 0 & 0 & 0 \\ 0 & 0 & 0 & 0 & C \\ 0 & 0 & 0 & 0 & 0 \\ 0 & 0 & 0 & 0 & 0 \end{bmatrix} \dot{x} = \begin{bmatrix} 0 & 0 & 0 & 0 & 0 \\ 0 & 0 & 1 & 0 & 0 \\ 0 & 0 & 0 & 1 & 0 \\ 0 & 1 & 0 & 0 & 0 \\ 0 & 0 & 1 & 0 & 0 \end{bmatrix} x.$$

It is easily seen that the corresponding consistency projectors Π_p, $p = 1, 2, 3$, are given by

$$\Pi_p = \begin{bmatrix} 1 & 0 \\ 0 & \overline{\Pi}_p \end{bmatrix},$$

where $\overline{\Pi}_p$ is the consistency projector from Example 6.5.4. Hence, the condition (6.8) can easily be checked for each mode change:

	$q = 1$	$q = 2$	$q = 3$
$p = 1$	$E_q(I - \Pi_q)\Pi_p = 0$	$E_q(I - \Pi_q)\Pi_p = 0$	$E_q(I - \Pi_q)\Pi_p \neq 0$
$p = 2$	$E_q(I - \Pi_q)\Pi_p = 0$	$E_q(I - \Pi_q)\Pi_p = 0$	$E_q(I - \Pi_q)\Pi_p \neq 0$
$p = 3$	$E_q(I - \Pi_q)\Pi_p = 0$	$E_q(I - \Pi_q)\Pi_p = 0$	$E_q(I - \Pi_q)\Pi_p = 0$

This is in correspondence to the results obtained in Example 6.5.4 where it was apparent that only a switch to mode three can produce impulses in the solution. However, the method based on Theorem 6.6.2 is simpler (for example, the transformation matrices S_p, $p \in \{1, \dots, \bar{p}\}$ are not needed) and it also takes into account which mode was active before the switch. For the considered circuit this doesn't make a difference, but in general the mode before the switch might prohibit certain initial values so that the switch does not produce impulses although it might produce switches for a general initial value. Actually, this is the case when (6.8) holds but $E_q(I - \Pi_q) \neq 0$. For a more complex example, see also [7].

6.6.2 Lyapunov Functions for Non-switched DAEs

In order to study stability of the switched DAE (6.7), the stability properties of the non-switched DAE

$$E\dot{x} = Ax \tag{6.9}$$

with regular matrix pair $(E, A) \in \mathbb{R}^{n\times n} \times \mathbb{R}^{n\times n}$ are studied first. From the quasi-Weierstrass form (6.3) of the pair (E, A) it is immediately clear that (6.9) is asymptotically stable if and only if the matrix J in (6.3) is Hurwitz. A direct method (i.e. without calculating the full quasi-Weierstrass form first) is given by the following result.

Theorem 6.6.4 (Generalised Lyapunov equation, [27])[2] *The DAE (6.9) is asymptotically stable if and only if there exists a positive definite symmetric $P = P^\top \in \mathbb{R}^{n\times n}$ and a symmetric $Q = Q^\top \in \mathbb{R}^{n\times n}$ which is positive definite on $\mathcal{V}^*$ (the limit of the first Wong sequence) such that*

$$A^\top PE + E^\top PA = -Q. \tag{6.10}$$

This result makes it possible to define a *Lyapunov function* as

$$V(x) = (Ex)^\top PEx \tag{6.11}$$

because for all (classical) solutions $x : \mathbb{R} \to \mathcal{V}^* \subseteq \mathbb{R}^n$ of (6.9) it follows that

$$\begin{aligned}\frac{\mathrm{d}}{\mathrm{d}t}V\big(x(t)\big) &= \big(E\dot{x}(t)\big)^\top PEx(t) + \big(Ex(t)\big)^\top PE\dot{x}(t)\\ &= x(t)^\top\big(A^\top PE + E^\top PA\big)x(t)\\ &= -x(t)Qx(t) < 0.\end{aligned}$$

---

[2]In [27], only the complex-valued case is studied. However, in the real-valued case it is easily seen that the real part of the complex solutions P and Q are also solutions of the Lyapunov equation.

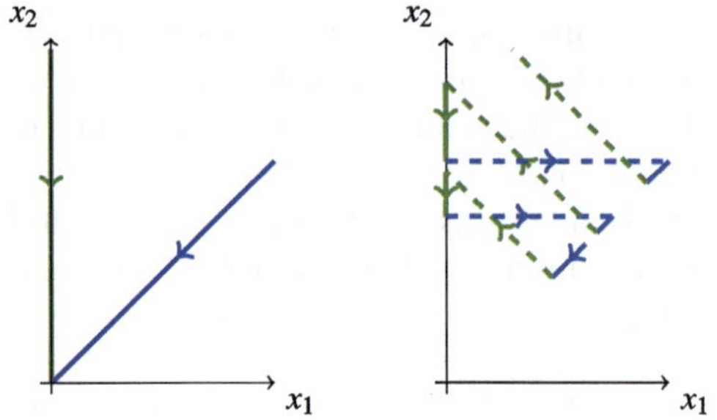

Fig. 6.4 Solutions of the switched DAE from Example 6.6.5. (*Left*): Dynamics without switching, (*Right*): Unstable solution behaviour for some switching signal; the dashed lines denote the jumps induced by the switching event

6.6.3 Asymptotic Stability of Switched DAEs

Consider the switched DAE (6.7)

$$E_\sigma \dot{x} = A_\sigma x.$$

Clearly, if one mode $E_p\dot{x} = A_p x$, $p \in \{1, \dots, \bar{p}\}$ is not asymptotically stable then (6.7) cannot be asymptotically stable for arbitrary switching (just chose the constant switching signal $\sigma(t) = p$ for all $t \in \mathbb{R}$). Hence for studying asymptotic stability for arbitrary switching signals or at least for slow switching it has to be assumed that each mode is asymptotically stable and hence permits a Lyapunov function (6.11). From the theory of switched ODEs (see e.g. [20]), it is known that switching between stable modes can lead to an unstable overall system, hence one might expect a similar behaviour for switched DAEs. This unstable behaviour cannot happen in the switched ODE case when there exists a common Lyapunov function. Surprisingly, this condition is not sufficient any more for the switched DAE case because the possible jumps must be incorporated in the right way, too. The following example shows that the induced jumps are important for the stability (for more examples, see [21]).

Example 6.6.5 (Unstable switched DAE with the same Lyapunov function for each subsystem) Consider the switched DAE (6.7) with two subsystems given by

$$(E_1, A_1) = \left(\begin{bmatrix} 0 & 0 \\ 0 & 1 \end{bmatrix}, \begin{bmatrix} 1 & -1 \\ 0 & -1 \end{bmatrix} \right), \qquad (E_2, A_2) = \left(\begin{bmatrix} 0 & 0 \\ 1 & 1 \end{bmatrix}, \begin{bmatrix} -1 & 0 \\ 0 & -1 \end{bmatrix} \right).$$

The consistency projectors are given by

$$\Pi_1 = \begin{bmatrix} 0 & 1 \\ 0 & 1 \end{bmatrix}, \qquad \Pi_2 = \begin{bmatrix} 0 & 0 \\ 1 & 1 \end{bmatrix}.$$

A possible solution behaviour of this switched DAE is shown in Fig. 6.4. Clearly, each subsystem is asymptotically stable. Furthermore, the solutions decrease along $V(x) = x_1^2 + x_2^2$ for both subsystems. However, when switches occur, the consistency projectors yield jumps which can destabilise the overall system.

In general, the jumps must be "compatible" with the Lyapunov function. Furthermore, impulse-freeness must be ensured additionally. Altogether, the following result holds.

Theorem 6.6.6 (Asymptotic stability under arbitrary switching) *Consider the switched DAE* (6.7) *with regular matrix pairs* (E_p, A_p), $p \in \{1, \dots, \bar{p}\}$. *Let* $\Pi_p := \Pi_{(E_p,A_p)}$ *be the consistency projector as given by Definition* 6.4.1. *If, for all* $p \in \{1, \dots, \bar{p}\}$,

(V$_p$) $E_p\dot{x} = A_p x$ *is asymptotically stable with Lyapunov function* V_p,
(IC) $E_q(I - \Pi_q)\Pi_p = 0$ *for all* $q \in \{1, \dots, \bar{p}\}$ *and*
(JC) $V_q(\Pi_q x) \le V_p(x)$ *for all* $x \in \mathcal{V}_p^* = \operatorname{im} \Pi_p$

then the switched DAE (6.7) *is asymptotically stable for all switching signals.*

Proof Attractivity is shown in [21] and stability (as well as attractivity in a nonlinear setting) is shown in [22]. The key idea is to consider the common Lyapunov function

$$V : \mathbb{R}^n \to \mathbb{R}, \qquad x \mapsto \begin{cases} V_p(x) & \text{if } x \in \mathcal{V}_p^*, \\ 0 & \text{otherwise,} \end{cases}$$

which is well-defined because for $x \in \mathcal{V}_p^* \cap \mathcal{V}_q^*$ it follows that $\Pi_p x = x = \Pi_q x$, and therefore

$$V_q(x) = V_q(\Pi_q x) \le V_p(x) = V_p(\Pi_p x) \le V_q(x),$$

hence $V_p(x) = V_q(x)$ for all $x \in \mathcal{V}_p^* \cap \mathcal{V}_q^*$. It then follows that V is decreasing along solutions and $V(x(t)) \to 0$ as $t \to \infty$. Finally, positive definiteness of V on each V_p implies $x(t) \to 0$ as $t \to \infty$. □

As highlighted in the proof, the jump condition (JC) is a generalisation of the common Lyapunov condition for switched ODEs, in fact, if this condition is applied to a switched ODE it reads as $V_p(x) \le V_q(x)$ for all p, q hence $V_p = V_q$, i.e. it is equivalent to the existence of a common Lyapunov function.

It is known from the switched ODEs theory that the destabilisation of a switched system can only be achieved by sufficiently fast switching. This result also holds for switched DAEs. Slow switching is characterised here by an average dwell time [18]. Therefore, denote with $N_\sigma(t_1, t_2)$ the number of switches of σ within the interval $[t_1, t_2)$. The class of switching signals with average dwell time $\tau_a > 0$ is then given by

$$\Sigma_{\tau_a} := \left\{ \sigma \in \Sigma \;\middle|\; \exists N_0 > 0 \; \forall t \in \mathbb{R} \; \forall \Delta t > 0 : \; N_\sigma(t, t + \Delta t) < N_0 + \frac{\Delta t}{\tau_a} \right\}.$$

The number $N_0 > 0$ in the definition of Σ_{τ_a} is the so-called *chatter bound* and is an upper bound for the number of switches within an interval of length smaller than τ_a.

For each switching signal $\sigma \in \Sigma_{\tau_a}$, this chatter bound is finite but it is not bounded for the whole class Σ_{τ_a}. The class of switching signals $\sigma \in \Sigma_{\tau_a}$ with chatter bound $N_0 = 1$ is exactly the class of switching signals with dwell time $\tau_d = \tau_a$.

Theorem 6.6.7 (Stability under slow switching, [22]) *Consider the switched DAE* (6.7) *satisfying conditions* (V_p) *and the impulse condition* (IC). *Then there exists* $\tau_a > 0$ *such that* (6.7) *is asymptotically stable if* $\sigma \in \Sigma_{\tau_a}$. *In fact, let* P_p, Q_p *be the solutions of the Lyapunov equation* (6.10) *corresponding to* (E_p, A_p), *let* O_p *be an orthogonal basis matrix of* $\mathcal{V}_p^* = \operatorname{im} \Pi_p$ *and let*

$$\mu_{p,q} := \frac{\lambda_{\max}(O_p^\top \Pi_q^\top E_q^\top P_q E_q \Pi_q O_p)}{\lambda_{\min}(O_p^\top E_p^\top P_p E_p O_p)} > 0,$$

$$\lambda_p := \frac{\lambda_{\min}(O_p^\top Q_p O_p)}{\lambda_{\max}(O_p^\top E_p^\top P_p E_p O_p)} > 0,$$

where $\lambda_{\min}(\cdot)$ *and* $\lambda_{\max}(\cdot)$ *denote the minimal and maximal eigenvalue of a symmetric matrix, respectively. Then an average dwell time of*

$$\tau_a > \frac{\max_{p,q} \ln \mu_{p,q}}{\min_p \lambda_p}$$

guarantees asymptotic stability of (6.7).

Example 6.6.8 (Example 6.6.5 continued) Consider the switched DAE from Example 6.6.5. It was already highlighted there that both systems share the same Lyapunov function $V(x) = x_1^2 + x_2^2$, hence condition (V_p) holds. Furthermore, it is not difficult to check that the impulse-freeness condition (IC) holds. Since the jump condition (JC) does not hold, Theorem 6.6.6 is not applicable, but Theorem 6.6.7 yields that for switching signals with sufficiently large average dwell time asymptotic stability is guaranteed. In order to calculate a sufficient average dwell time, the Lyapunov functions must be rewritten as $V(x) = (Ex)^\top PEx$, where P solves the generalised Lyapunov equation (6.10). This is achieved by choosing

$$P_1 = \begin{bmatrix} 0 & 0 \\ 0 & 2 \end{bmatrix}, \qquad Q_1 = \begin{bmatrix} 0 & 0 \\ 0 & 4 \end{bmatrix}$$

and

$$P_1 = \begin{bmatrix} 0 & 0 \\ 0 & 2 \end{bmatrix}, \qquad Q_2 = \begin{bmatrix} 0 & 1 \\ 1 & 2 \end{bmatrix}.$$

The associated Lyapunov functions are $V_1(x) = 2x_2^2$ and $V_2(x) = (x_1 + x_2)^2$ which coincide with $V(x) = x_1^2 + x_2^2$ on the corresponding consistency spaces. As orthonormal basis of $\mathcal{V}_1^*$ and $\mathcal{V}_2^*$ choose

$$O_1 = \frac{1}{2}\begin{bmatrix} \sqrt{2} \\ \sqrt{2} \end{bmatrix} \quad \text{and} \quad O_2 = \begin{bmatrix} 0 \\ 1 \end{bmatrix}.$$

Then

$$O_1^\top E_1^\top P_1 E_1 O_1 = 1, \qquad O_2^\top E_2^\top P_2 E_2 O_2 = 1,$$
$$O_1^\top \Pi_2^\top E_2^\top P_2^\top E_2 \Pi_2 O_1 = 2, \qquad O_2^\top \Pi_1^\top E_1^\top P_1^\top E_1 \Pi_1 O_2 = 2,$$
$$O_1^\top Q_1 O_1 = 2, \qquad O_2^\top Q_2 O_2 = 2,$$

hence $\mu := \max_{p,q} \mu_{p,q} = 2$ and $\lambda := \min_p \lambda_p = 2$. Therefore, the corresponding switched DAE is asymptotically stable for all switching signals $\sigma \in \Sigma_\tau$ with $\tau_a > \ln 2/2$. For this case, the bound is actually sharp because it is easily seen that for a periodic switching signal with periodicity $\ln 2/2$ the corresponding solution is also periodic (cf. [21]).

6.6.4 Commutativity and Asymptotic Stability

For switched ODEs with asymptotically stable subsystems, it is well known [26] that commuting A-matrices imply asymptotic stability for arbitrary switching signal and also guarantee a common quadratic Lyapunov function. The aim of this section is to generalise this result to switched DAEs (6.1). Example 6.6.5 shows that commutativity of the A-matrix is not the right condition to guarantee asymptotic stability for arbitrary switching. In particular, for switched DAEs instability can be induced by the jumps, hence one would expect that the consistency projectors play a prominent role again. Surprisingly, this is not the case as the following result shows.

Theorem 6.6.9 (Commutativity of A^{diff}-matrices implies stability, [23]) *Consider a switched DAE* (6.7) *with corresponding matrices* A_p^{diff} *as defined in Theorem* 6.4.4. *Assume asymptotic stability of each subsystem, i.e.* (V_p), *and impulse-freeness, i.e.* (IC), *then*

$$\left[A_p^{\text{diff}}, A_q^{\text{diff}}\right] := A_p^{\text{diff}} A_q^{\text{diff}} - A_q^{\text{diff}} A_p^{\text{diff}} = 0 \quad \forall p, q \in \{1, 2, \dots, \bar{p}\}$$

implies asymptotic stability of the switched DAE (6.7) *for arbitrary switching signals.*

Proof The key observation is that $[A_p^{\text{diff}}, A_q^{\text{diff}}] = 0$ implies commutativity also with the consistency projectors:

$$\left[\Pi_p, A_p^{\text{diff}}\right] = 0, \qquad \left[\Pi_p, A_q^{\text{diff}}\right] = 0, \qquad [\Pi_p, \Pi_q] = 0.$$

Hence the flow commutes and asymptotic stability easily follows, see [23] for details. There it is also shown that a common quadratic Lyapunov function exists which is compatible with the jumps. □

6.7 Bibliographical Notes

The theory of DAEs gained popularity in the 1980s, see e.g. the textbooks [3, 4, 6, 16], and was already discussed by Gantmacher in the 1950s [12]. That DAEs are especially suitable to model electrical circuits was highlighted by Verghese et al. [35]. There impulsive behaviour was also discussed and motivated; however, the underlying distributional solution space was not formalised which is also true for the above mentioned textbooks. Later on, the focus of research was on numerics of DAEs and (smoothly) time-varying DAEs, see e.g. the textbook [19] and the references therein.

The existence of impulsive solutions was a reoccurring topic in the literature, see e.g. [5, 30]; for a more detailed overview, see the introduction of [34]. Apart from the author's own work (presented in this chapter), there is not much work on switched DAEs available [14, 15, 24, 25, 38] and none of these works resolve or even discuss the problem of multiplying a piecewise-constant function with a distribution which naturally occurs when studying switched differential algebraic equations of the form (6.1). Another approach to model electrical circuits with switches uses the complementarity framework and is discussed in the next chapter. Combining both approaches seems fruitful and is a topic of ongoing research.

6.8 Summary

In this chapter, it was motivated that switched DAEs are a suitable modelling framework for electrical circuits with switches. The space of piecewise-smooth distributions was introduced as an underlying solution space. Explicit solution formulas were given in particular for the impulsive parts induced by the inconsistent initial values. A simple matrix condition was given which makes it possible to exclude impulsive solutions for arbitrary switching. Finally, stability of switched DAEs was studied. With the help of Lyapunov functions, a sufficient condition for asymptotic stability under arbitrary and slow switching was presented. The theoretical results were illustrated with an examples stemming from a simple electrical circuit.

Acknowledgements This work was supported by the DFG grant Wi1458/10-1. Many thanks to Roman Geiselhart for giving valuable comments on the manuscript of this book chapter.

References

1. Armentano, V.A.: The pencil $(sE - A)$ and controllability-observability for generalized linear systems: a geometric approach. SIAM J. Control Optim. **24**, 616–638 (1986)
2. Berger, T., Ilchmann, A., Trenn, S.: The quasi-Weierstraß form for regular matrix pencils. Linear Algebra Appl. (2010). doi:10.1016/j.laa.2009.12.036. Preprint available online, Institute for Mathematics, Ilmenau University of Technology, Preprint Number 09-21
3. Campbell, S.L.: Singular Systems of Differential Equations I. Pitman, New York (1980)

4. Campbell, S.L.: Singular Systems of Differential Equations II. Pitman, New York (1982)
5. Cobb, J.D.: Controllability, observability and duality in singular systems. IEEE Trans. Autom. Control **AC-29**, 1076–1082 (1984)
6. Dai, L.: Singular Control Systems. Lecture Notes in Control and Information Sciences, vol. 118. Springer, Berlin (1989)
7. Domínguez-García, A.D., Trenn, S.: Detection of impulsive effects in switched DAEs with applications to power electronics reliability analysis. In: Proc. of the IEEE Conference on Decision and Control, Atlanta, Georgia, USA, pp. 5662–5667 (2010)
8. Ferreira, J.C.: Introduction to the Theory of Distributions. Pitman Monographs and Surveys in Pure and Applied Mathematics, vol. 87. Wesley, Harlow (1997). Translated by J. Sousa Pinto and R. F. Hoskins
9. Frasca, R., Çamlıbel, M.K., Goknar, I.C., Iannelli, L., Vasca, F.: Linear passive networks with ideal switches: Consistent initial conditions and state discontinuities. IEEE Trans. Circuits Syst. I, Regul. Papers **57**(12), 3138–3151 (2010)
10. Fuchssteiner, B.: Eine assoziative Algebra über einen Unterraum der Distributionen. Math. Ann. **178**, 302–314 (1968)
11. Fuchssteiner, B.: Algebraic foundation of some distribution algebras. Stud. Math. **76**, 439–453 (1984)
12. Gantmacher, F.R.: The Theory of Matrices, vols. I & II. Chelsea, New York (1959)
13. Geerts, A.H.W.: Solvability conditions, consistency and weak consistency for linear differential-algebraic equations and time-invariant linear systems: The general case. Linear Algebra Appl. **181**, 111–130 (1993)
14. Geerts, A.H.W., Schumacher, J.M.: Impulsive-smooth behavior in multimode systems. Part I: State-space and polynomial representations. Automatica **32**(5), 747–758 (1996)
15. Geerts, A.H.W., Schumacher, J.M.: Impulsive-smooth behavior in multimode systems. Part II: Minimality and equivalence. Automatica **32**(6), 819–832 (1996)
16. Griepentrog, E., März, R.: Differential-Algebraic Equations and Their Numerical Treatment. Teubner-Texte zur Mathematik, vol. 88. Teubner, Leipzig (1986)
17. Hautus, M.L.J., Silverman, L.M.: System structure and singular control. Linear Algebra Appl. **50**, 369–402 (1983)
18. Hespanha, J.P., Morse, A.S.: Stability of switched systems with average dwell-time. In: Proc. of the IEEE Conference on Decision and Control, Phoenix, Arizona, USA, pp. 2655–2660 (1999)
19. Kunkel, P., Mehrmann, V.: Differential-Algebraic Equations. Analysis and Numerical Solution. EMS Publishing House, Zürich (2006)
20. Liberzon, D.: Switching in Systems and Control. Systems and Control: Foundations and Applications. Birkhäuser, Boston (2003)
21. Liberzon, D., Trenn, S.: On stability of linear switched differential algebraic equations. In: Proc. of the IEEE Conference on Decision and Control, Shanghai, China, pp. 2156–2161 (2009)
22. Liberzon, D., Trenn, S.: Switched nonlinear differential algebraic equations: Solution theory, Lyapunov functions, and stability. Automatica (2011). doi:10.1016/j.automatica.2012.02.041. Preprint available from the websites of the authors
23. Liberzon, D., Trenn, S., Wirth, F.: Commutativity and asymptotic stability for linear switched DAEs. In: Proc. of 50th IEEE Conf. on Decision and Control and European Control Conference 2011, Orlando, USA, pp. 417–422 (2011)
24. Meng, B.: Observability conditions of switched linear singular systems. In: Proc. of the Chinese Control Conference, Harbin, Heilongjiang, China, pp. 1032–1037 (2006)
25. Meng, B., Zhang, J.F.: Reachability conditions for switched linear singular systems. IEEE Trans. Autom. Control **51**(3), 482–488 (2006)
26. Narendra, K.S., Balakrishnan, J.: A common Lyapunov function for stable LTI systems with commuting A-matrices. IEEE Trans. Autom. Control **29**(12), 2469–2471 (1994)
27. Owens, D.H., Debeljkovic, D.L.: Consistency and Liapunov stability of linear descriptor systems: A geometric analysis. IMA J. Math. Control Inf. **2**(2), 139–151 (1985)

28. Rabier, P.J., Rheinboldt, W.C.: Classical and generalized solutions of time-dependent linear differential-algebraic equations. Linear Algebra Appl. **245**, 259–293 (1996)
29. Rabier, P.J., Rheinboldt, W.C.: Time-dependent linear DAEs with discontinuous inputs. Linear Algebra Appl. **247**, 1–29 (1996)
30. Rabier, P.J., Rheinboldt, W.C.: Theoretical and numerical analysis of differential-algebraic equations. In: Ciarlet, P.G., Lions, J.L. (eds.) Handbook of Numerical Analysis, vol. VIII, pp. 183–537. Elsevier, Amsterdam (2002)
31. Schwartz, L.: Théorie des Distributions I, II. Publications de l'institut de mathématique de l'Universite de Strasbourg, vols. IX, X. Hermann, Paris (1950/1951)
32. Tanwani, A., Trenn, S.: On observability of switched differential-algebraic equations. In: Proc. of the IEEE Conference on Decision and Control, Atlanta, Georgia, USA, pp. 5656–5661 (2010)
33. Trenn, S.: Distributional differential algebraic equations. PhD thesis, Institut für Mathematik, Technische Universität Ilmenau, Universitätsverlag Ilmenau, Ilmenau, Germany (2009). URL http://www.db-thueringen.de/servlets/DocumentServlet?id=13581
34. Trenn, S.: Regularity of distributional differential algebraic equations. Math. Control Signals Syst. **21**(3), 229–264 (2009)
35. Verghese, G.C., Levy, B.C., Kailath, T.: A generalized state-space for singular systems. IEEE Trans. Autom. Control **AC-26**(4), 811–831 (1981)
36. Weierstraß, K.: Zur Theorie der bilinearen und quadratischen Formen. Monatsh. Akad. Wiss. 310–338 (1868)
37. Wong, K.T.: The eigenvalue problem $\lambda Tx + Sx$. Int. J. Differ. Equ. **16**, 270–280 (1974)
38. Wunderlich, L.: Analysis and numerical solution of structured and switched differential-algebraic systems. PhD thesis, Fakultät II Mathematik und Naturwissenschaften, Technische Universität Berlin, Berlin, Germany (2008)

The design of a PBC in the BM framework follows the same principle as for Euler–Lagrange systems [19]. This means that the controller dynamics is determined by making a copy of (8.32)–(8.33) in terms of the controller state ξ and add damping in terms of the difference (the error) between z and ξ, i.e.

$$Q\frac{\mathrm{d}}{\mathrm{d}t}\xi = \nabla P^{\mu}(\xi) + \nabla_{\xi} P_a(z-\xi). \tag{8.34}$$

Here $P_a := P_{R_a} - P_{G_a}$ represents the added damping through the injection of additional content and co-content functions. The damping injection process can be viewed as shaping the mixed-potential function.

At this point, one is usually tempted to add damping on those states that do not contain any damping a priori. For the boost converter, this would mean that, since the output capacitor voltage already contains a damping term given by the load resistor, damping should be added on the current coordinate. However, Theorem 8.7.1 and Theorem 8.7.2 suggest that there are two natural ways of injecting additional damping: *series damping injection* (current feedback) and *parallel damping injection* (voltage feedback), respectively. Series damping PBC forces the closed-loop dynamics to act as if there are resistors in series with the inductances, while parallel damping PBC corresponds to virtually 'injecting' resistors in parallel with the capacitors. Indeed, let $e := z - \xi$ and suppose that we set $P_{R_a}(e_1) = \frac{1}{2}R_s e_1^2$ and $P_{G_a} = 0$. Subtracting (8.34) from (8.33) yields the closed-loop error dynamics

$$Q\frac{\mathrm{d}}{\mathrm{d}t}e = \nabla P_e^{\mu}(e),$$

with closed-loop error mixed-potential

$$P_e^{\mu}(e) = -\frac{e_2^2}{2R_o} - \mu e_1 e_2 + \frac{R_s}{2}e_1^2.$$

Clearly, these equations corresponds to the circuit depicted in Fig. 8.9a.

On the other hand, setting $P_{R_a} = 0$ and $P_{G_a}(e_2) = \frac{1}{2}R_p^{-1}e_2^2$, yields a closed-loop error mixed-potential of the form

$$P_e^{\mu}(e) = -\frac{e_2^2}{2(R_o + R_p)} - \mu e_1 e_2,$$

which corresponds to the circuit depicted in Fig. 8.9b.

Besides the type of injected damping, Theorem 8.7.1 and Theorem 8.7.2 also provide information on the lower and upper bounds for tuning the control parameters. For a circuit with linear (or linearized) dynamics it can be shown that if one the theorems is satisfied, the corresponding eigenvalues of the (linearized) system matrix lie either on the real axis (away from the origin) or on a circle in the left-half plane. The radius of this circle can be made arbitrary small with δ, which may therefore be considered as a fine tuning parameter. Note that for the PBC controlled boost converter example, the theorems are satisfied if either $R_s > (1-\delta)^{-1}\mu\sqrt{L/C}$

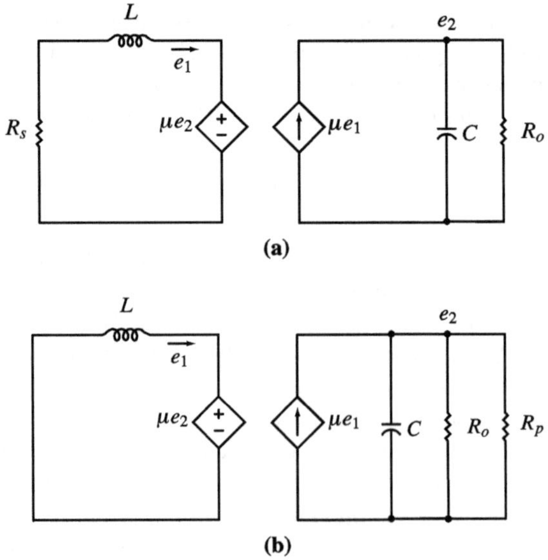

Fig. 8.9 Closed-loop interpretation of a (**a**) series and (**b**) parallel damping PBC of a boost converter

for series damping injection, or $R_p < (1-\delta)\mu^{-1}\sqrt{L/C} - R_o$ for parallel damping injection. These fairly sharp criteria form a systematic and straightforward tool for solving the tuning problem for a general class of PBC controlled power converters. The interested reader is referred to [13] for more details.

8.8 Summary

This chapter treats a power based modelling framework for switching electrical circuits. The Brayton–Moser framework is discussed, and the advantages for stability analysis, circuit design and control design are treated. In particular, for controller design, the new passivity property treated in this chapter is not yet exploited for controller design. It is expected that such exploitation in combination with the BM structure will result in clear tuning rules for the controllers to achieve better performance. Furthermore, passivity-based tracking control, particularly relevant for DC–AC conversion, needs to be further studied and developed. The canonical transformation framework that so far has been mainly studied for tracking control of mechanical systems in [3, 4, 8] is expected to be instrumental.

References

1. Brayton, R.K., Moser, J.K.: A theory of nonlinear networks I. Q. Appl. Math. **22**(1), 1–33 (1964)
2. Desoer, C.A., Kuh, E.S.: Basic Circuit Theory. McGraw-Hill, New York (1969)

3. Dirksz, D.A.: Robust energy- and power-based control design: Port-Hamiltonian and Brayton–Moser systems. Ph.D. thesis, University of Groningen (2011)
4. Dirksz, D.A., Scherpen, J.M.A.: Power-based control: Canonical coordinate transformations, integral and adaptive control. Automatica (2012). doi:10.1016/j.automatica.2012.03.003
5. Duindam, V., Macchelli, A., Stramigioli, S., Bruyninckx, H. (eds.): Modeling and Control of Complex Physical Systems: The Port-Hamiltonian Approach. Springer, Berlin (2009)
6. Escobar, G., Van der Schaft, A.J., Ortega, R.: A Hamiltonian viewpoint in the modeling of switching power converters. Automatica **35**, 445–452 (1999)
7. Favache, A., Dochain, D.: Power-shaping control of reaction systems: The CSTR case. Automatica **46**, 1877–1883 (2010)
8. Fujimoto, K., Sugie, T.: Canonical transformation and stabilization of generalized Hamiltonian systems. Syst. Control Lett. **42**(3), 217–227 (2001)
9. García-Canseco, E., Jeltsema, D., Ortega, R., Scherpen, J.M.A.: Power shaping control of physical systems. Automatica **46**, 127–132 (2010)
10. Hernandez-Gomez, M., Ortega, R., Lamnabhi-Lagarrigue, F., Escobar, G.: Adaptive PI stabilization of switched power converters. IEEE Trans. Control Syst. Technol. **18**(3), 688–698 (2010)
11. Hill, D., Moylan, P.: The stability of nonlinear dissipative systems. IEEE Trans. Autom. Control **21**(5), 708–711 (1976)
12. Hiti, S., Borojević, D.: Control of front-end three-phase Boost rectifier. In: Proc. of the Applied Power Electronics Conference, Orlando, Florida, USA, pp. 927–933 (1994)
13. Jeltsema, D., Scherpen, J.M.A.: Tuning of passivity-preserving controllers for switched-mode power converters. IEEE Trans. Autom. Control **49**(8), 1333–1344 (2004)
14. Jeltsema, D., Scherpen, J.M.A.: On Brayton and Moser's missing stability theorem. IEEE Trans. Circuits Syst. II **52**(9), 550–552 (2005)
15. Kazmierkowski, M., Krishnan, R., Blaabjerg, F. (eds.): Control in Power Electronics: Selected Problems. Academic Press, San Diego (2002)
16. Leonhard, W.: Control of Electrical Drives, 3rd edn. Springer, Berlin (2001)
17. Moser, J.K.: Bistable systems of differential equations with applications to tunnel diode circuits. IBM J. Res. Dev. **5**, 226–240 (1960)
18. Ortega, R., Jeltsema, D., Scherpen, J.M.A.: Power shaping: A new paradigm for stabilization of nonlinear RLC circuits. IEEE Trans. Autom. Control **48**(10), 1762–1767 (2003)
19. Ortega, R., Loría, A., Nicklasson, P.J., Sira-Ramírez, H.: Passivity-Based Control of Euler-Lagrange Systems; Mechanical, Electrical and Electromechanical Applications. Springer, Berlin (1998)
20. Ortega, R., van der Schaft, A.J., Mareels, I., Maschke, B.M.: Putting energy back in control. IEEE Control Syst. Mag. **21**(2), 18–33 (2001)
21. Savant, C.J. Jr., Roden, M.S., Carpenter, G.L.: Electronic Design: Circuits and Systems, 2nd edn. Benjamin–Cummings Publishing Company, San Francisco (1991)
22. Scherpen, J.M.A., Jeltsema, D., Klaassens, J.B.: Lagrangian modeling of switching electrical networks. Syst. Control Lett. **48**(5), 365–374 (2003)
23. Sira-Ramírez, H., de Nieto, M.D.: A Lagrangian approach to average modeling of pulse width-modulation controlled DC-to-DC power converters. IEEE Trans. Circuits Syst. I **43**(5), 427 (1996)
24. Stöhr, A.: Über gewisse nich notwendig lineare $(n+1)$-pole. Arch. Electron. Übertr. Tech. **7**, 546–548 (1953)
25. Van der Schaft, A.J.: $\mathscr{L}_2$-Gain and Passivity Techniques in Nonlinear Control. Springer, Berlin (2000)
26. Weiss, L., Mathis, W., Trajkovic, L.: A generalization of Brayton–Moser's mixed-potential function. IEEE Trans. Circuits Syst. I **45**(4), 423–427 (1998)
27. Wells, D.A.: A power function for the determination of Lagrangian generalized forces. J. Appl. Phys. **16**(9), 535–538 (1945)
28. Willems, J.C.: Dissipative dynamical system part i: General theory. Arch. Ration. Mech. Anal. **45**(5), 321–351 (1972)

Part III
Advanced Control Strategies for Power Converters

Chapter 9
Averaging Results Pertaining to the Implementation of Hybrid Feedback via PWM Control

Wei Wang, Andrew R. Teel, and Dragan Nešić

9.1 Introduction

Electrical power systems are an indispensable technology that continues to impact every aspect of the modern society. While these systems are prevalent in many branches of technology, manufacturing, transportation, and so on, the tools for their analysis and design are still underdeveloped; this is partially due to the inherent switching behaviour of their key elements—electronic power converters—that is difficult to analyse [20, 24].

Modern electronic power converters have a very high efficiency and power density and can operate at very high frequencies due to efficient power semi-conductors that allow for high frequency switching with minimal losses. Indeed, this allows for an efficient implementation of various pulse-width modulation (PWM) techniques that are ubiquitous in power converter systems.

Pulse-width modulation is a technique in which the width of a train of voltage (or current) pulses is adjusted (modulated) by rapidly tuning the switch between the supply and load ON and OFF. This technology is used extensively in power electronics and finds wide applications in industry [11, 20, 22, 28, 34, 35]. The net effect of the modulated voltage pulse train on a load can be shown to be equal to the average voltage of the pulse train; while this observation can be proved in many situations, rigorous averaging techniques for general pulse-width modulated systems are still

W. Wang (✉) · D. Nešić
Department of Electrical and Electronic Engineering, The University of Melbourne, Parkville, 3010 Victoria, Australia
e-mail: wewang@ee.unimelb.edu.au

D. Nešić
e-mail: dnesic@unimelb.edu.au

A.R. Teel
Department of Electrical Engineering, University of California Santa Barbara, Santa Barbara, CA 93106, USA
e-mail: teel@ece.ucsb.edu

F. Vasca, L. Iannelli (eds.), *Dynamics and Control of Switched Electronic Systems*, Advances in Industrial Control,
DOI 10.1007/978-1-4471-2885-4_9,

not fully developed. Moreover, refining the existing analysis techniques may lead to performance improvements in electronic power converters. Hence, averaging techniques for PWM systems continue to attract research attention [21, 22, 26].

A useful modelling tool based on an averaging technique for power converters is provided in [26]; however, the results in [26] are heuristic in nature, and the relationships between the average models and the actual system are unclear. A more rigorous use of averaging theory for electronic power systems was proposed in [21] to model the open-loop converter systems; this provided a technique for recovering oscillatory effects, such as the ripple, from the averaged model. In [22], stability of the closed-loop PWM power electronic systems is analysed via averaging. In particular, it was demonstrated that if a continuous feedback controller is designed to stabilise the averaged system, then the same controller would stabilise the actual system in an appropriate sense if the switching frequency is high enough.

Results in [22] are tailored for electronic power converters that are controlled with a continuous feedback controller, such as a linear state feedback controller. Note that there are situations when certain closed-loop performance specifications can not be achieved with any linear feedback controller whereas they are achievable with a hybrid controller, see [3]. For instance, a switched controller designed via Lyapunov approach in [6] is employed to control a power converter system in [23], and it was shown to provide better performance on transient and steady dynamics than continuous proportional–integral–derivative (PID) controllers. More details can be found in the survey of hybrid control techniques for power converter systems [10, 25]. This observation provides a partial motivation for developing averaging techniques for hybrid systems that can be used to analyse a class of hybrid PWM systems in general and power converter systems in particular.

The main purpose of this chapter is to provide a rigorous theory of averaging for a class of hybrid dynamical systems with disturbances. This theory can be used as a framework for a systematic design of hybrid controllers for electronic power systems, such as power converters. In particular, our results can be used to justify PWM implementation of arbitrary hybrid controllers.

The averaging method is an old research topic that employs a time-scale separation between the time variations of the state of a dynamical system and the time variations of the derivative of that state. This important tool has been developed for the analysis of time-varying continuous-time systems [1, 5, 27, 31], discrete-time systems [2, 4, 17, 33] and differential inclusions [12–14]. The averaging results are also extended for special classes of hybrid systems. In [29], exponential stability of a class of linear switched systems is considered via averaging. Averaging can also be applied to systems with dither[1] to infer practical stability of the system [18, 19].

For dynamical systems with exogenous disturbances, there has not been a suitable version of averaging stability theory until recently, see [27, 37]. With the definitions of strong average and weak average from [27, 37], robustness to disturbances is investigated respectively for switched systems and discrete-time systems in [39]

[1] Dither is a high-frequency signal introduced into system with the object of modifying its nonlinear characteristics.

and [40] with the assumption that their averages are input-to-state stable (ISS). The results in [27, 37] have also been extended to consider PWM control systems in [38], where the nonsmooth nature is accommodated by working with upper semicontinuous set-valued maps. Finite $\mathscr{L}_2$ gain property of rapidly switching linear systems with disturbances via the averaging method was considered in [30].

Note that all aforementioned results on averaging for hybrid systems approximate the time-varying hybrid systems by a non-hybrid average system. On the other hand, it is sometimes appropriate to average the time-varying hybrid system by a time-invariant hybrid system. For instance, systems controlled with hybrid feedback controllers that are implemented by PWM require this type of results.

We consider stability analysis of a class of hybrid dynamical systems that cover an important class of systems controlled with PWM hybrid feedback controllers. Applying definitions of strong and weak averages introduced in [27], we first consider closeness of solutions between the average system and solutions of the actual hybrid system on compact time domains. We also show that ISS of the average system implies some weaker ISS properties for the actual system. For the special case of zero inputs, where the strong and the weak average coincide, we show that uniform global asymptotic stability of the average system implies semi-global practical stability properties of the actual system.

Moreover, we show that our results provide a framework for hybrid feedback control design for power electronic systems. We first illustrate how to model a typical electronic converter that is controlled with a hybrid controller as a hybrid dynamical system; as the hybrid controller is implemented using PWM, the closed-loop system is modelled as a time-varying hybrid system. Then, we introduce an average time-invariant hybrid system that is used to analyse the properties of the original time-varying hybrid system. We show that if the hybrid controller stabilises the average system, then it would also stabilise the original system for sufficiently high PWM frequency. We also present results for systems with disturbances and investigate input-to-state stability of time-varying systems via averaging.

The chapter is organised as follows. Models of a boost converter controlled with a continuous linear and a hybrid controller are given in Sect. 9.2. Some useful definitions in the hybrid setting are given in Sect. 9.3. A class of hybrid dynamical systems generalised from the PWM hybrid feedback control model in Sect. 9.2 are introduced in Sect. 9.4. Section 9.5 contains the main results, and the PWM hybrid feedback control example is revisited to illustrate how our main results apply in this particular case. Conclusions are provided in the last section. All proofs are omitted for space reasons and can be found in the cited references.

9.2 Modelling of PWM Hybrid Feedback Control Systems

Consider a typical single-rate PWM boost power converter in Fig. 9.1 as an example. We illustrate how to represent closed-loop dynamics of a general PWM hybrid feedback control system as a hybrid dynamical system. Moreover, noting that the open-loop power converter system is time-varying due to PWM, we show that one

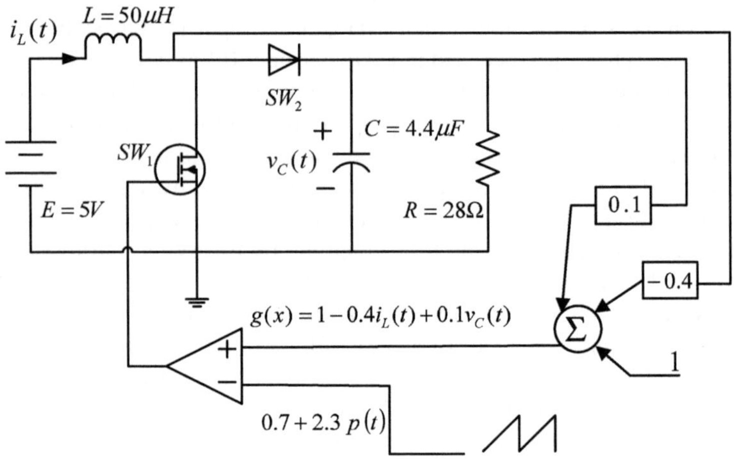

Fig. 9.1 Continuous-time feedback control boost converter [22]

can design hybrid feedbacks for its time-invariant average system for some specific control aim and use the main results of this chapter to obtain similar properties for the actual system under the same hybrid controller.

For the single-rate PWM power converter example (see Fig. 9.1), its open-loop model was considered in [21], and the closed-loop model with a continuous feedback controller was given in [22, Sect. 4]. We first present its open-loop and closed-loop models as given in [21] and [22], respectively. Moreover, a closed-loop model of this converter with general hybrid controllers is also given. After that, we consider a general continuous-time plant with hybrid feedbacks that are implemented via multi-rate PWM. Our main results are then developed for a class of hybrid models that includes this general multi-rate PWM hybrid system as a special case.

Suppose that the boost converter in Fig. 9.1 operates in the continuous conduction mode; in this case there are two configuration modes for the converter system corresponding to the ON and OFF states of the switches. Namely, the mode q_1 corresponds to the switch SW_1 ON and SW_2 OFF, and the mode q_2 corresponds to SW_1 OFF and SW_2 ON. A PWM produces a signal in the form of a pulse train (e.g. that takes values only at 0 and 1) of a constant frequency but with pulses that have a varying width; the pulse width is modulated using the measurements from the plant to determine the duty ratio, i.e., the ratio of time spent at 0 to the time spent at 1 or SW_1 ON to OFF for the converter system in Fig. 9.1.

Let ξ_1 denote the instantaneous value of the inductor current i_L, and $\xi_2 := v_C$ be the capacitor voltage. Let $\xi := (\xi_1, \xi_2)$. Considering the circuit in Fig. 9.1, we have that dynamics of states ξ agree with $\dot{\xi} = A_{q_i}\xi + B_{q_i}$ on the q_i configuration for $i = 1, 2$, where

$$A_{q_1} := \begin{pmatrix} 0 & 0 \\ 0 & -\frac{1}{RC} \end{pmatrix}, \qquad A_{q_2} := \begin{pmatrix} 0 & -\frac{1}{L} \\ \frac{1}{C} & -\frac{1}{RC} \end{pmatrix}, \qquad B_{q_1} = B_{q_2} := \begin{pmatrix} \frac{E}{L} \\ 0 \end{pmatrix}.$$

Noting that the point of equilibrium of the converter system can be moved to the origin using a coordinate transformation, one can consider the stabilisation of the origin for the closed-loop converter system. For the converter example in [22], note that the triangle switched signal in Fig. 9.1 is denoted by $0.7 + 2.3\, p(t)$, where $p(t)$ is periodic in t satisfying $p(t) = t/T$ for $t \in [0, T)$ and $T > 0$. Then, we have the closed-loop model of the converter system [22]:

$$\dot{\xi} = A_{q_2}\xi + B_{q_1} + (A_{q_1}\xi - A_{q_2}\xi)u\big(d(\xi) - p(\tau)\big), \tag{9.1}$$

where the duty ratio function $d(\xi) := (g(\xi) - 0.7)/(3 - 0.7)$ with $g(\xi) = 1 - 0.4\xi_1 + 0.1\xi_2$ is scaled using the minimum and the maximum values of the triangle signal so that $d(\xi)$ takes values in $[0, 1]$; $u : \mathbb{R} \to [0, 1]$ is the unit step function with $u(s) = 1$ for $s \geq 0$ and $u(s) = 0$ for $s < 0$.

The open-loop model for this converter system is given in [21]:

$$\dot{\xi} = A_{q_2}\xi + B_{q_1} + (A_{q_1}\xi - A_{q_2}\xi)u\big(d - p(\tau)\big), \tag{9.2}$$

where $d \in [0, 1]$ is the duty cycle for the open-loop PWM operation.

We next consider the same converter, but instead of the continuous controller $g(\xi)$ in Fig. 9.1 we want to apply a hybrid feedback controller, see Fig. 9.2. Suppose that hybrid controller $U : \bar{C} \times \bar{D} \to [0, 1]$ that was designed to satisfy given performance specifications is given as:

$$\begin{aligned} \dot{\eta} &= R(\eta, \xi), && (\xi, \eta) \in \bar{C}, \\ \eta^+ &\in S(\eta, \xi), && (\xi, \eta) \in \bar{D}, \\ U &:= U(\eta, \xi), \end{aligned}$$

where $\eta \in \mathbb{R}^n$, $\bar{C}$, $\bar{D}$ are the constraint sets that allow flows and jumps for η, $R : \bar{C} \to \mathbb{R}^n$ is a flow mapping, and $S : \mathbb{R}^n \rightrightarrows \mathbb{R}^n$ is a set-valued mapping. Note that states η may include physical variables together with logic variables or operation modes that are used to describe the hybrid feedback control law.

Applying this $U(\eta, \xi)$ to the open-loop converter system (9.2), we have that the closed-loop model of the converter system in Fig. 9.2 is:

$$\begin{aligned} &\begin{pmatrix} \dot{\xi} \\ \dot{\eta} \end{pmatrix} = \begin{pmatrix} A_{q_2}\xi + B_{q_1} \\ R(\eta, \xi) \end{pmatrix} + \begin{pmatrix} A_{q_1}\xi - A_{q_2}\xi \\ 0 \end{pmatrix} u\big(U(\eta, \xi) - p(\tau)\big), \quad (\xi, \eta) \in \bar{C}, \\ &\begin{pmatrix} \xi^+ \\ \eta^+ \end{pmatrix} \in \begin{pmatrix} \xi \\ S(\eta, \xi) \end{pmatrix}, \quad (\xi, \eta) \in \bar{D}. \end{aligned} \tag{9.3}$$

Averaging results from [22] can be used to analyse the model (9.1) but not the model (9.3). It is the main purpose of this chapter to give averaging results that can be used to analyse models of the form (9.3). With these analysis results, one can design hybrid controllers $U(\eta, \xi)$ based on the simpler average model to obtain similar properties of the actual PWM closed-loop system under the same U.

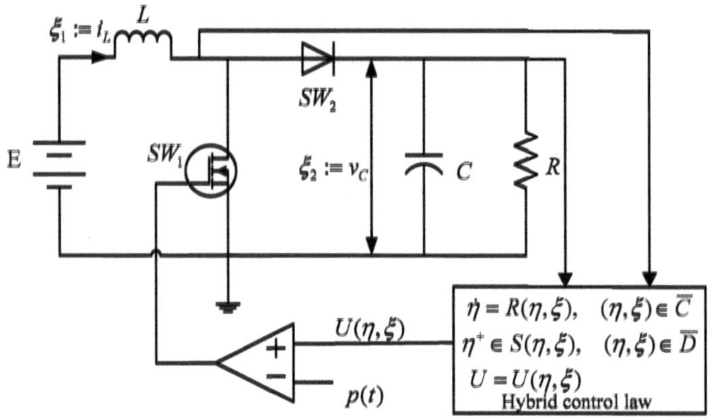

Fig. 9.2 Hybrid feedback control boost converter

Note that the averaged model of open-loop converter system (9.2) is given in [21]:

$$\dot{\xi} = A_{q_2}\xi + B_{q_1} + d(A_{q_1} - A_{q_2})\xi, \tag{9.4}$$

where the duty cycle $d \in [0, 1]$ can be taken as a control signal. Suppose that a controller $U : \bar{C} \cup \bar{D} \to [0, 1]$ is designed to stabilise the open-loop averaged system (9.4) by letting $d := U$. Then, our results in Sect. 9.5 show that the stability properties of the PWM converter system (9.3) can be implied by stability of the closed-loop of system (9.4) under the same controller U.

Remark 9.2.1 Suppose that we need a controller $\tilde{U} = \tilde{U}(\xi, \eta)$ that takes values in $[a, b]$ to stabilise the plant and achieve appropriate performance. To implement this controller via PWM, we need to get an average ranging from a to b using a step function $\tilde{u}(\cdot)$ that satisfies $\tilde{u}(s) = a$ for $s < 0$ and $\tilde{u}(s) = b$ for $s \geq 0$. Suppose also that we want to use a triangle signal $\hat{p}(\cdot) = c + kp(\cdot) \in [c, c + k]$, with $k > 0$ to implement this controller, where $p(\cdot)$ is the triangle wave defined earlier. Then, we need a duty cycle function $\hat{U}(\xi, \eta)$, generated from $\tilde{U}(\xi, \eta)$, but taking values in $[c, c + k]$. In particular, we take $\hat{U}(\xi, \eta) := c + k(\tilde{U}(\xi, \eta) - a)/(b - a)$. The PWM control that we need to implement is then

$$\tilde{u}\big(\hat{U}(\xi, \eta) - \hat{p}(\tau)\big),$$

which can be written in other ways, and it is also equal to

$$a + (b - a)u\big(U(\xi, \eta) - p(\tau)\big),$$

where

$$U(\xi, \eta) := \frac{\tilde{U}(\xi, \eta) - a}{b - a},$$

$$u(s) := \frac{\tilde{u}(s) - a}{b - a},$$

$$p(\tau) := \frac{\hat{p}(\tau) - c}{k}.$$

The averaging results given in Sect. 9.5 pertain to a more general class of PWM systems that is presented next.

We consider a general continuous-time plant with disturbances controlled by hybrid feedbacks that are implemented via multi-rate PWM. Consider a continuous-time plant with states $\xi \in \mathbb{R}^n$, disturbances $w \in \mathscr{W} \subset \mathbb{R}^m$ and outputs $y \in \mathbb{R}^l$:

$$\begin{aligned} \dot{\xi} &= O(\xi, w) + \sum_{l}^{k} P_l(\xi, w) U_l, \\ y &= Q(\xi, w). \end{aligned} \tag{9.5}$$

For this continuous-time plant, the hybrid feedback controllers U_l are given through the following auxiliary hybrid system with states $\eta \in \mathbb{R}^h$:

$$\begin{aligned} \dot{\eta} &= R(\eta, y), & (\eta, y) &\in C_1, \\ \eta^+ &\in S(\eta, y), & (\eta, y) &\in D_1, \\ U_l &= U_l(\eta, y), \end{aligned}$$

where C_1, $D_1 \in \mathbb{R}^h \times \mathbb{R}^l$ are the constraint sets that allow flows and jumps for η; $S : \mathbb{R}^h \rightrightarrows \mathbb{R}^h$ is a set-valued mapping that is outer semi-continuous, locally bounded, and for each $(\eta, y) \in D_1$, $S(\eta, y)$ is nonempty; functions $O : \mathbb{R}^n \times \mathscr{W} \to \mathbb{R}^n$ and $R : C_1 \to \mathbb{R}^h$ are continuous, while $P_l : \mathbb{R}^n \times \mathscr{W} \to \mathbb{R}^n$, $Q : \mathbb{R}^n \times \mathscr{W} \to \mathbb{R}^l$ and $U_l : \mathbb{R}^h \times \mathbb{R}^l \to [0, 1]$ are locally Lipschitz.
Let $x := [\xi, \eta]^T$, $h_l(x, w) := U_l(\eta, Q(\xi, w))$, $\psi_l(x, w) := [P_l(\xi, w), 0]^T$, $C := C_1 \times \mathscr{W}$, $D := D_1 \times \mathscr{W}$,

$$\mathscr{F}_0(x, w) := \begin{pmatrix} O(\xi, w) \\ R(\eta, Q(\xi, w)) \end{pmatrix} \quad \text{and} \quad \mathscr{G}(x, w) := \begin{pmatrix} \xi \\ S(\eta, Q(\xi, w)) \end{pmatrix}.$$

In the case that feedback controllers U_l are implemented by multi-rate PWM, the closed-loop of system (9.5) becomes:

$$\begin{aligned} &\left.\begin{aligned} \dot{x} &= \mathscr{F}(x, w, \tau) \\ \dot{\tau} &= \tfrac{1}{\varepsilon} \end{aligned}\right\}, && ((x, w), \tau) \in C \times \mathbb{R}_{\geq 0}, \\ &\left.\begin{aligned} x^+ &\in \mathscr{G}(x, w) \\ \tau^+ &= \tau \end{aligned}\right\}, && ((x, w), \tau) \in D \times \mathbb{R}_{\geq 0}, \end{aligned} \tag{9.6}$$

with

$$\mathscr{F}(x, w, \tau) := \mathscr{F}_0(x, w) + \sum_{i=1}^{m} \psi_i(x, w) u\big(h_i(x, w) - p_i(\tau)\big), \qquad (9.7)$$

where the second term of $\mathscr{F}$ in (9.7) is used to model a multi-rate implementation of a PWM feedback. Note that $p_i(\tau)$ are the only time-varying terms, the small parameter $\varepsilon > 0$ in (9.6) is used to guarantee that the switching signals p_i change fast compared with state ξ and so the effect of p_i can be averaged.

In the next section, we give some useful definitions under the hybrid dynamical systems framework in [7–9, 15, 16] that are needed in the sequel.

9.3 Preliminaries

$\mathbb{R}_{\geq 0} := [0, +\infty)$, $\mathbb{Z}_{\geq 0} := \{0, 1, 2, \ldots\}$, $\mathbb{B}$ is the closed unit ball in an Euclidean space, the dimension of which should be clear from the context, and $|\cdot|$ refers to the Euclidean norm. A set-valued mapping $M : \mathbb{R}^n \rightrightarrows \mathbb{R}^n$ is outer semi-continuous at $x \in \mathbb{R}^n$ if for all sequences $x_i \to x$ and $y_i \in M(x_i)$ such that $y_i \to y$, we have $y \in M(x)$, and M is outer semi-continuous if it is outer semi-continuous at each $x \in \mathbb{R}^n$. A set-valued mapping $M : \mathbb{R}^n \rightrightarrows \mathbb{R}^n$ is locally bounded if for any compact set $A \subset \mathbb{R}^n$, there exists $r > 0$ such that $M(A) := \bigcup_{x \in A} M(x) \subset r\mathbb{B}$; if M is outer semi-continuous and locally bounded, then $M(A)$ is compact for any compact set A. A function $x : \mathbb{R}_{\geq 0} \to \mathbb{R}^n$ is locally absolutely continuous if its derivative is defined almost everywhere and we have $x(t) - x(t_0) = \int_{t_0}^{t} \dot{x}(s)\, ds$ for all $t \geq t_0 \geq 0$. Given a set S, $\overline{\operatorname{con}}\, S$ denotes its convex hull. Given a compact set $A \subset \mathbb{R}^n$ and an $x \in \mathbb{R}^n$, define $|x|_A := \min_{y \in A} |x - y|$. A continuous function $\sigma : \mathbb{R}_{\geq 0} \to \mathbb{R}_{\geq 0}$ is of class-$\mathscr{L}$ if it is non-increasing and converging to zero as its argument grows unbounded. A continuous function $\gamma : \mathbb{R}_{\geq 0} \to \mathbb{R}_{\geq 0}$ is of class-$\mathscr{G}$ if it is zero at zero and non-decreasing. It is of class-$\mathscr{K}$ if it is of class-$\mathscr{G}$ and strictly increasing. A continuous function $\beta : \mathbb{R}_{\geq 0} \times \mathbb{R}_{\geq 0} \to \mathbb{R}_{\geq 0}$ is of class-$\mathscr{K}\mathscr{L}$ if it is of class-$\mathscr{K}$ in its first argument and class of $\mathscr{L}$ in its second argument. A set $S \subset \mathbb{R}_{\geq 0} \times \mathbb{Z}_{\geq 0}$ is called a compact hybrid time domain if $S = \bigcup_{j=0}^{J-1}([t_j, t_{j+1}], j)$ for some finite sequence of times $0 = t_0 \leq t_1 \leq t_2 \leq \cdots \leq t_J$. The set S is a hybrid time domain if for all $(T, J) \in S$, $S \cap ([0, T] \times \{0, 1, \ldots, J\})$ is a compact hybrid time domain.

Consider a time-invariant hybrid system with $\xi \in \mathbb{R}^n$ and $w \in \mathscr{W} \subset \mathbb{R}^m$ specified as:

$$H \quad \begin{aligned} \dot{\xi} &= F(\xi, w), \quad && (\xi, w) \in C, \\ \xi^+ &\in G(\xi, w), \quad && (\xi, w) \in D. \end{aligned} \tag{9.8}$$

Solutions to the hybrid system H are given by functions, called *hybrid arcs*, defined on hybrid time domains and satisfying the dynamics and the constraints given by the data of system H in (9.8).

Definition 9.3.1 A hybrid signal is a function defined on a hybrid time domain. $w : \operatorname{dom} w \to \mathscr{W}$ is called a hybrid input if $w(\cdot, j)$ is Lebesgue measurable and locally essentially bounded for each j. A hybrid signal $\xi : \operatorname{dom} \xi \mapsto \mathbb{R}^n$ is called a hybrid arc if $\xi(\cdot, j)$ is locally absolutely continuous for each j. A hybrid arc $\xi : \operatorname{dom} \xi \mapsto \mathbb{R}^n$ and a hybrid input $w : \operatorname{dom} w \mapsto \mathscr{W}$ form a solution pair to system H in (9.8) if $\operatorname{dom} \xi = \operatorname{dom} w$, $(\xi(0,0), w(0,0)) \in C \cup D$ and

- for all $j \in \mathbb{Z}_{\geq 0}$ and almost all t such that $(t, j) \in \operatorname{dom} \xi$, $(\xi(t, j), w(t, j)) \in C$ and $\dot{\xi}(t, j) = F(\xi(t, j), w(t, j))$;
- for all $(t, j) \in \operatorname{dom} \xi$ such that $(t, j+1) \in \operatorname{dom} \xi$, $(\xi(t, j), w(t, j)) \in D$ and $\xi(t, j+1) \in G(\xi(t, j), w(t, j))$.

A solution pair is maximal if it cannot be extended.

In the case of 0-input for system H, the definition of solution can be obtained with letting $\operatorname{dom} w := \operatorname{dom} \xi$ and $w(t, j) = 0$ for all $(t, j) \in \operatorname{dom} w$.

We next give some definitions on properties of the input signals. Given any hybrid signal $w : \operatorname{dom} w \mapsto \mathscr{W}$, let $\Gamma(w)$ denote the set of $(t, j) \in \operatorname{dom} w$ such that $(t, j+1) \in \operatorname{dom} w$, and define

$$|w| := \max\Big\{ \operatorname*{ess\,sup}_{(t,j)\in \operatorname{dom} w \setminus \Gamma(w)} \big|w(t, j)\big|, \ \sup_{(t,j)\in\Gamma(w)} \big|w(t, j)\big| \Big\}. \tag{9.9}$$

Note that for each j, the set $\{t : (t, j) \in \Gamma(w)\} \cap I_j$ with $I_j := \{t \in \mathbb{R} : (t, j) \in \operatorname{dom} w\}$ has one-dimensional zero measure, and thus we can define $\dot{w}(t, j) = 0$ on this set for all $(t, j) \in \operatorname{dom} w$ without affecting w. With this convention, the definition $|\dot{w}(t, j)|$ reduces to

$$|\dot{w}| := \operatorname*{ess\,sup}_{(t,j)\in \operatorname{dom} w} \big|\dot{w}(t, j)\big|.$$

Let $\mathscr{L}_{\mathscr{W}}$ be a given subset of hybrid signals $w : \operatorname{dom} w \to \mathscr{W}$. We need equi-essential boundedness and local equi-uniform Lipschitz continuity of $\mathscr{L}_{\mathscr{W}}$ to present the main results for a set of disturbances with specific properties. The set $\mathcal{L}_{\mathcal{W}}$ is called equi-essentially bounded if there exists $\Omega > 0$ such that $|w| \leq \Omega$ for all $w \in \mathcal{L}_{\mathcal{W}}$. It is called locally equi-uniformly Lipschitz continuous if there exists $L > 0$ such that, for all $w \in \mathcal{L}_{\mathcal{W}}$ and $(t, j), (s, j) \in \operatorname{dom} w$,

$$\left|w(t, j) - w(s, j)\right| \leq L|t - s|.$$

A sufficient condition for $\mathcal{L}_{\mathcal{W}}$ to be locally equi-uniformly Lipschitz continuous is that there exists a strictly positive real number Ω_1 such that, for each $w \in \mathcal{L}_{\mathcal{W}}$, $w(\cdot, j)$ is locally absolutely continuous for each j and for all $(t, j) \in \operatorname{dom} w$ such that $|\dot{w}| \leq \Omega_1$.

In order to exploit recent results in the literature on robustness for hybrid systems, for any $\Omega \geq 0$, let

$$F_\Omega(\xi) := \left\{v \in \mathbb{R}^n : v = F(\xi, w), w \in \mathcal{W} \cap \Omega\mathbb{B} \text{ and } (\xi, w) \in C\right\},$$
$$C_\Omega := \left\{\xi : \exists w \in \mathcal{W} \cap \Omega\mathbb{B} \text{ such that } (\xi, w) \in C\right\},$$

and assume the following basic regular conditions, except the convexity condition for F_Ω, that combine what is typically assumed in continuous-time and in discrete-time systems.

Assumption 1 The sets $C \subset \mathbb{R}^n \times \mathbb{R}^m$, $D \subset \mathbb{R}^n \times \mathbb{R}^m$ and $\mathcal{W} \subset \mathbb{R}^m$ are closed; $F : C \to \mathbb{R}^n$ is continuous, and for each $\Omega \geq 0$ and $\xi \in C_\Omega$, the set $F_\Omega(\xi)$ is convex; $G : \mathbb{R}^n \times \mathbb{R}^m \rightrightarrows \mathbb{R}^n$ is outer semi-continuous and locally bounded, and for each $(\xi, w) \in D$, $G(\xi, w)$ is nonempty.

Note that the convexity condition for F_Ω in Assumption 1 is not typically assumed for continuous-time systems, but it is motivated by [7, Remark 3], where an example illustrates that the convexity condition is used to guarantee robustness to disturbances for hybrid systems, and the results presented in this chapter are based on the robustness properties of hybrid dynamical systems. In particular, consider a hybrid system H_δ with $\bar{x} \in \mathbb{R}^n$, $w \in \mathcal{W} \subset \mathbb{R}^m$ that is inflated from system H by a small parameter $\delta > 0$:

$$H_\delta \quad \begin{aligned} \dot{\bar{x}} &\in F_\delta(\bar{x}, w), & (\bar{x}, w) &\in C_\delta, \\ \bar{x}^+ &\in G_\delta(\bar{x}, w), & (\bar{x}, w) &\in D_\delta, \end{aligned} \tag{9.10}$$

where the data $\{F_\delta, G_\delta, C_\delta, D_\delta\}$ are defined as

$$F_\delta(\bar{x}, w) := \overline{\operatorname{con}}\, F\big((\bar{x} + \delta\mathbb{B}, w) \cap C\big) + \delta\mathbb{B},$$
$$G_\delta(\bar{x}, w) := G\big((\bar{x} + \delta\mathbb{B}, w) \cap D\big) + \delta\mathbb{B},$$
$$C_\delta := \left\{(\bar{x}, w) : (\bar{x} + \delta\mathbb{B}, w) \cap C \neq \emptyset\right\},$$
$$D_\delta := \left\{(\bar{x}, w) : (\bar{x} + \delta\mathbb{B}, w) \cap D \neq \emptyset\right\}.$$

The preliminary results on robustness of hybrid dynamical systems include that forward pre-completeness from a compact set K_0 (Definition 9.3.2) for system H, with disturbances coming from an equi-essentially bounded set, implies that the reachable set of solutions starting from the set K_0 for system H is compact. Moreover, for each solution $\bar{x}$ starting from the set K_0 of its inflated system H_δ, there exists a solution ξ of system H such that they are (T, ρ)-close (Definition 9.3.3), see [41, Propositions 1–2]. They also include the results on robust stability properties of the inflated system H_δ with the assumption that the actual system H is input-to-state stable (Definition 9.3.4) in [41, Proposition 3]. Without the convex condition for the flow mapping F_Ω with respect to disturbances in Assumption 1, [7, Remark 3] shows that the above preliminary results may not be preserved.

The definitions of forward pre-completeness and (T, ρ)-closeness of hybrid signals are first given.

Definition 9.3.2 (Forward completeness) A hybrid solution pair is said to be forward complete if its domain is unbounded. A hybrid solution pair is said to be forward pre-complete if its domain is compact or unbounded. System H is said to be forward pre-complete from a compact set $K_0 \subset \mathbb{R}^n$ with a disturbance bound $\Omega \geq 0$ if all maximal solution pairs (ξ, w) with $\xi(0,0) \in K_0$ and w with $|w| \leq \Omega$ are forward pre-complete.

Definition 9.3.3 (Closeness of hybrid signals) Two hybrid signals $\xi_1 : \operatorname{dom} \xi_1 \mapsto \mathbb{R}^n$ and $\xi_2 : \operatorname{dom} \xi_2 \mapsto \mathbb{R}^n$ are said to be (T, ρ)-close if:

(a) for each $(t, j) \in \operatorname{dom} \xi_1$ with $t + j \leq T$, there exists s such that $(s, j) \in \operatorname{dom} \xi_2$, with $|t - s| \leq \rho$ and $|\xi_1(t, j) - \xi_2(s, j)| \leq \rho$,
(b) for each $(t, j) \in \operatorname{dom} \xi_2$ with $t + j \leq T$, there exists s such that $(s, j) \in \operatorname{dom} \xi_1$, with $|t - s| \leq \rho$ and $|\xi_2(t, j) - \xi_1(s, j)| \leq \rho$.

For pure continuous-time systems, forward completeness guarantees the existence of solutions and that no finite-time escape occurs. On the other hand, forward pre-completeness of hybrid systems, without requiring the domain of its solutions to be unbounded, shows that solutions are contained in a compact set in a finite time domain. Different from pure continuous-time systems or discrete-time systems, the concept of (T, ρ)-closeness provides an equivalent characterisation of graphical convergence of hybrid arcs, which implies that closeness of hybrid solutions not only depends on initial conditions but also on their time domains.

For a given compact set $A \subset \mathbb{R}^n$, a function $\chi : \mathbb{R}^n \to \mathbb{R}_{\geq 0}$ is said to be a proper indicator function for A on $\mathbb{R}^n$ if χ is continuous, $\chi(x) = 0$ if and only if $x \in A$, and $\chi(x) \to \infty$ when $|x| \to \infty$.

The following definitions are input-to-state stability (ISS) for system H and two kinds of ISS-like properties: semi-globally practically ISS (SGP-ISS) and semi-globally practically derivative ISS (SGP-DISS) for system H_μ, which depends on a small parameter $\mu > 0$ with states $\xi \in \mathbb{R}^n$ and inputs $w \in \mathscr{W} \subset \mathbb{R}^m$:

$$H_\mu \quad \begin{aligned} \dot{\xi} &= F_\mu(\xi, w), \quad && (\xi, w) \in C_\mu, \\ \xi^+ &\in G_\mu(\xi, w), \quad && (\xi, w) \in D_\mu. \end{aligned} \tag{9.11}$$

For convenience of studying these stability concepts with respect to a certain measure instead of a vector norm, we let $A \subset \mathbb{R}^n$ be nonempty and compact and let $\chi : \mathbb{R}^n \to \mathbb{R}_{\geq 0}$ be a proper indicator for A.

Definition 9.3.4 System H in (9.8) is called ISS with respect to (χ, β, γ) with $\beta \in \mathscr{KL}$ and $\gamma \in \mathscr{G}$ if each solution pair (ξ, w) satisfies

$$\chi\big(\xi(t, j)\big) \leq \max\big\{\beta\big(\chi(\xi_0), t + j\big), \gamma\big(|w|\big)\big\} \quad \forall\, (t, j) \in \operatorname{dom} \xi.$$

Definition 9.3.5 System H_μ in (9.11) is called SGP-ISS with respect to (χ, β, γ) with $\beta \in \mathscr{KL}$ and $\gamma \in \mathscr{G}$ if, for each compact set $K_0 \subset \mathbb{R}^n$ and any positive real numbers Ω and ν, there exists $\mu^* > 0$ such that for each $\mu \in (0, \mu^*]$, each solution pair (x, w) with $\xi_0 := \xi(0, 0) \in K_0$ and $|w| \leq \Omega$ satisfies

$$\chi\big(\xi(t, j)\big) \leq \max\big\{\beta\big(\chi(\xi_0), t + j\big), \gamma\big(|w|\big)\big\} + \nu \quad \forall\, (t, j) \in \operatorname{dom} \xi.$$

Definition 9.3.6 System H_μ in (9.11) is called SGP-DISS with respect to (χ, β, γ) with $\beta \in \mathscr{KL}$ and $\gamma \in \mathscr{G}$ if, for each compact set $K_0 \subset \mathbb{R}^n$ and each triple of positive real numbers (Ω, Ω_1, ν), there exists $\mu^* > 0$ such that for each $\mu \in (0, \mu^*]$, each solution pair (ξ, w), where $w(\cdot, j)$ is locally absolutely continuous, with $\xi_0 := \xi(0, 0) \in K_0$, $|w| \leq \Omega$ and $|\dot{w}| \leq \Omega_1$ satisfies:

$$\chi\big(\xi(t, j)\big) \leq \max\big\{\beta\big(\chi(\xi_0), t + j\big), \gamma\big(|w|\big)\big\} + \nu \quad \forall\, (t, j) \in \operatorname{dom} \xi.$$

We can see in the above definitions that the bounded-input bounded-output property holds globally for ISS, while for SGP-ISS and SGP-DISS, the bounds of the solution of hybrid systems are not only determined by the properties of disturbances signals but also by the parameter μ. Moreover, by tuning the parameter μ, which depends on the bound of norm of input signals for SGP-ISS and depends on the bound of norm of input signals with their derivatives for SGP-DISS, the estimate of the domain of attraction for system H_μ can be arbitrarily enlarged.

9.4 Strong and Weak Averages

A class of time-varying hybrid dynamical systems H_ε that are generalised from PWM hybrid control systems are considered in this section. With the strong and weak average definitions for functions, the simplified average systems for system H_ε are given. We also revisit the PWM hybrid control systems (9.6) and power converter systems (9.3) to discuss their averages.

Consider a class of time-varying hybrid systems H_ε that depend on a small parameter $\varepsilon > 0$:

$$H_\varepsilon \quad \begin{array}{ll} \left.\begin{array}{l} \dot{x} = f_\varepsilon(x, w, \tau) \\ \dot{\tau} = \frac{1}{\varepsilon} \end{array}\right\}, & \big((x, w), \tau\big) \in C \times \mathbb{R}_{\geq 0}, \\ \left.\begin{array}{l} x^+ \in G(x, w) \\ \tau^+ \in H(x, w, \tau) \end{array}\right\}, & \big((x, w), \tau\big) \in D \times \mathbb{R}_{\geq 0}, \end{array} \tag{9.12}$$

where $x \in \mathbb{R}^n$, $w \in \mathbb{R}^m$, $f_\varepsilon : \mathbb{R}^n \times \mathbb{R}^m \times \mathbb{R}_{\geq 0} \to \mathbb{R}^n$, $G : \mathbb{R}^n \times \mathbb{R}^m \rightrightarrows \mathbb{R}^n$ and $H : \mathbb{R}^n \times \mathbb{R}^m \times \mathbb{R}_{\geq 0} \rightrightarrows \mathbb{R}_{\geq 0}$, based on the following mathematical assumptions.

Assumption 2 The sets $C \subset \mathbb{R}^n \times \mathbb{R}^m$, $D \subset \mathbb{R}^n \times \mathbb{R}^m$ and $\mathscr{W} \subseteq \mathbb{R}^m$ are closed; $G : \mathbb{R}^n \times \mathbb{R}^m \rightrightarrows \mathbb{R}^n$ is outer semi-continuous and locally bounded, and for each $(\xi, w) \in D$, $G(\xi, w)$ is nonempty; $\tau \mapsto f_\varepsilon(x, w, \tau)$ is measurable for each $(x, w) \in C$; and for each $\delta > 0$ and compact set $K \subset \mathbb{R}^n \times \mathbb{R}^m$ there exist $M(K) > 0$ and $\varepsilon^*(K, \delta) > 0$ such that

$$\begin{aligned} \big|f_0(x, w, \tau)\big| &\leq M \quad \forall\big((x, w), \tau\big) \in (C \cap K) \times \mathbb{R}_{\geq 0}, \\ \big|f_\varepsilon(x, w, \tau) - f_0(x, w, \tau)\big| &\leq \frac{\delta}{3} \quad \forall\big((x, w), \tau, \varepsilon\big) \in (C \cap K) \times \mathbb{R}_{\geq 0} \times (0, \varepsilon^*]. \end{aligned} \tag{9.13}$$

Note that only local boundedness but no continuity condition is needed for the flow mapping f_0 that holds for PWM hybrid feedback control systems, for which equations with a discontinuous right-hand side are used to describe flow dynamics. The assumptions for (G, C, D) are also mild, which guarantee that the sets of solutions of H have good sequential compactness properties [32].

Let $\tilde{\mathscr{L}}_{\mathscr{W}}$ be a subset of signals $\tilde{w} : \mathbb{R}_{\geq 0} \to \mathscr{W}$. The following definitions of weak and strong averages for nonlinear functions come from [27].

Definition 9.4.1 (Weak Average) For a function $f_0 : C \times \mathbb{R}_{\geq 0} \to \mathbb{R}^n$, the function $f_{\text{wa}} : C \to \mathbb{R}^n$ is said to be a *weak average* on C if for each compact set $K \subset \mathbb{R}^n \times \mathbb{R}^m$, there exists a class-$\mathscr{L}$ function σ_K such that, for all

$((x, w), \tau, T) \in (C \cap K) \times \mathbb{R}_{\geq 0} \times \mathbb{R}_{\geq 0}$,

$$\left| \frac{1}{T} \int_{\tau}^{\tau+T} \left[f_{\mathrm{wa}}(x, w) - f_0(x, w, s) \right] \mathrm{d}s \right| \leq \sigma_K(T). \tag{9.14}$$

Definition 9.4.2 (Strong Average) For a function $f_0 : C_1 \times \mathcal{W} \times \mathbb{R}_{\geq 0} \to \mathbb{R}^n$, the function $f_{\mathrm{sa}} : C_1 \times \mathcal{W} \to \mathbb{R}^n$ is said to be a *strong average* on $C_1 \times \mathcal{W}$ if for each compact set $K \subset \mathbb{R}^n \times \mathbb{R}^m$, there exists a class-$\mathcal{L}$ function σ_K such that, for all $\tilde{w} \in \tilde{\mathcal{L}}_{\mathcal{W}}$ with $((x, \tilde{w}(s)), \tau, T) \in ((C_1 \times \mathcal{W}) \cap K) \times \mathbb{R}_{\geq 0} \times \mathbb{R}_{\geq 0}$ for all $s \geq 0$, the following holds:

$$\left| \frac{1}{T} \int_{\tau}^{\tau+T} \left[f_{\mathrm{sa}}(x, \tilde{w}(s)) - f_0(x, \tilde{w}(s), s) \right] \mathrm{d}s \right| \leq \sigma_K(T). \tag{9.15}$$

Note that the main difference between the weak and the strong average is that in the definition of weak average the disturbance is kept constant in (9.14), whereas for strong average, inequality (9.15) needs to hold for all disturbances $\tilde{w} \in \tilde{\mathcal{L}}_{\mathcal{W}}$.

Remark 9.4.3 For 0-input case, note that the strong and the weak averages in Definitions 9.4.1 and 9.4.2 coincide. Let f_{av} denote the average of f_0 in this case. A complete characterisation of strong averages for continuous-time systems was given in [27]; it is shown that any $f(x, w, \tau)$ that is periodic in τ has a strong average if and only if the function f has the structure

$$f(x, w, \tau) = \tilde{f}(x, \tau) + \tilde{g}(x, w), \tag{9.16}$$

where $\tilde{f}(x, \tau)$ has a well-defined average $\tilde{f}_{\mathrm{av}}$ in the sense of Definition 9.4.1 or 9.4.2. Moreover, we have $f_{\mathrm{sa}}(x, w) = \tilde{f}_{\mathrm{av}}(x) + \tilde{g}(x, w)$.

Letting f_{wa} come from Definition 9.4.1 and (G, C, D) from (9.12), the weak average H_{wa} of system H_ε is

$$H_{\mathrm{wa}} \quad \begin{aligned} \dot{\xi} &= f_{\mathrm{wa}}(\xi, w), \quad (\xi, w) \in C, \\ \xi^+ &\in G(\xi, w), \quad (\xi, w) \in D. \end{aligned} \tag{9.17}$$

Similarly, for the case where $C = C_1 \times \mathcal{W}$, the strong average H_{sa} of system H_ε is

$$H_{\mathrm{sa}} \quad \begin{aligned} \dot{\xi} &= f_{\mathrm{sa}}(\xi, w), \quad (\xi, w) \in C, \\ \xi^+ &\in G(\xi, w), \quad (\xi, w) \in D, \end{aligned} \tag{9.18}$$

where f_{sa} comes from Definition 9.4.2.

To present some useful results for hybrid systems without disturbances, we also define an average system H_{av} for the 0-input case, which is formed as (9.17) or (9.18) with the data $\{f_{\mathrm{av}}, G, C, D\}$ and letting $w(t, j) = 0$.

With the definitions of weak and strong averages, we revisit the PWM hybrid feedback control system (9.6) introduced in Sect. 9.2 to show how to get its weak and strong averages in the following example.

Example 9.4.4 We first show that there exists a weak average for the function $\mathscr{F}: C \times \mathbb{R}_{\geq 0} \to \mathbb{R}^n$ in (9.7). Given $T > \max\{T_1, \dots, T_n\}$, let $k_l = k_l(T) \in \mathbb{Z}_{\geq 0}$ and $\tilde{T}_l \in [0, T_l)$ satisfying $T = k_l T_l + \tilde{T}_l$. Note that $k_l(T) \to \infty$ as the given T approaches infinity. We have

$$\begin{aligned}
&\frac{1}{T}\int_{\tau}^{\tau+T}\sum_{l=1}^{m}\psi_l(x,w)u\big(h_l(x,w)-p_l(s)\big)\,ds \\
&= \frac{1}{T}\sum_{l=1}^{m}\psi_l(x,w)\bigg\{\int_{\tau}^{\tau+k_lT_l}u\big(h_l(x,w)-p_l(s)\big)\,ds \\
&+ \int_{\tau+k_lT_l}^{\tau+k_lT_l+\tilde{T}_l}u\big(h_l(x,w)-p_l(s)\big)\,ds\bigg\}, \\
&= \sum_{l=1}^{m}\psi_l(x,w)\bigg(\frac{k_lT_l}{k_lT_l+\tilde{T}_l}h_l(x,w)+\frac{v_l(x,w,\tau,\tilde{T}_l)}{k_lT_l+\tilde{T}_l}\bigg),
\end{aligned}$$

where $v_l(x, w, \tilde{T}_l) := \int_{\tau}^{\tau+\tilde{T}_l} u(h_l(x, w) - p_l(s))\,ds$ satisfies $|v_l(x, w, \tilde{T}_l)| \leq \tilde{T}_l$. Let

$$f_{\mathrm{wa}}(x,w) := \mathscr{F}_0(x,w) + \sum_{l=1}^{m}\psi_l(x,w)h_l(x,w). \tag{9.19}$$

Then, for any compact set $K \subset \mathbb{R}^n \times \mathbb{R}^m$, there exists $r > 0$ such that, for all $(x, w) \in C \cap K$,

$$\begin{aligned}
\bigg|f_{\mathrm{wa}}(x,w) - \frac{1}{T}\int_{\tau}^{\tau+T}\mathscr{F}(x,w,s)\,ds\bigg| &\leq \frac{1}{T}\sum_{l=1}^{m}\big|\psi_l(x,w)v_l(x,w,\tilde{T}_l)\big| \\
&\leq \frac{r}{T+1} := \sigma_K(T),
\end{aligned}$$

which shows that f_{wa} agrees with Definition 9.4.1. Let $\mathscr{G}, C, D$ come from (9.6). Then, the hybrid system H_{wa} with data $\{f_{\mathrm{wa}}, \mathscr{G}, C, D\}$ formed as (9.17) is the weak average for the PWM closed-loop control system in (9.6).

Noting the structure of f in (9.16) that allows for strong average in Remark 9.4.3, there exists a strong average if $\psi_l(x, w)$ and $h_l(x, w)$ are independent of w, i.e., $\psi_l(x, w) := \psi_l(x)$ and $h_l(x, w) := h_l(x)$. In this case, following the calculations used to establish the weak average, we get that

$$f_{\mathrm{sa}}(x,w) := \mathscr{F}_0(x,w) + \sum_{l=1}^{m}\psi_l(x)h_l(x) \tag{9.20}$$

on the set C, at least when C has the form $C = C_1 \times \mathscr{W}$, and the strong average of system (9.6) is system H_{sa} in (9.18) with data $\{f_{\mathrm{sa}}, \mathscr{G}, C, D\}$.

Applying the results obtained in Example 9.4.4, we revisit the power converter example in Fig. 9.2.

Example 9.4.5 Note that the single-rate PWM power converter system in Fig. 9.2 is a special case of PWM hybrid feedback control systems and its hybrid feedback closed-loop (9.3) can be formed in (9.6). We can directly apply the results on strong or weak average obtained in Example 9.4.4, noting that strong and weak averages coincide for the 0-input case, to get that the average for the closed-loop converter system (9.3) is:

$$\begin{aligned} \begin{pmatrix} \dot{\xi} \\ \dot{\eta} \end{pmatrix} &= \begin{pmatrix} A_{q2}\xi + B_{q1} \\ R(\eta,\xi) \end{pmatrix} + \begin{pmatrix} A_{q1}\xi - A_{q2}\xi \\ 0 \end{pmatrix} U(\eta,\xi), \quad (\xi,\eta) \in \bar{C}, \\ \begin{pmatrix} \xi^+ \\ \eta^+ \end{pmatrix} &\in \begin{pmatrix} \xi \\ S(\eta,\xi) \end{pmatrix}, \qquad\qquad\qquad\qquad\qquad (\xi,\eta) \in \bar{D}, \end{aligned} \tag{9.21}$$

which is consistent to the model by applying the hybrid controller $U(\eta, \xi)$ to the open-looped average converter system (9.4), where $U : \bar{C} \times \bar{D} \to [0, 1]$ acts as a duty ratio function.

To relate the actual hybrid system H_ε with its weak average H_{wa} or strong average H_{sa}, we define the following functions η_{wa} and η_{sa} that facilitate the averaging technique in a coordinate transformation, and present conditions on local Lipschitz continuity of these functions to guarantee the main results. Let f_{wa} and f_{sa} come from the definitions of weak average and strong average. For each $((x, w), \tau, \mu) \in C \times \mathbb{R}_{\geq 0} \times \mathbb{R}_{\geq 0}$ and $\tau_0 \in [0, \tau]$, let

$$\eta_{\mathrm{wa}}(x, w, \tau, \tau_0, \mu) := \int_{\tau_0}^{\tau} \exp\bigl(\mu(s-\tau)\bigr)\bigl[f_0(x, w, s) - f_{\mathrm{wa}}(x, w)\bigr]\,\mathrm{d}s. \tag{9.22}$$

Let $0 \leq \tau_0 \leq \tau_1$ and $\tilde{w} : [\tau_0, \tau_1] \to \mathcal{W}$ be given. For each $\tau \in [\tau_0, \tau_1]$ and $(x, \mu) \in C_1 \times \mathbb{R}_{\geq 0}$, let

$$\eta_{\mathrm{sa}}(x, \tilde{w}, \tau, \tau_0, \mu) := \int_{\tau_0}^{\tau} \exp\bigl(\mu(s-\tau)\bigr)\bigl[f_0\bigl(x, \tilde{w}(s), s\bigr) - f_{\mathrm{sa}}\bigl(x, \tilde{w}(s)\bigr)\bigr]\,\mathrm{d}s. \tag{9.23}$$

Let $\bar{N} := \{1, \ldots, n\}$. For each $i \in \bar{N}$, η^i_{sa} represents the ith component of η_{sa}, and similarly for η^i_{wa}. We assume the following assumptions.

Assumption 3 For a function f_0 defined on $C \times \mathbb{R}_{\geq 0}$, f_{wa} is a continuous function that is a weak average of f_0 on C, and, for each compact set $K \subset \mathbb{R}^n \times \mathbb{R}^m$, there exists $L(K)$ such that, for all $i \in \bar{N}$, $((x_1, w_1), \tau_a), ((x_2, w_2), \tau_b) \in (C \cap K) \times \mathbb{R}_{\geq 0}$ and $\tau_0 \in [0, \min\{\tau_a, \tau_b\}]$,

$$\begin{aligned} &\bigl|\eta^i_{\mathrm{wa}}(x_1, w_1, \tau_a, \tau_0, 0) - \eta^i_{\mathrm{wa}}(x_2, w_2, \tau_b, \tau_0, 0)\bigr| \\ &\quad \leq L\bigl(|x_1 - x_2| + |w_1 - w_2| + |\tau_a - \tau_b|\bigr). \end{aligned}$$

Assumption 4 For a function f_0 defined on $C_1 \times \mathcal{W} \times \mathbb{R}_{\geq 0}$, where $C \subset C_1 \times \mathcal{W}$, f_{sa} is a continuous function that is a strong average of f_0 on $C_1 \times \mathcal{W}$, and, for each compact set $K \subset \mathbb{R}^n \times \mathbb{R}^m$, there exists $L(K)$ such that, for all $i \in \bar{N}$, $0 \leq \tau_0 \leq \tau_1$, $\tilde{w} : [\tau_0, \tau_1] \to \mathcal{W}$ and $((x_1, \tilde{w}(s)), \tau_a), ((x_2, \tilde{w}(s)), \tau_b) \in ((C_1 \times \mathcal{W}) \cap K) \times [\tau_0, \tau_1]$ for all $s \in [\tau_0, \tau_1]$:

$$\left|\eta_{\mathrm{sa}}^i(x_1, \tilde{w}, \tau_a, \tau_0, 0) - \eta_{\mathrm{sa}}^i(x_2, \tilde{w}, \tau_b, \tau_0, 0)\right| \leq L\big(|x_1 - x_2| + |\tau_a - \tau_b|\big).$$

Assumptions 3 and 4 may hold even when f is not periodic in τ nor continuous in (x, w). We will revisit PWM hybrid feedback control system (9.6) in Sect. 9.5 to illustrate this situation. Now we are ready to present the main results. The proofs are all omitted due to the space limitation, and the details can be found in [41].

9.5 Main Results

Compared to strong averages, weak averages exist for a larger class of systems. For instance, an example in [27] illustrates that existence of the strong average always implies existence of weak average, but the opposite does not hold. However, we can only state weaker results with the weak average definition. In particular, we show in Theorem 9.5.2 that solutions of the actual system can be arbitrarily close to solutions of its strong average in compact time domains for input signals with bounded norm. For the weak average case, the results on closeness of solutions in Theorem 9.5.1 pertain to a smaller class of input signals for which the norm of derivative is also bounded. Moreover, we show that ISS of the strong average implies SGP-ISS of the actual system in Theorem 9.5.5, whereas we can only obtain SGP-DISS of the actual system from ISS of its weak average in Theorem 9.5.4.

Theorems 9.5.1 and 9.5.2, demonstrate that for each solution x of the hybrid system H_ε in (9.12) under disturbances w, there exists some solution ξ of its strong or weak average under similar disturbances w_1 such that x and ξ can be made arbitrarily close on compact time domains if the parameter ε is sufficiently small.

Theorem 9.5.1 (Weak average) *Suppose that the set $\mathcal{L}_\mathcal{W}$ is equi-essentially bounded and locally equi-uniformly Lipschitz continuous, system H_ε in* (9.12) *satisfies Assumptions* 2 *and* 3, *and its weak average system H_{wa} satisfies Assumption* 1 *and is forward pre-complete from a compact set $K_0 \subset \mathbb{R}^n$ with a disturbance bound $\Omega \geq 0$. Then, for each $T \geq 0$ and $\rho > 0$, there exists $\varepsilon^* > 0$ such that, for all $\varepsilon \in (0, \varepsilon^*]$ and $w \in \mathcal{L}_\mathcal{W}$, and for each solution pair (x, w) to H_ε with $x(0, 0) \in K_0$, there exists some solution pair (ξ, w_1) to H_{wa} with $\xi(0, 0) \in K_0$ and $|w_1| \leq |w|$ such that x and ξ are (T, ρ)-close.*

Theorem 9.5.2 (Strong average) *Suppose that the set $\mathcal{L}_\mathcal{W}$ is equi-essentially bounded, system H_ε in* (9.12) *satisfies Assumptions* 2 *and* 4, *and its strong* *average system H_{sa} satisfies Assumption 1 and is forward pre-complete from a compact set $K_0 \subset \mathbb{R}^n$ with a disturbance bound $\Omega \geq 0$. Then, for each $T \geq 0$ and $\rho > 0$, there exists $\varepsilon^* > 0$ such that, for all $\varepsilon \in (0, \varepsilon^*]$ and $w \in \mathscr{L}_\mathscr{W}$, and for each solution pair (x, w) to H_ε with $x(0,0) \in K_0$, there exists some solution pair (ξ, w_1) to H_{sa} with $\xi(0,0) \in K_0$ and $|w_1| \leq |w|$ such that x and ξ are (T, ρ)-close.*

For 0-input case, we also give the result on closeness of solutions of the hybrid system H_ε and its average system H_{av}, see [36, Theorem 1]. In this case, assumptions and definitions for system H_ε with disturbances can be directly used with letting $w = 0 \subset \mathbb{R}^m$ and $\operatorname{dom} w := \operatorname{dom} \xi$.

Corollary 9.5.3 *Suppose that system H_ε, where $w = 0 \subset \mathbb{R}^m$, in* (9.12) *satisfies Assumptions* 2 *and* 3, *and its average system H_{av} satisfies Assumption* 1 *and is forward pre-complete from a compact set $K_0 \subset \mathbb{R}^n$. Then, for each $T \geq 0$ and $\rho > 0$, there exists $\varepsilon^* > 0$ such that, for all $\varepsilon \in (0, \varepsilon^*]$ and each solution x to H_ε with $x(0,0) \in K_0$, there exists some solution ξ to H_{av} with $\xi(0,0) \in K_0$ such that x and ξ are (T, ρ)-close.*

Note that the above results on closeness of solutions hold for arbitrarily large compact hybrid time domains and use forward pre-completeness, properties that are weaker than stability, of the average system. We can get that strong/weak averages approximate well solutions of the original system on compact time domains, where disturbances are required to be Lipschitz and bounded for weak averages whereas only bounded for the strong averages.

In order to apply the results of Theorem 9.5.1 (Theorem 9.5.2) for PWM hybrid control systems (9.6), we have to check if Assumption 3 (Assumption 4) is satisfied, which is not straightforward. We revisit Example 9.4.4 to show that Assumptions 3 and 4 hold for the PWM hybrid control system.

Example 9.4.4 (continued) We first check if Assumption 3 holds for η_{wa}. Considering the definitions of η_{wa} in (9.22) and f_{wa} in (9.19), it follows for each $\tau \in (0, \min_l\{T_l\})$ and $\tau_0 \in [0, \tau]$ that

$$\begin{aligned}\eta_{wa}(x, w, \tau, \tau_0, 0) &= \int_{\tau_0}^{\tau} \big[\mathscr{F}(x, w, s) - f_{wa}(x, w)\big]\,ds,\\ &= \sum_{l=1}^{m} \psi_l(x, w) \int_{\tau_0}^{\tau} \big[u\big(h_l(x, w) - p_l(s)\big) - h_l(x, w)\big]\,ds,\\ &= \sum_{l=1}^{m} \psi_l(x, w)\big(\min\{(\tau - \tau_0), h_l(x, w)\}\\ &\quad - (\tau - \tau_0)h_l(x, w)\big),\end{aligned} \tag{9.24}$$

which is bounded for any (x, w) in a compact set, and locally Lipschitz as functions ψ_i and h_i are assumed to be locally Lipschitz.

For the strong average case, considering the definitions of η_{sa} in (9.23) and f_{sa} in (9.20), we have

$$\begin{aligned}\eta_{sa}(x, \tilde{w}, \tau, \tau_0, 0) &= \int_{\tau_0}^{\tau} \left[\mathscr{F}\left(x, \tilde{w}(s), s\right) - f_{sa}\left(x, \tilde{w}(s)\right)\right] ds, \\ &= \sum_{i=1}^{m} \psi_i(x) \int_{\tau_0}^{\tau} \left(u\left(h_i(x) - p_i(s)\right) - h_i(x)\right) ds.\end{aligned}$$

Noting (9.24), it follows that Assumption 4 holds for the function $\eta_{sa}(x, w, \tau, \tau_0, 0)$.

Now, we study *robustness to disturbances* for the class of time-varying hybrid systems H_ε with the assumption that its strong or weak average system is ISS.

Theorem 9.5.4 *Suppose that the set $\mathscr{L}_{\mathscr{W}}$ is equi-essentially bounded and locally equi-uniformly Lipschitz continuous, system H_ε in* (9.12) *satisfies Assumptions* 2 *and* 3 *and its weak average system H_{wa} satisfies Assumptions* 1 *and is ISS with respect to (χ, β, γ). Then, system H_ε is SGP-DISS with respect to (χ, β, γ).*

Theorem 9.5.5 *Suppose that the set $\mathscr{L}_{\mathscr{W}}$ is equi-essentially bounded, system H_ε in* (9.12) *satisfies Assumptions* 2 *and* 4 *and its strong average system H_{sa} satisfies Assumption* 1 *and is ISS with respect to (χ, β, γ). Then, system H_ε is SGP-ISS with respect to (χ, β, γ).*

We next apply the main results on the PWM hybrid feedback control system in (9.6) and give two corollaries on robust stability properties of system (9.6) based on its weak and strong average systems are ISS.

Example 9.4.4 (continued) Let $\mathscr{G}, C, D$ come from (9.6). Let the hybrid system $\tilde{H}_{wa}$ and $\tilde{H}_{sa}$ be formed in (9.17) with data $\{f_{wa}, \mathscr{G}, C, D\}$ and $\{f_{sa}, \mathscr{G}, C, D\}$, respectively, where f_{wa} comes from (9.19) and f_{sa} from (9.20). Then, noting the result in Example 9.4.4, $\tilde{H}_{wa}$ is the weak average, and $\tilde{H}_{sa}$ is the strong average for the PWM closed-loop control system in (9.6).

Note that Assumptions 3 and 4 had been checked that they hold for the PWM hybrid control system (9.6) when Example 9.4.4 was revisited last time. We can directly apply the results in Theorems 9.5.4–9.5.5 to get the following corollaries.

Corollary 9.5.6 *Suppose that the set $\mathscr{L}_{\mathscr{W}}$ is equi-essentially bounded and locally equi-uniformly Lipschitz continuous, the PWM hybrid control system in* (9.6) *satisfies Assumptions* 2 *and its weak average system $\tilde{H}_{wa}$ satisfies Assumption* 1 *and is ISS with respect to (χ, β, γ). Then, the PWM hybrid control system in* (9.6) *is SGP-DISS with respect to (χ, β, γ).*

Corollary 9.5.7 *Suppose that the set $\mathscr{L}_{\mathscr{W}}$ is equi-essentially bounded, the PWM hybrid control system in* (9.6) *satisfies Assumption* 2 *and its strong average system* $\tilde{H}_{\mathrm{sa}}$ *satisfies Assumption* 1 *and is ISS with respect to* (χ, β, γ). *Then, the PWM hybrid control system in* (9.6) *is SGP-ISS with respect to* (χ, β, γ).

We also consider asymptotic stability properties for the hybrid system H_ε without influence of disturbances in the following corollary, see also [36, Theorem 4]. The system H in (9.8) is called globally asymptotically stable with respect to (χ, β) with letting $w \equiv 0$ in the definition of ISS. System H_μ in (9.11) is called semi-globally asymptotically stable (SGP-AS) with respect to (χ, β) with letting $w \equiv 0$ in the definition of SGP-ISS.

Corollary 9.5.8 *Suppose that system* H_ε *in* (9.12), *where* $w = 0$, *satisfies Assumptions* 2 *and* 3 *and its average system* H_{av} *satisfies Assumption* 1 *and is globally asymptotically stable with respect to* (χ, β). *Then, system* H_ε *is SGP-AS with respect to* (χ, β).

Example 9.4.5 (continued) For the PWM hybrid feedback power converter example in (9.3) and its average system in (9.21), Assumptions 1–3 hold since the local boundedness and Lipschitz condition are naturally satisfied for linear systems noting that the auxiliary system generating hybrid feedback controllers for linear plant is usually linear. Then, if we design hybrid feedbacks such that some specifications on its response are achieved for its average model (9.21), Corollary 9.5.3 shows that (9.21) can approximate solutions of the actual system on arbitrarily large compact time domains. Moreover, if the average system (9.21) is globally asymptotically stable under a hybrid feedback, we can get that the actual converter system (9.3) under the same hybrid controller is SGP-AS.

The main results can still be applied for the case where the controllers can only locally stabilise the average system. To illustrate this, suppose that a controller U locally asymptotically stabilises system (9.21) with the domain of attraction $K \subset \bar{C} \cup \bar{D}$. Let $C_1 := \bar{C} \cap K$ and $D_1 := \bar{D} \cap K$. Then, we can get that the hybrid average system (9.21), where the constraint sets $\bar{C}$ and $\bar{D}$ are replaced by C_1 and D_1, respectively, is globally asymptotically stable, and we can apply the results in the above example to consider the stability properties for the actual converter system.

9.6 Summary

We presented stability and robustness to disturbances analysis results for a class of hybrid dynamical systems via the averaging method. Using the notions of strong and weak average, the time-varying hybrid system that includes PWM hybrid feedback control system as a special case is approximated by a time-invariant hybrid system. The results showed that the solutions of the actual time-varying hybrid system and its weak or strong average can be made arbitrarily close on compact time domains by reducing the parameter ε if the average system is forward pre-complete. Our results also showed that ISS of the strong (weak) average implies SGP-ISS (SGP-DISS) of the actual system and asymptotic stability of the average system implies SGP-asymptotic stability of the actual system in the case of no disturbances. Through a PWM power converter example, we showed that the main results can be taken as a framework for a systematic design of hybrid controllers for electronic power systems.

Acknowledgements Supported by the Australian Research Council under the Discovery Project, Future Fellow program, AFOSR (Grant FA9550-09-1-0203), NSF (Grant ECCS-0925637) and NSF (CNS-0720842).

References

1. Aeyels, D., Peuteman, J.: On exponential stability of nonlinear time-varying differential equations. Automatica **35**(6), 1091–1100 (1999)
2. Bai, E.W., Fu, L.C., Sastry, S.S.: Averaging analysis for discrete time and sampled data adaptive systems. IEEE Trans. Circuits Syst. **35**(2), 137–148 (1988)
3. Beker, O., Hollot, C.V., Chait, Y.: Plant with integrator: an example of reset control overcoming limitations of linear feedback. IEEE Trans. Autom. Control **46**(11), 1797–1799 (2002)
4. Bitmead, R.R., Johnson, C.R. Jr.: Discrete averaging principles and robust adaptive identification. In: Advances in Control and Dynamic Systems, vol. 25, pp. 237–271. Academic Press, New York (1987)
5. Bogoliubov, N.N., Mitropolsky, Y.A.: Asymptotic Methods in the Theory of Nonlinear Oscillations. Gordon and Breach, New York (1961)
6. Bolzem, P., Spinellin, W.: Quadratic stabilization of a switched affine system about a nonequilibrium point. In: Proc. of the American Control Conference, Boston, Massachusetts, USA, pp. 3890–3895 (2004)
7. Cai, C., Teel, A.R.: Results on input-to-state stability for hybrid systems. In: Proc. of the IEEE Conference on Decision and Control and European Control Conference, Seville, Spain, pp. 5403–5406 (2005)
8. Cai, C., Teel, A.R., Goebel, R.: Smooth Lyapunov functions for hybrid systems, part I: Existence is equivalent to robustness. IEEE Trans. Autom. Control **52**(7), 1264–1277 (2007)
9. Cai, C., Teel, A.R., Goebel, R.: Smooth Lyapunov functions for hybrid systems, part II: (pre)asymptotically stable compact sets. IEEE Trans. Autom. Control **53**(3), 734–748 (2008)
10. Cervanters, I., Perez-Pinal, F.J., Torres, A.M.: Hybrid control of DC–DC power converters. In: Renewable Energy, pp. 173–196. InTech, Rijeka (2009)
11. Czarkowski, D., Pujara, L.R., Kazimierczuk, M.K.: Robust stability of state-feedback control of PWM DC–DC push-pull converter. IEEE Trans. Ind. Electron. **42**(1), 108–111 (1995)
12. Donchev, T.: Functional differential inclusions involving dissipative and compact multifunctions. Glas. Mat. **33**(53), 51–60 (1998)
13. Donchev, T., Grammel, G.: Averaging of functional differential inclusions in Banach spaces. J. Math. Anal. Appl. **311**(2), 402–416 (2005)
14. Donchev, T., Slavov, I.: Averaging method for one-sided Lipschitz differential inclusions with generalized solutions. SIAM J. Control Optim. **37**(5), 1600–1613 (1999)
15. Goebel, R., Teel, A.R.: Solutions to hybrid inclusions via set and graphical convergence with stability theory applications. Automatica **42**(4), 573–587 (2006)

16. Goebel, R., Sanfelice, R.G., Teel, A.R.: Hybrid dynamical systems. IEEE Control Syst. Mag. **29**(2), 28–93 (2009)
17. Grammel, G.: Limits of nonlinear discrete-time control systems with fast subsystems. Syst. Control Lett. **36**(4), 277–283 (1999)
18. Iannelli, L., Johansson, K.H., Jonsson, U.T., Vasca, F.: Dither shape in the averaging of switched systems. In: Proc. of the American Control Conference, Boston, Massachusetts, USA, pp. 5812–5817 (2004)
19. Iannelli, L., Johansson, K.H., Jonsson, U.T., Vasca, F.: Averaging of nonsmooth systems using dither. Automatica **42**(4), 669–676 (2006)
20. Kassakian, J.G., Schlecht, M.F., Verghese, G.C.: Principles of Power Electronics. Addison Wesley, New York (1991)
21. Krein, P.T., Bentsman, J., Bass, R.M., Lesieutre, B.L.: On the use of averaging for the analysis of power electronic systems. IEEE Trans. Power Electron. **5**(2), 182–190 (1990)
22. Lehman, B., Bass, R.M.: Extensions of averaging theory for power electronics systems. IEEE Trans. Power Electron. **11**(4), 542–553 (1996)
23. Liu, Y.M., Huang, X.F., Zhang, B., Yin, L.Y.: Hybrid feedback switching control in a buck converter. In: Proc. of the IEEE International Conference on Automation and Logistics, Qingdao, China, pp. 207–210 (2008)
24. Machowski, J., Bialek, J., Bumby, J., James, R.: Power System Dynamics—Stability and Control, 2nd edn. Wiley, Chichester (2008)
25. Mariéthoz, S., Almér, S., et al.: Comparison of hybrid control techniques for buck and boost DC–DC converters. IEEE Trans. Control Syst. Technol. **18**(5), 1126–1145 (2010)
26. Middlebrook, R.D., Ćuk, S.: A general unified approach to modeling switching-converter power stages. In: Proc. of the IEEE Power Electronics Specialists Conference, Cleveland, Ohio, USA, pp. 18–34 (1976)
27. Nešić, D., Teel, A.R.: Input-to-state stability for nonlinear time-varying systems via averaging. Math. Control Signals Syst. **14**(3), 257–280 (2001)
28. Papafotiou, G., Geyer, T., Morari, M.: Hybrid modelling and optimal control of switched-mode DC–DC converters. In: Proc. of the IEEE Workshop on Computers in Power Electronics, Zhangjiajie, China, pp. 148–155 (2004)
29. Porfiri, M., Roberson, D.G., Stilwell, D.J.: Fast switching analysis of linear switched systems using exponential splitting. SIAM J. Control Optim. **47**(5), 2582–2597 (2008)
30. Roberson, D.G., Stilwell, D.J.: $\mathscr{L}_2$ gain performance analysis of linear switched systems: fast switching behavior. In: Proc. of the American Control Conference, New York, USA, pp. 2084–2089 (2007)
31. Sanders, J.A., Verhulst, F.: Averaging Methods in Nonlinear Dynamical Systems. Springer, New York (1985)
32. Sanfelice, R.G., Teel, A.R., Goebel, R., Prieur, C.: On the robustness to measurement noise and unmodeled dynamics if stability in hybrid systems. In: Proc. of the American Control Conference, Minneapolis, Minnesota, USA, pp. 4061–4066 (2006)
33. Solo, V., Kong, X.: Adaptive Signal Processing Algorithms. Prentice Hall, Englewood Cliffs (1995)
34. Sreekumar, C., Agarwal, V.: A hybrid control algorithm for voltage regulation in DC–DC boost converter. IEEE Trans. Ind. Electron. **55**(6), 2530–2538 (2008)
35. Sun, J., Mitchell, D.M., Greuel, M.F., Krein, P.T., Bass, R.M.: Averaged modeling of PWM converters operating in discontinuous conduction mode. IEEE Trans. Power Electron. **16**(4), 482–492 (2001)
36. Teel, A.R., Nešić, D.: Averaging theory for a class of hybrid systems. Dyn. Contin. Discrete Impuls. Syst. **17**, 829–851 (2010)
37. Teel, A.R., Moreau, L., Nešić, D.: A unified framework for input-to-state stability in systems with two time scales. IEEE Trans. Autom. Control **48**(9), 1526–1544 (2003)
38. Teel, A.R., Moreau, L., Nešić, D.: Input-to-state set stability for pulse width modulated control systems with disturbances. Syst. Control Lett. **51**(1), 23–32 (2004)

39. Wang, W., Nešić, D.: Input-to-state stability analysis via averaging for parameterized discrete-time systems. Dyn. Contin. Discrete Impuls. Syst. **17**, 765–787 (2010)
40. Wang, W., Nešić, D.: Input-to-state stability and averaging of linear fast switching systems. IEEE Trans. Autom. Control **55**(5), 1274–1279 (2010)
41. Wang, W., Nešić, D., Teel, A.R.: Input-to-state stability for a class of hybrid systems via averaging. Math. Control Signals Syst. **23**(4), 223–256 (2012)

Chapter 10
Sliding Modes

Domingo Biel and Enric Fossas

10.1 Introduction

In 1974, J. R. Wood [26] wrote a report for the National Aerospace and Space Administration about power conversion in electrical networks. Although there is no reference there to sliding mode control, the author analysed ideal DC–DC power converters using Filippov[1] methods [6], hence designing discontinuous control laws and obtaining ideal sliding dynamics. It is worth reproducing here a few sentences:

The line $x + \beta \dot{x} = 0$ is called switching line, and we need to consider carefully what happens at this line. ... However, because of the chattering behaviour along switching lines which is observed in practice, ... In addition, the designer will want to include either a small fixed delay or a small amount of hysteresis, in order to limit the switching frequency of the power transistor because of efficiency considerations ...

The key concepts of discontinuous analysis and design for DC–DC conversion were already in Wood's paper, as well as the drawbacks this discontinuous design involves.

Among the first papers dealing with DC–DC conversion in a sliding mode framework, we must quote [2] and [25]. The English version of the book *Sliding Modes and their Application in Variable Structure Systems* by Utkin had been published by MIR in 1978 [24] and, from 1983 to 1985, a young Yugoslavian professor, A. Sabanoviç, visited Caltech when R. Venkataramanan worked on his thesis. In [25], DC–DC converters were explored in the light of sliding mode and equivalent control

[1]The seminal paper by Filippov had been published by AMS Translations ten years before.

D. Biel (✉) · E. Fossas
Institute of Industrial and Control Engineering, Universitat Politècnica de Catalunya,
Avda. Diagonal, 647, pl. 11, 08028 Barcelona, Spain
e-mail: biel@eel.upc.edu

E. Fossas
e-mail: enric.fossas@upc.edu

F. Vasca, L. Iannelli (eds.), *Dynamics and Control of Switched Electronic Systems*, Advances in Industrial Control,
DOI 10.1007/978-1-4471-2885-4_10, © Springer-Verlag London Limited 2012

concepts to arrive at large-signal and small-signal models; and experimental results were presented as well. The paper includes a sliding mode equivalent representation of DC–DC duty ratio controllers that allows an easy and elegant analysis; the ramp generator defining the switching frequency was incorporated as an integrator plus unitary feedback. Linear (buck) and nonlinear elementary converters (boost, buck–boost) were analysed. A switching line based on a proportional–derivative (PD) design in the output-error were synthesised for the buck converter, while for bilinear converters, two nested loops were reported: an inner current sliding mode programmed loop and an outer proportional–integral (PI) to tune the appropriate current reference to regulate the output voltage. Practical considerations were also reported.

From this time too there are Hebertt Sira-Ramírez's papers [17, 18, 20] where DC–DC conversion was reformulated as an application of variable structure theory and sliding regimes to study energy transfers between dynamic storage elements in lossless bilinear networks. In Europe (Genoa and Toulouse), DC–AC conversion was analysed in sliding domain at the end of the 1980s by Carpita et al. [5] and Boudjema et al. [3]. The implementation of these designs entails considering two problems already reported by Wood and Venkataramanan et al., namely fixed switching frequency and chattering reduction.

Several techniques are reported in the literature to obtain a fixed switching frequency. For instance, processing the equivalent control through a pulse-width modulation (PWM) [18] and the addition of a hysteresis cycle to the sliding mode control comparator [2, 4] and [13]. Several approaches [10, 15] consider a variable bandwidth hysteresis cycle, the implementation of which depends on the system parameters and is complex. Other electronic implementations of quasi-sliding controls are reported in [14, 16], where the fixed switching frequency is synchronised by an external signal defined by a periodic bipolar pulse train. Finally, in [7], the duty cycle is defined so that the sliding surface average is zero in each commutation period. Readers interested in a more in depth description of these methods are referred to [1] and references therein.

Chattering reduction was also considered in [7] and more recently in [9] by V. Utkin and co-workers. They proposed a new challenge in multi-phase converter systems that allows reducing chattering to desired level under any given switching frequency. This can be achieved by providing an appropriate phase shift to implement the so-called "ripple cancellation" or "harmonic cancellation" method.

There are several papers dealing with PWM-based SMC [8, 12, 21, 22] and deserving specific mention. They are based on a result by H. Sira-Ramírez [19] which states that in the limit, i.e. when the switching period tends to zero, the equivalent control is equal to the duty cycle. However, in most of these papers, the switch reference signal appears to be designed in a classic linear control framework and SMC is not used in implementation issues.

The remaining part of this chapter is organised as follows: in the next section, the designing of a sliding mode controller for a boost converter [25] is presented, while a design based on the averaging technique is presented in Sect. 10.3. The new chattering reduction method by Utkin is described in Sect. 10.4 and experiments are

Buck converter

$$\begin{cases} L\dfrac{\mathrm{d}i}{\mathrm{d}t} = -v + Eu \\ C\dfrac{\mathrm{d}v}{\mathrm{d}t} = i - \dfrac{v}{R} \end{cases}$$

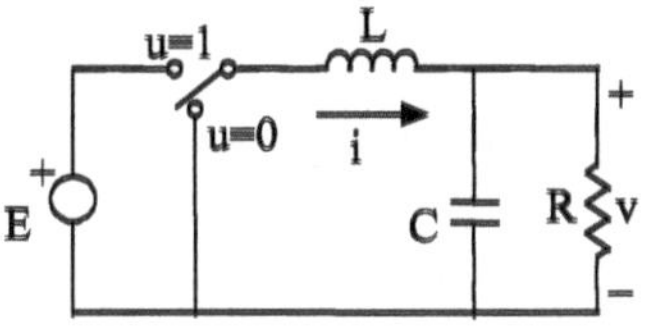

Fig. 10.1 Buck converter topology

reported as well. The chapter closes by drawing some conclusions from the work presented.

10.2 Sliding Mode Control for DC–DC Power Converters

The aim of DC–DC power conversion is to obtain a regulated, continuous voltage (or current) at the load terminals. The power regulator consists of a power stage composed of semiconductors, inductors and capacitors, and a control stage commonly based on the processing of an error signal (the difference between a reference and an output voltage) and a voltage–time conversion through a PWM. The control objective is to achieve a regulated robust output voltage with good dynamic performance.

10.2.1 Electrical and State Space Models

The ideal[2] buck, boost and buck–boost topologies feeding a resistive load are depicted in Figs. 10.1 and 10.2. The converter dynamics is modelled by two state variables, i.e. i, the inductor current, and v, the capacitor voltage[3], and by the control input $u \in \{0, 1\}$, which describes the position of a bidirectional switch. The state equations of the converter are listed below, where E is the DC-input voltage, L and C, the inductor and the capacitor value, respectively, and R, the resistive load. Note that, in the case of the buck–boost converter, there exists an output voltage polarity inversion with respect to the input voltage.

[2]Without semiconductor, capacitor and inductor losses.

[3]The capacitor voltage coincides with the output voltage because the resistor losses associated to the capacitor are not considered.

Boost converter

$$\begin{cases} L\dfrac{di}{dt} = -v(1-u)+E \\ C\dfrac{dv}{dt} = i(1-u)-\dfrac{v}{R} \end{cases}$$

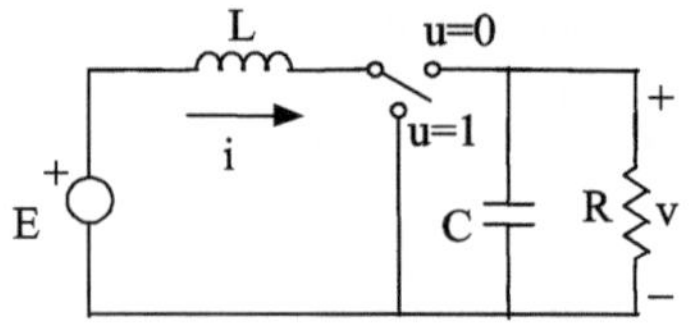

Buck–boost converter

$$\begin{cases} L\dfrac{di}{dt} = -(1-u)v+Eu \\ C\dfrac{dv}{dt} = (1-u)i-\dfrac{v}{R} \end{cases}$$

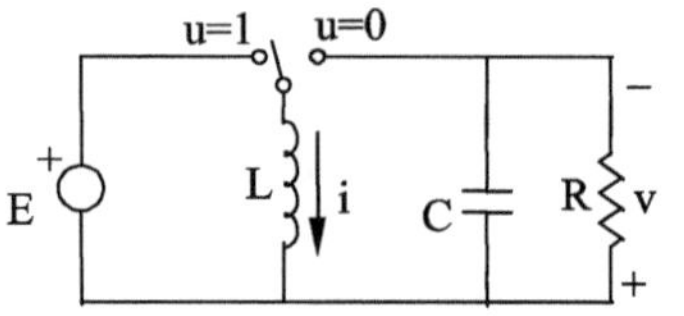

Fig. 10.2 Boost and buck–boost converter topologies

A general model for buck, boost and buck–boost converters can be written as

$$\frac{\mathrm{d}}{\mathrm{d}t}\begin{pmatrix} Li \\ Cv \end{pmatrix} = \begin{pmatrix} 0 & -1+\lambda u \\ 1-\lambda u & -\frac{1}{R} \end{pmatrix}\begin{pmatrix} i \\ v \end{pmatrix} + \begin{pmatrix} E(1+\gamma(u-1)) \\ 0 \end{pmatrix}. \qquad (10.1)$$

Specific models can be obtained by selecting the parameters λ and γ as follows: $\lambda = 0$, $\gamma = 1$ for the buck converter, $\lambda = 1$, $\gamma = 0$ for the boost converter and $\lambda = 1$, $\gamma = 1$ for the buck–boost converter:

Therefore, DC–DC switching converters can be modelled as bilinear systems, and represented as variable structure systems.

10.2.2 Sliding Mode Control Analysis and Design

Output voltage regulation is the general control objective in DC–DC power conversion. A naive approach would be to design the action of the switch, the control action, based uniquely on the output voltage error (direct control). This approach will not be successful in general. An indirect approach, based on both the output voltage and the inductor current, is needed to achieve robust regulation.

SMC strategies for the DC–DC conversion problem via direct and indirect control will be considered here. Starting from a switching surface, the transversality condition is checked and the equivalent control is derived. The latter is used to obtain the ideal sliding dynamics and, when the ideal sliding dynamics are stable, to deduce the sliding domain. For further details on the subject, the reader is referred to [17, 23, 25].

10.2.2.1 Direct Output Voltage Control

The direct output voltage control corresponds to the use of the switching surface

$$\sigma_v = v - V_{\text{ref}}, \tag{10.2}$$

where $V_{\text{ref}} > 0$ is a constant voltage reference.

Note that the transversality condition is not fulfilled in the buck converter case ($\lambda = 0$). For the other cases, ($\lambda = 1$ and $i \neq 0$), the equivalent control and the ideal sliding dynamics are given by

$$u_{\text{eq}} = \frac{i - \frac{V_{\text{ref}}}{R}}{i}, \tag{10.3}$$

$$\begin{cases} v = V_{\text{ref}}, \\ \frac{\mathrm{d}}{\mathrm{d}t} i = \frac{1}{Li}\left(Ei - \frac{V_{\text{ref}}}{R}(E\gamma + V_{\text{ref}})\right). \end{cases} \tag{10.4}$$

The ideal sliding dynamics has an equilibrium point at (i^*, V_{ref}), with

$$i^* = \frac{(E\gamma + V_{\text{ref}})V_{\text{ref}}}{RE}.$$

Its stability is analysed by the first linear approximation, namely

$$\frac{\mathrm{d}}{\mathrm{d}t}\hat{i} = \frac{R}{L} \frac{E^2}{V_{\text{ref}}(E\gamma + V_{\text{ref}})} \hat{i}, \tag{10.5}$$

where $\hat{i} = i - i^*$. Since

$$\frac{R}{L} \frac{E^2}{V_{\text{ref}}(E\gamma + V_{\text{ref}})} > 0,$$

the equilibrium point is unstable; hence direct voltage regulation results in instability of the inductor current.

10.2.2.2 Indirect Output Voltage Control

With the so-called indirect output voltage control, the switching surface is chosen as

$$\sigma_i = i - I_{\text{ref}}, \tag{10.6}$$

where I_{ref} denotes a constant inductor current reference.

Then, the equivalent control and the ideal sliding dynamics are given by

$$u_{\mathrm{eq}} = \frac{v - E(1-\gamma)}{E\gamma + \lambda v}, \tag{10.7}$$

$$\begin{cases} i = I_{\mathrm{ref}}, \\ \frac{\mathrm{d}}{\mathrm{d}t} v = \frac{1}{C}\left(I_{\mathrm{ref}} - \lambda I_{\mathrm{ref}}\left(\frac{v-E(1-\gamma)}{E\gamma+\lambda v}\right) - \frac{v}{R}\right). \end{cases} \tag{10.8}$$

The geometric locus defined by the equilibrium points is described in coordinates (I_{ref}, v^*) with

$$v^{*2} + E\gamma v^* - E I_{\mathrm{ref}} R = 0$$

for $\lambda = 1$, and

$$v^* = I_{\mathrm{ref}} R$$

for $\lambda = 0$. Linearising (10.8) around the equilibrium point (I_{ref}, v^*) yields

$$\frac{\mathrm{d}}{\mathrm{d}t}\hat{v} = \frac{1}{C}\left(-I_{\mathrm{ref}} E\left(\frac{\lambda}{E\gamma + \lambda v^*}\right)^2 - \frac{1}{R}\right)\hat{v}, \tag{10.9}$$

where $\hat{v} = v - v^*$. Thus, the indirect control results in a stable ideal sliding dynamics. The sliding domain on $i = I_{\mathrm{ref}}$ resulting from $0 < u_{\mathrm{eq}} < 1$, assuming $E > 0$, gives the following sliding domain converter characteristics:

- Buck, $i = I_{\mathrm{ref}}$ and $0 < v < E$;
- Boost, $i = I_{\mathrm{ref}}$ and $0 < E < v$;
- Buck–boost, $i = I_{\mathrm{ref}}$ and $0 < v$, $0 < E$.

Finally, the switching strategy is defined so that σ_i^2 qualifies as a Lyapunov function. From Eq. (10.7),

$$\frac{\mathrm{d}}{\mathrm{d}t}\sigma_i^2 = \sigma_i \frac{\mathrm{d}}{\mathrm{d}t}\sigma_i = \sigma_i(\lambda v + E\gamma)(u - u_{\mathrm{eq}}). \tag{10.10}$$

Then, since $0 < u_{\mathrm{eq}} < 1$ is assumed,

$$u = \begin{cases} 0 & \text{if } \sigma_i(\lambda v + E\gamma) > 0, \\ 1 & \text{if } \sigma_i(\lambda v + E\gamma) < 0. \end{cases} \tag{10.11}$$

In summary, the indirect output voltage control provides output voltage regulation presuming the converter states meet the sliding domain conditions. However, the output voltage depends on the load resistance; therefore, these controllers do not produce systems that are robust with respect to load variations.

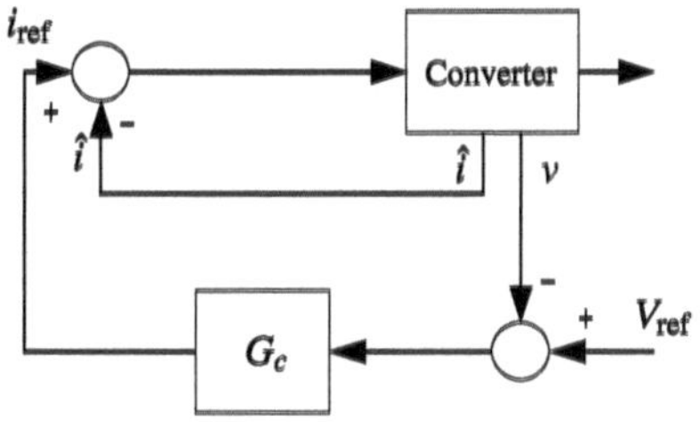

Fig. 10.3 Classical two loops DC–DC regulator diagram

10.2.2.3 Robustness

Two strategies to robustify indirect output voltage control are given here. The first is a specific method for linear systems of relative degree greater than 1 and the second is a PI-type strategy.

As for linear systems, by adding higher derivatives of the error, the relative degree decreases. For the buck converter, see [25] and [23], a suitable switching surface to be considered is

$$\sigma_{vr} = e + k\frac{\mathrm{d}}{\mathrm{d}t}e, \tag{10.12}$$

where $e = V_{\mathrm{ref}} - v$.

This strategy cannot be applied to the boost or to the buck–boost converters because these systems have relative degree 1 and the derivative of the control input would appear in the expression of the equivalent control. Note, in addition, that the derivative of the output voltage (a discontinuous signal) should be processed in the nonlinear converters case. This makes it impossible for designers to use this switching scheme. To avoid processing discontinuities, a high frequency filtering (averaging) is used in [11]. A linearising process can alternatively be used, as in [20].

The switching surface

$$\sigma_{ir} = i + k_e e + k_{v_a} v_a, \tag{10.13}$$

where $e = V_{\mathrm{ref}} - v$ and v_a satisfies $\frac{\mathrm{d}}{\mathrm{d}t} v_a = e$, is a robust alternative.

Both switching surfaces σ_{vr} and σ_{ir} can be represented by the block diagram in Fig. 10.3 which highlights two control loops: a fast inner control loop corresponding to the current dynamics and a slow outer control loop which processes the output voltage error. The G_c block is a P (resp., a PI) controller and $\hat{\imath}$ is the capacitor (resp., inductor) current for σ_{vr} (resp., for σ_{ir}). The analysis of the controlled systems (buck, boost and buck–boost), namely checking the transversality condition, computing the equivalent control, the sliding domain, the equilibrium points and their stability is left to the reader.

10.3 Averaging and Pulse-Width Modulation Based Sliding Modes

SMC theory presumes an infinite switching frequency when the system operates in sliding mode, however actual switches cannot commute at infinite frequency. At any rate, higher switching frequencies become harmful in some of the applications (the higher the switching frequency, the higher the losses in the converter). Consequently, actual sliding mode controls operate at high, finite, possibly variable frequency, which results in a *chattering* around the sliding surface. This problem has been tackled through fixed and variable bandwidth hysteresis comparators, by the addition of an external synchronous signal, by the use of equivalent control as duty cycle and through a zero average dynamics (ZAD) control strategy. Readers interested in SMC implementation in power converters are referred to [1].

Most of the works devoted to PWM based sliding mode controllers take benefit from a result by H. Sira-Ramírez [19], proving that in the limit, when the switching period tends to zero, the duty cycle is equal to the equivalent control. As for the ZAD strategy [7], the duty cycle is defined so that the average of the sliding surface is zero in each commutation period; that is to say, the controller guarantees

$$\langle\sigma(x)\rangle = \frac{1}{T}\int_{kT}^{(k+1)T} \sigma\big(x(\tau)\big)\,\mathrm{d}\tau = 0. \tag{10.14}$$

The ZAD-based duty-cycle design yields chattering reduction and a fixed switching frequency. The ZAD-based duty cycle is defined according to the sampled values of σ and $\frac{\mathrm{d}}{\mathrm{d}t}\sigma$. In particular, consider the case $\sigma(x(t_k), t_k) \geq 0$. Denoting by $\{u^+, u^-\}$ the set of discrete values the switch can take (in the next example, $u^- = 1$ and $u^+ = 0$), we then choose $u(t_k) = u^+$, getting the following implications:

$$\sigma\big(x(t_k), t_k\big) + \frac{T}{2}\frac{\mathrm{d}}{\mathrm{d}t}\sigma\bigg|_{(k,u^+)} \geq 0 \Rightarrow d_k = 1,$$

$$\sigma\big(x(t_k), t_k\big) + \frac{T}{2}\frac{\mathrm{d}}{\mathrm{d}t}\sigma|_{(k,u^+)} < 0 \Rightarrow d_k = 1 - \sqrt{\frac{\left|\frac{\mathrm{d}}{\mathrm{d}t}\sigma|_{(k,u^+)}\right| - 2\frac{|\sigma(x(t_k),t_k)|}{T}}{\left|\frac{\mathrm{d}}{\mathrm{d}t}\sigma|_{(k,u^+)}\right| + \left|\frac{\mathrm{d}}{\mathrm{d}t}\sigma|_{(k,u^-)}\right|}}.$$

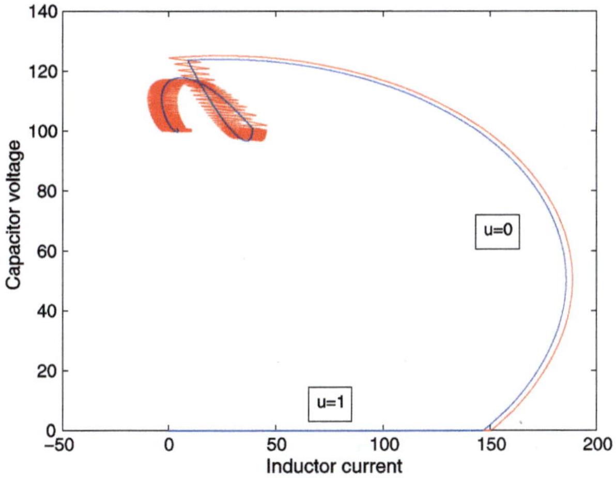

Fig. 10.4 Projection of the three dimensional dynamics on the (i, v)-plane

On the other hand, in the case $\sigma(x(t_k), t_k) \leq 0$, we have $u(t_k) = u^-$ and the following implications:

$$\sigma\big(x(t_k), t_k\big) + \frac{T}{2}\frac{\mathrm{d}}{\mathrm{d}t}\sigma\bigg|_{(k,u^-)} \leq 0 \Rightarrow d_k = 1,$$

$$\sigma\big(x(t_k), t_k\big) + \frac{T}{2}\frac{\mathrm{d}}{\mathrm{d}t}\sigma\bigg|_{(k,u^+)} > 0 \Rightarrow d_k = 1 - \sqrt{\frac{\left|\frac{\mathrm{d}}{\mathrm{d}t}\sigma|_{(k,u^-)}\right| - 2\frac{|\sigma(x(t_k),t_k)|}{T}}{\left|\frac{\mathrm{d}}{\mathrm{d}t}\sigma|_{(k,u^+)}\right| + \left|\frac{\mathrm{d}}{\mathrm{d}t}\sigma|_{(k,u^-)}\right|}}.$$

This methodology is applied here to a boost converter. In order to guarantee robustness with respect to line and load variations, the dynamics are extended by adding the integral of the voltage error, and a switching surface like σ_{ir} in (10.13) is designed taking $k_e = -1.2$ and $k_{v_a} = -4500$. The converter parameters are $E = 50\,\mathrm{V}$, $L = 20\,\mu\mathrm{H}$ and $C = 100\,\mu\mathrm{F}$. The load resistance varies from $5\,\Omega$ to $40\,\Omega$ and the output voltage reference is set to $100\,\mathrm{V}$.

The projection of the state trajectory on the (i, v)-plane is depicted in Fig. 10.4. The trajectory in *red* corresponds to the ZAD-based duty cycle with a switching frequency of 100 kHz, while the one in *blue* corresponds to the ideal sliding controller. The initial load value of $40\,\Omega$ changes to $5\,\Omega$ when trajectory reaches the equilibrium point. Note that the average of the dynamics in *red* results in the dynamics in *blue*. Nevertheless, the two trajectories show some differences. Namely, the ZAD-based entails a delay because of the synchrony execution of the algorithm and it shows the chattering typical of switched circuits. The ideal sliding dynamics, i.e. the dynamics on the switching surface, is zoomed in Fig. 10.5, *blue* trajectory.

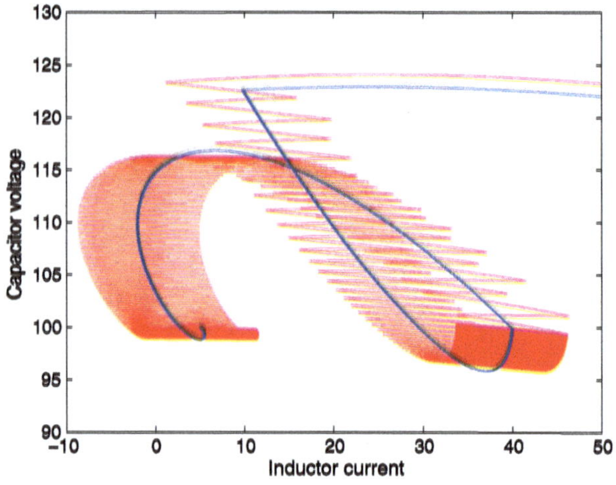

Fig. 10.5 Ideal sliding and ZAD dynamics

10.4 Multiphase Systems and Interleaving

The chattering reduction problem has been recently considered using *interleaving*, a strategy well known in electronics. V. Utkin and co-workers proposed a new methodology in SCM framework that allows reducing chattering to desired level under any given switching frequency [9]. This can be achieved by providing an appropriate phase shift to implement the so-called "ripple cancellation" or "harmonic cancellation" method. In this section, this new challenge is implemented on a four-phases, half-bridge buck converter prototype.

10.4.1 Analogue Interleaving and Chattering Reduction

Let us consider a system with an inner loop which is in charge of regulating an inner output y by means of a relay as in Fig. 10.6 where the 'Phase-1' block corresponds to a relative degree 1, first order transfer function. Hence,

$$\frac{\mathrm{d}}{\mathrm{d}t}s = \frac{\mathrm{d}}{\mathrm{d}t}y_{\mathrm{ref}} - \frac{\mathrm{d}}{\mathrm{d}t}y = a - M\mathrm{sign}(s) \tag{10.15}$$

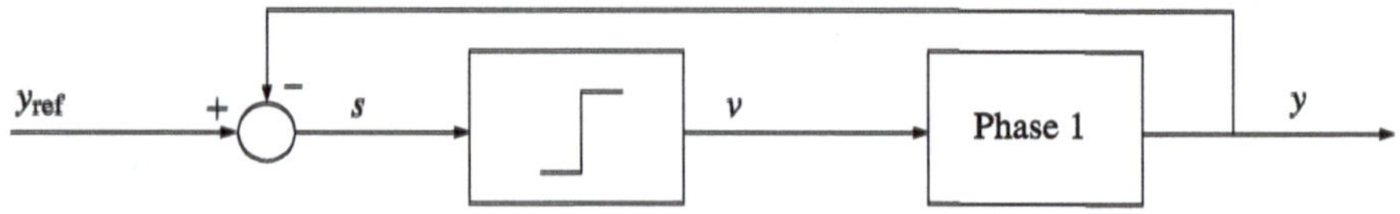

Fig. 10.6 Inner loop in a single-phase system

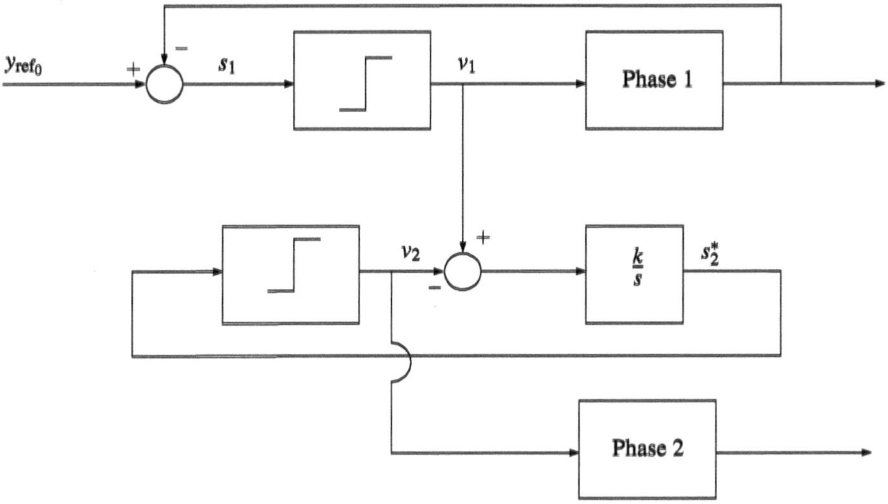

Fig. 10.7 A two-phase master–slave system

where $a = \frac{d}{dt}y_{ref} + y/\tau$ and $M = k_0/\tau$. Note that there is a sliding motion on $s = 0$ provided that $|a| < M$.

The chattering reduction method proposed in [9] is based on:

- A multiphase inner loop instead of a single-phase one,
- Modifying the reference appropriately, i.e. taking as a new reference $y_{ref_0} = y_{ref}/m$ and
- Taking benefit of some properties of the Fourier expansions.

Two possible schemes, called an m-phase interconnected system and an m-phase master–slave system, were proposed, and the experiments were performed in a four-phase master–slave DC–DC buck converter. Here we summarise the main results by considering for simplicity a two-phase master–slave model which is shown in Fig. 10.7.

As in the single phase case,

$$\frac{d}{dt}s_1 = a - M\,\text{sign}(s_1), \tag{10.16}$$

$$\frac{d}{dt}s_2^* = kM\big[\text{sign}(s_1) - \text{sign}(s_2^*)\big], \tag{10.17}$$

where

$$s_1 = y_{ref_0} - y_1, \tag{10.18}$$

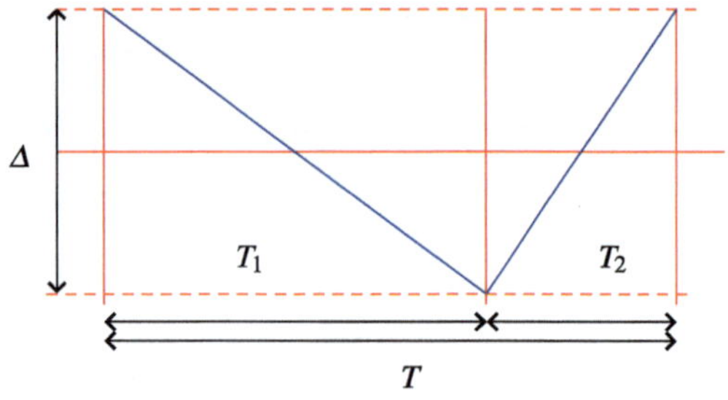

Fig. 10.8 Periodic s_1 dynamics

$$s_2^* = k \int M\big[\operatorname{sign}(s_1) - \operatorname{sign}(s_2^*)\big]\,\mathrm{d}t, \tag{10.19}$$

presuming that the relay gains and the dynamics in the two phases are identical.

In Fig. 10.8, the s_1 dynamics close to $s_1 = 0$ are sketched. Computing the period T from the figure yields

$$T = T_1 + T_2 = \frac{\Delta}{M - a} + \frac{\Delta}{M + a} = \frac{2\Delta M}{M^2 - a^2}. \tag{10.20}$$

It is presumed that $\frac{\mathrm{d}}{\mathrm{d}t} y_{\mathrm{ref}_0}$ and state variables are practically constant, i.e. the dominant term in $\frac{\mathrm{d}}{\mathrm{d}t} s_1$ is $M\operatorname{sign}(s_1)$.

In Fig. 10.9, the (s_1, s_2^*)-plane for $a > 0$ is sketched with the vector field values indicated in the corners. From the figure, the phase shift can be written as

$$T_\phi = \frac{\Delta}{2kM}, \tag{10.21}$$

which is equal to the time from changing s_2^* from point (2) to point (3).

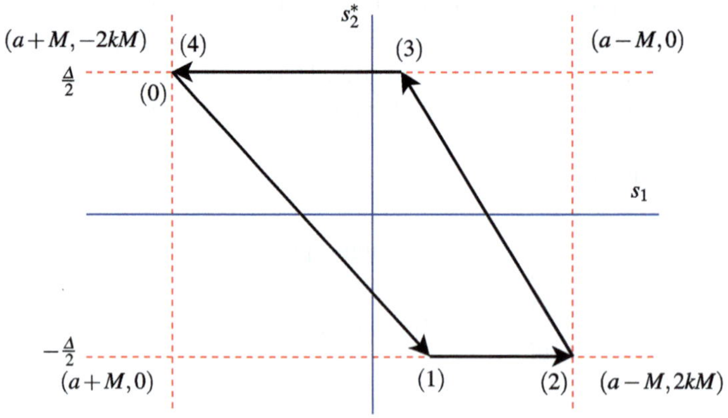

Fig. 10.9 Periodic (s_1, s_2^*) dynamics

The scheme sketched in Fig. 10.7 allows implementing a phase shift between phases 1 and 2 while the following inequality holds

$$\frac{\Delta}{2kM}\big(M+|a|\big)<\Delta, \tag{10.22}$$

which is equivalent to

$$M+|a|<2kM, \tag{10.23}$$

where Δ is the hysteresis width and k and M are the integral and relay gains, respectively; see [9] for details.

The ideal sliding dynamics in the interconnected system results in $y_1 = y_2 = y_{\text{ref}_0}$.

10.4.1.1 Selection of Phase Number

Suppose that a master–slave m-phase linear system is to be designed so that the period of chattering is the same in each phase, and two consecutive phases have the same phase-shift T/m. Since chattering is a periodic function, it can be represented using Fourier series with frequencies

$$\omega_n = \frac{2\pi n}{T}, \qquad n = 1, 2, \ldots, \infty. \tag{10.24}$$

The following well known lemmas are used for the design.

Lemma 10.4.1 *Let $f_1(t)$ be a periodic function, $f_2(t) \triangleq f_1(t+\tau)$, and let*

$$\mathscr{F}_1(t) = a_0 + \sum_n \big(a_n \cos(\omega_n t) + b_n \sin(\omega_n t)\big)$$

be the Fourier series of f_1. Then

$$\mathscr{F}_2(t) = \mathscr{F}_1(t+\tau).$$

Lemma 10.4.2 *Let a_n, b_n be the cosine and sine coefficients, respectively. If n/m is not an integer then $a_n = b_n = 0$.*

As a consequence of the above lemmas, all harmonics except for $n = lm$, $l \in \mathbb{Z}$, are suppressed in the output signal. As a result, the amplitude of chattering can be reduced to a desired level by increasing the number of phases. Additionally, since the sliding surface T-average is zero, there is no continuous component in the Fourier expansion.

10.4.2 A Four-Phase Parallel Buck Converter

In this section, we present an application of the SMC described above to a parallel buck converter. The dynamics of the converter can be modelled as

$$L\frac{\mathrm{d}}{\mathrm{d}t}i_1 = -R_L i_1 - v_o + E u_1, \tag{10.25}$$

$$\vdots \tag{10.26}$$

$$L\frac{\mathrm{d}}{\mathrm{d}t}i_m = -R_L i_m - v_o + E u_m, \tag{10.27}$$

$$C\frac{\mathrm{d}}{\mathrm{d}t}v_o = i_1 + \cdots + i_m - \frac{v_o}{R}, \tag{10.28}$$

where i_k refers to the current in the kth phase inductor, v_o refers to the output voltage and $u_k \in \{0,\ 1\}$ are the switches' control signals. Note that inductances L and losses resistors R_L are presumed to be the same for all phases.

Let us assume we deal with an m-phase buck converter which phases are shifted by $T_\phi = T/m$. Then from (10.20) and (10.21),

$$\frac{\Delta}{2kM} = \frac{1}{m}\frac{2\Delta M}{M^2 - a^2}, \tag{10.29}$$

hence

$$k = \frac{m(M^2 - a^2)}{4M^2}. \tag{10.30}$$

Finally, (10.23) results in

$$|a| < M\left(1 - \frac{2}{m}\right). \tag{10.31}$$

Particularising it for the half-bridge DC–DC buck converter gives

$$\frac{1}{L}\left|\frac{E}{2} - \left(\frac{R_L}{mR} + 1\right)v_o^*\right| < \frac{E}{2L}\left(1 - \frac{2}{m}\right) \tag{10.32}$$

which, in turn, taking $v_o^* = \alpha E$ and defining $\hat{\alpha} = \alpha(R_L/(mR) + 1)$ yields the necessary and sufficient conditions:

$$\hat{\alpha} > \frac{1}{m} \quad \text{if } \hat{\alpha} < 0.5 \tag{10.33}$$

and

$$\hat{\alpha} < 1 - \frac{1}{m} \quad \text{if } \hat{\alpha} > 0.5. \tag{10.34}$$

Table 10.1 shows the minimum and maximum $\hat{\alpha}$ values for $m = 3, 4, 5, 6$.

It is now interesting to make some considerations:

Table 10.1 Minimum and maximum values of $\hat{\alpha}$ as a function of m

m	$\hat{\alpha} < 0.5$	$0.5 < \hat{\alpha}$
3	0.3333	0.6666
4	0.25	0.75
5	0.20	0.8
6	0.1666	0.8333

- It is not necessary to exactly set the switching period. Taking Δ sufficiently small will be enough.
- T is linear with respect to Δ.
- Defining k properly yields the desired phase shift between consecutive currents.
- k depends on the input voltage and on the current reference.
- Differences in the losses among phases will result in differences in the current averaged values.

The controller is now implemented on a real DC–DC buck converter with the following parameters: $E = 10\,\text{V}$, $L = 22\,\mu\text{H}$, $C = 10\,\mu\text{F}$ and $R_L = 0.7\,\Omega$ (this includes $300\,\text{m}\Omega$ corresponding to semiconductor losses and $400\,\text{m}\Omega$ corresponding to inductor losses). The electronic switch is a metal–oxide–semiconductor field-effect transistor (MOSFET) that works at a frequency of 100 kHz. The system is controlled using two loops: a current inner loop, sliding mode controlled with a hysteresis band, and a voltage outer loop that defines the current reference through a proportional–integral (PI) controller.

The power converter feeds a $2\,\Omega$ load providing several output voltages, namely 3 V, 5 V, 7 V. Also, the specific output voltage value $v_o^* = 4.59\,\text{V}$ is selected so that the relation T_{on}/T, i.e. the resulting duty cycle, equals to 0.5. This case is particularly interesting because the Fourier coefficients corresponding to the harmonics that are multiples of four cancel. Thus the chattering reduction is really relevant. The cases $v_o^* = 3\,\text{V}$, $v_o^* = 7\,\text{V}$ show that the system does not work properly, as it is stated by the theoretical analysis.

Figure 10.10 shows the oscilloscope signals when the half-bridge, single-phase DC–DC converter provides an output voltage of 5 V. The triangular signal corresponds to the input current, the rectangular one corresponds to the control signal while the third one is the output voltage. The average current value is 2.44 A and the chattering width is 0.47 A. The output voltage average is 4.96 V.

Let us now consider the half-bridge, four-phase DC–DC converter. The four shifted currents are depicted in Fig. 10.11 and the MOSFET drain voltages are shown in Fig. 10.12. Both figures show that the master–slave algorithm operates properly with respect to current shifting. The current values of each phase are $i_{L_1} = 0.62 \pm 0.43\,\text{A}$, $i_{L_2} = 0.6 \pm 0.42\,\text{A}$, $i_{L_3} = 0.63 \pm 0.43\,\text{A}$ and $i_{L_4} = 0.66 \pm 0.42\,\text{A}$. The output current, one of the duty cycles and the output voltage are depicted in

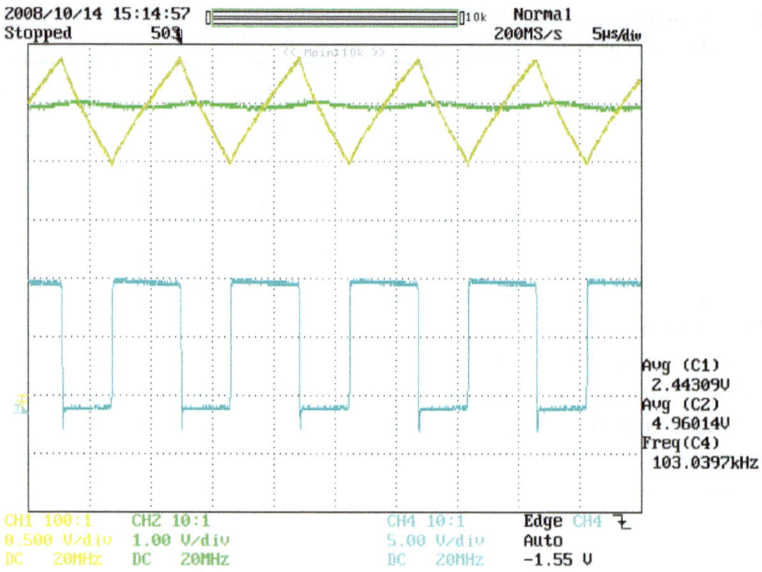

Fig. 10.10 Oscilloscope signals for single-phase and $v_o^* = 5\,\mathrm{V}$

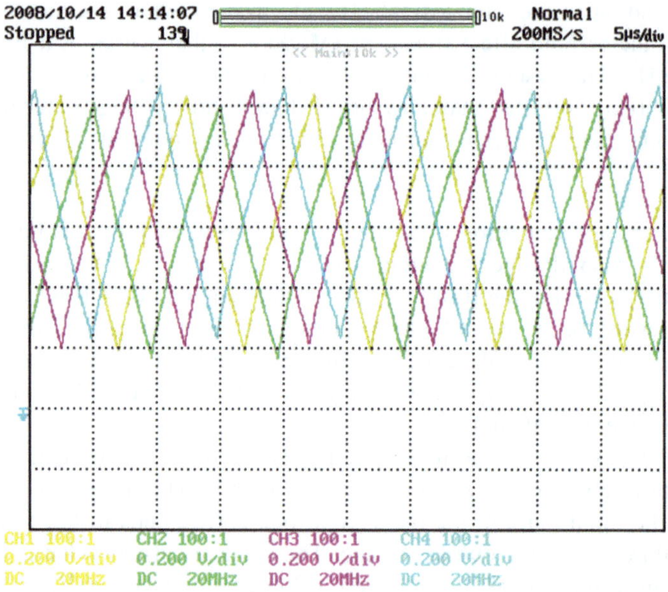

Fig. 10.11 Currents oscilloscope signals

Fig. 10.13, which is zoomed in Fig. 10.14 to show the voltage ripple and the current chattering. The average value of the output current is 2.48 A, with a chattering width of 0.095 A. Note the reduction in the current chattering from 0.47 A to 0.095 A.

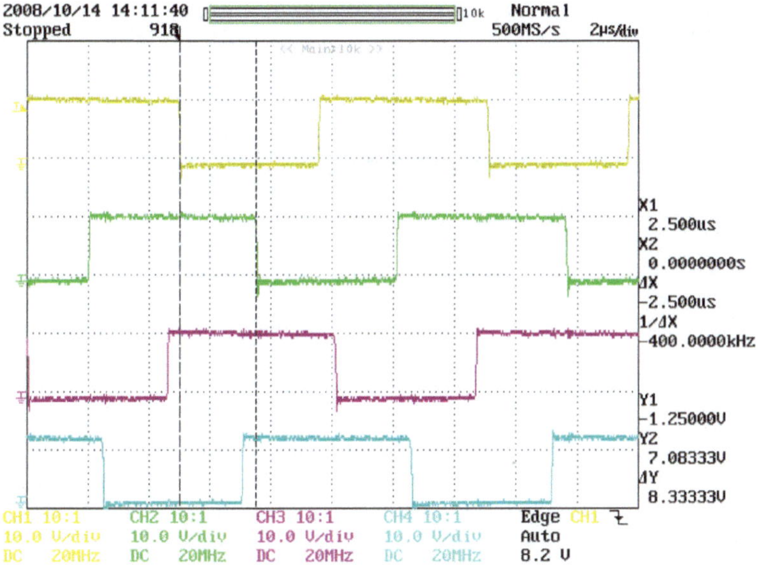

Fig. 10.12 MOSFET drain voltages oscilloscope signals

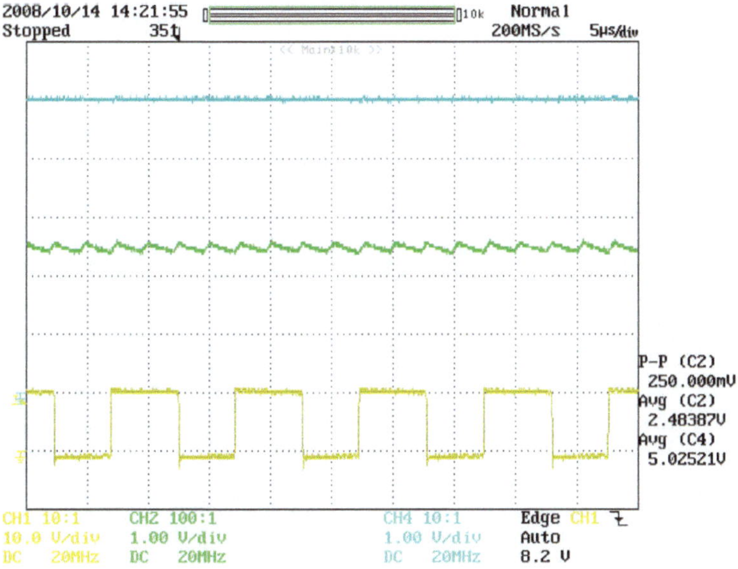

Fig. 10.13 Oscilloscope signals for four-phase and $v_o^* = 5\,\mathrm{V}$

Chattering reduction in the case of 4.59 V output voltage is amazing. This is because the four duty cycles are equal to 0.5 and the Fourier coefficients corresponding to the harmonics that are multiples of four cancel.

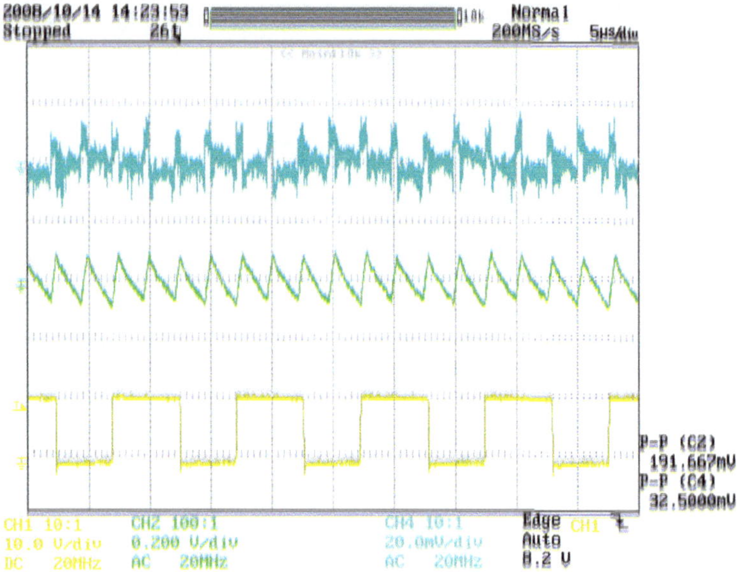

Fig. 10.14 Voltage ripple and current chattering oscilloscope signals

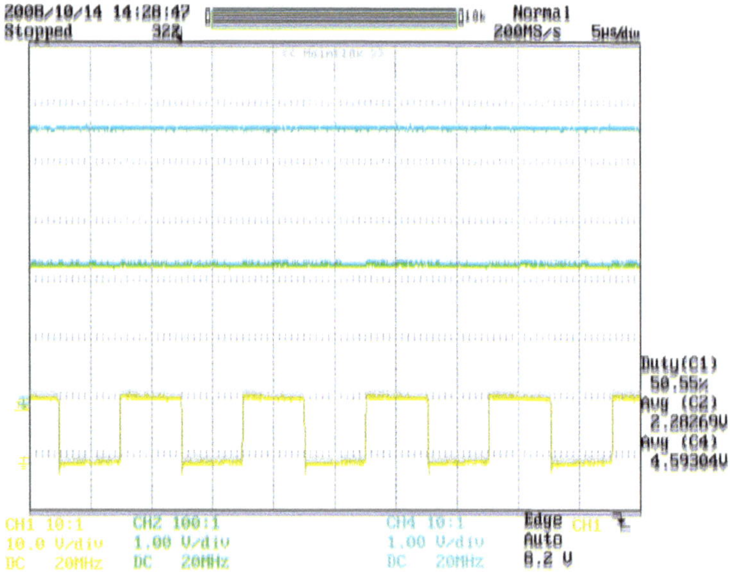

Fig. 10.15 Oscilloscope signals for four-phase and $v_o^* = 4.59$ V

In this case, the output current and voltage are shown in Fig. 10.15 and zoomed in Fig. 10.16. The average value of the output current is 2.28 A, with a chattering width of 0.033 A. It is difficult to distinguish this chattering from measurement noise. Note the reduction in the current chattering from 0.47 A to 0.033 A.

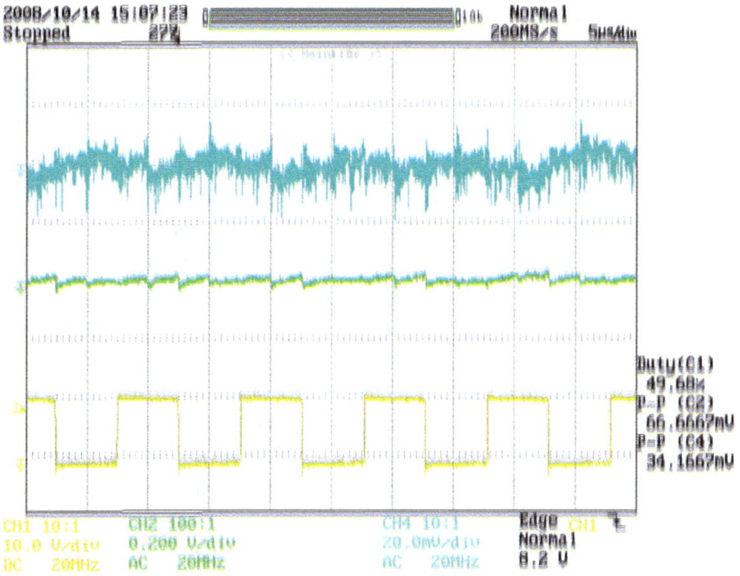

Fig. 10.16 Voltage ripple and current chattering oscilloscope signals

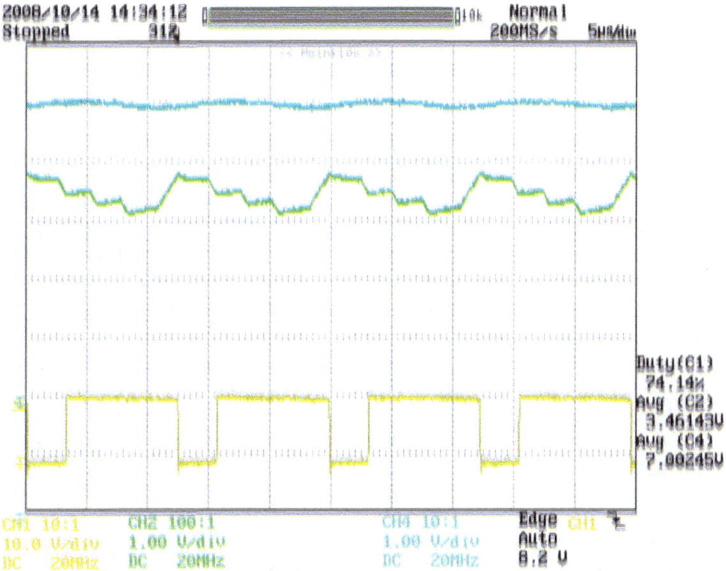

Fig. 10.17 Oscilloscope signals for four-phase and $v_o^* = 7\,\mathrm{V}$

Finally, Fig. 10.17 shows the output current and voltage for the four-phase, master–slave, DC–DC converter when a 7 V output voltage is demanded. Note that the system does not work properly. This is because our demand yields an $\hat{\alpha}$ which is not in the interval available using four phases as shown in Table 10.1. Actually, the $\hat{\alpha}$ value for $\alpha = 0.7$ is $0.76 \notin (0.25, 0.75)$.

10.5 Summary

In this chapter, the application of sliding mode control to DC–DC power converters has been reviewed. The successful theoretical and practical solution proposed in the pioneering work of Sabanović, Venkataramanan and Ćuk has been reported. The main drawbacks of sliding mode control for power converters are then analysed, focusing on the issues of non-constant switching frequency and chattering, and some of the possible corresponding solutions proposed in the literature have been revisited. The most recent results of V. Utkin for getting a fixed switching frequency and a great chattering reduction are presented in the final part of the chapter. All the theory is particularised to ideal single-phase and multi-phase DC–DC converters.

Acknowledgements D. Biel was partially supported by the Spanish government research projects DPI2009-14713-C03-03 and DPI2010-15110, while E. Fossas was partially supported by the Spanish government research projects DPI2008-01408 and DPI2010-15110. The authors would like to express their gratitude to Prof. Vadim I. Utkin, for most of what they know on this subject is because of him.

References

1. Biel, D., Fossas, E.: SMC applications to power electronics, pp. 265–294. The Institution of Electrical Engineers, London, United Kingdom (2004)
2. Bilalović, F., Mušić, O., Šabanović, A.: Buck converter regulator operating in the sliding mode. In: Proc. of the Power Conversion International Conference, pp. 331–340 (1983)
3. Boudjema, F., Boscardin, M., Bidan, P., Marpinard, J., Valentin, M., Abatut, J.: Vss approach to a full-bridge buck converter used for ac sine voltage generation. In: Proc. of the IEEE Industrial Electronics Society Annual Conference, Philadelphia, Pennsylvania, USA, pp. 82–89 (1989)
4. Bühler, H.: Réglage Par Mode de Glissement. Presses Polytechniques Romandes, Lousanne (1986)
5. Carpita, M., Marchesioni, M., Oberti, M., Puguisi, L.: Power conditioning system using sliding mode control. In: Proc. of the IEEE Power Electronics Specialist Conference, Kyoto, Japan, pp. 623–633 (1988)
6. Filippov, A.: Differential equations with discontinuous right-hand side. Transl. Am. Math. Soc. **42**, 199–231 (1964)
7. Fossas, E., Griñó, R., Biel, D.: Quasi-Sliding Control Based on Pulse Width Modulation, Zero Averaged Dynamics and the L2 Norm, pp. 335–344. World Scientific, Singapore (2001)
8. Lai, Y.M., Tan, S.C., Wu, C.K.: Design of a PWM based sliding mode controlled buck-boost converter in continuous-conduction-mode. ECTI Trans. Electr. Eng. Electron. Commun. **5**(1), 129–133 (2007)
9. Lee, H., Malinin, A., Utkin, V.: Chattering suppression in multiphase power converters. Int. J. Control **82**(9), 1720–1737 (2009)
10. Malesani, L., Rossetto, L., Spiazzi, G., Zuccato, A.: An AC power supply with sliding-mode control. IEEE Ind. Appl. Mag. **2**(5), 32–38 (1996)

11. Mattavelli, P., Rossetto, L., Spiazzi, G.: Small-signal analysis of DC-to-DC converters with sliding mode control. IEEE Trans. Power Electron. **12**(1), 96–102 (1997)
12. Nguyen, V., Lee, C.: Tracking control of buck converter using sliding-mode with adaptive hysteresis. In: Proc. of the IEEE Power Electronics Specialist Conference, Atlanta, Georgia, USA, pp. 1086–1093 (1995)
13. Nicolas, B., Fadel, M., Chéron, Y.: Sliding mode control of DC-to-DC converters with input filter based on the Lyapunov-function approach. In: Proc. of the European Power Electronics Conference, Seville, Spain, pp. 1338–1343 (1995)
14. Pinheiro, H., Martins, A., Pinheiro, J.: A sliding mode controller in single phase voltage source inverters. In: Proc. of the IEEE Industrial Electronics Society Annual Conference, Bologna, Italy, pp. 394–398 (1994)
15. Ruiz, J., Lorenzo, S., Lobo, I., Amigo, J.: Minimal ups structure with sliding mode control and adaptive hysteresis band. In: Proc. of the IEEE Industrial Electronics Society Annual Conference, Pacific Grove, California, USA, pp. 1063–1067 (1990)
16. Silva, J., Paulo, S.: Fixed frequency sliding modulator for current mode PWM inverters. In: Proc. of the IEEE Power Electronics Specialist Conference, Seattle, Washington, USA, pp. 623–629 (1993)
17. Sira-Ramírez, H.: Sliding motions in bilinear switched networks. IEEE Trans. Circuits Syst. **34**(8), 919–933 (1987)
18. Sira-Ramírez, H.: Differential geometric methods in variable structure control. Int. J. Control **48**(4), 1359–1390 (1988)
19. Sira-Ramírez, H.: A geometric approach to pulse-width modulated control in nonlinear dynamical systems. IEEE Trans. Autom. Control **34**(2), 184–187 (1989)
20. Sira-Ramírez, H., Ilic, M.: A geometric approach to the feedback control of switch mode DC-to-DC power supplies. IEEE Trans. Circuits Syst. **35**(10), 1291–1298 (1988)
21. Tan, S.C., Lai, Y.M., Tse, C.K.: General design issues of sliding-mode controllers in DC–DC converters. IEEE Trans. Ind. Electron. **55**(3), 1160–1173 (2008)
22. Tan, S.C., Lai, Y.M., Tse, C.K., Cheung, M.K.H.: A fixed-frequency pulse-width-modulation-based quasi-sliding-mode controller for buck converters. IEEE Trans. Power Electron. **20**(6), 1379–1392 (2005)
23. Utkin, V., Guldner, J., Shi, J.: Sliding Mode Control in Electromechanical Systems. Taylor and Francis, London (1999)
24. Utkin, V.I.: Sliding Modes and Their Application in Variable Structure Systems. MIR, Moscow (1978)
25. Venkataramanan, R., Sabanović, A., Ćuk, S.: Sliding mode control of DC-to-DC converters. In: Proc. of the IEEE Industrial Electronics Society Annual Conference, San Francisco, California, USA, pp. 251–258 (1985)
26. Wood, J.R.: Power Conversion in Electrical Networks. NASA Rep. no. CR-120830, Harvard University, Division of Engineering and Applied Physics, Cambridge, Massachusetts, USA (1974)

Chapter 11
Model Predictive Control

Sébastien Mariéthoz and Stefan Almér

11.1 Introduction

Model predictive control (MPC) is mostly known from the process industry where it has become the preferred tool of many practicing control engineers [9, 13, 16]. MPC provides a systematic and intuitively clear method of controlling multi-variable systems subject to input and state constraints. As a consequence, MPC often provides satisfactory performance in complex systems where traditional techniques like proportional–integral–derivative (PID) controllers fall short. Furthermore, MPC allows the plant to be operated closer to the limitations of actuators and state. This can be favourable since a process is often most profitable when certain variables are kept close to the boundaries. Traditionally, MPC has been reserved for applications with relatively slow time scales. The reason for this is that MPC requires an optimisation problem to be solved at each sampling instant in order to obtain the control input. The optimisation problem is often large, and obtaining the solution is time consuming, thus limiting the rate at which the control input can be updated. However, recent advances in both computing hardware and optimisation algorithms have allowed one to consider MPC as an option in many new applications. In particular, MPC has been applied successfully in power electronics where the control frequency ranges from a few hundred hertz up to a few megahertz [11, 12].

The main challenge of implementing MPC consists in solving the associated optimisation problem in a bounded time. The level of difficulty depends on the problem class, which is essentially determined by the system model and cost criterion. The two main approaches that are employed to deal with the hybrid dynamics of power electronics systems are distinguished by the way the switches are controlled. The switches can be controlled through pulse-width modulation (PWM), or they

S. Mariéthoz (✉) · S. Almér
Automatic Control Laboratory, ETH Zürich, Physikstrasse 3, ETL CH, 8092 Zürich, Switzerland
e-mail: sebastien.mariethoz@control.ee.ethz.ch

S. Almér
e-mail: almers@control.ee.ethz.ch

F. Vasca, L. Iannelli (eds.), *Dynamics and Control of Switched Electronic Systems*, Advances in Industrial Control,
DOI 10.1007/978-1-4471-2885-4_11,

can be controlled directly by choosing between the ON and OFF states at discrete-time instants. Employing the PWM approach, together with a suitable model of the dynamics, results in an optimisation problem which is a linear, quadratic or mixed-integer linear or quadratic problem. The direct approach results in a pure integer problem. The solution to linear and quadratic problems are piecewise affine functions of the initial state. For systems of limited complexity, this allows the solution to be computed explicitly off-line. The solution can then be implemented in extremely short time on-line. In the sections below, the modelling and solution approaches are explained and illustrated through a few practical examples.

11.2 MPC Principle

Three distinct features characterise the model predictive control approach:

1. An explicit system model is used to describe the behaviour of the plant over a finite time horizon.
2. A cost criterion is used to describe the desired closed-loop behaviour of the plant.
3. A receding horizon procedure is employed to control the closed-loop system.

These three principles of operation are combined as follows. Assume a discrete-time setting and let

$$x_{l+1} = f(x_l, u_l) \tag{11.1}$$

be the system model describing the plant behaviour, where $x_{l+1} \in \mathbb{R}^n$ is the (predicted) state, and $u_l \in \mathbb{R}^m$ is the control input. At each time instant k, the state $\hat{x}_k$ of the plant is measured or estimated. The plant behaviour is predicted over a time horizon stretching from the current time k to the time $k+N$, where $N \geq 1$ is the prediction horizon. The prediction is given by the response x_l, $l = k+1, \ldots, k+N$, of (11.1) to the control input u_l, $l = k, \ldots, k+N-1$, with the measured or estimated state $\hat{x}_k$ as initial condition.

A cost criterion defined as a function of the prediction x_k is optimised to obtain the most suitable control input u^*. Only the first component of the optimiser u^* is implemented; at the next time instant (time $k+1$) the state of the plant is measured again, and the procedure is repeated.

The MPC approach can be summarised as follows. At each time instant k

1. Measure or estimate the current state $\hat{x}_k$
2. Solve the finite-time-horizon optimal control problem

$$\text{minimise} \quad \sum_{l=k}^{k+N-1} J(x_{l+1}, u_l, l) \tag{11.2a}$$

$$\text{subject to:} \quad x_{l+1} = f(x_l, u_l), \quad l = k, \ldots, k+N-1 \tag{11.2b}$$

$$g(x_{l+1}, u_l) \le 0, \qquad l = k, \dots, k+N-1 \tag{11.2c}$$

$$x_k = \hat{x}_k \tag{11.2d}$$

3. Apply the first component u_k^*.

In the optimisation problem above, (11.2a) formulates the cost criterion, (11.2b) the dynamic model equality constraints, (11.2c) the system admissible state and input, which are specified as inequality constraints.

The optimisation problem initial condition (11.2d) will be omitted in the sequel for brevity. The main challenge of implementing MPC consists in solving (11.2a)–(11.2d) in a bounded time. The level of difficulty depends on the problem class, which is itself mainly determined by the model class and employed cost criterion. In the sections below, we discuss the main problem classes that have been considered in power electronics applications so far.

11.3 MPC Model Classes and Approaches

The switches of a power electronics system can do transitions either in a forced way imposed by the control law or spontaneously based on the system states. Spontaneous commutations can for instance be due to a current through a diode going to zero. From the point of view of deriving a control model to synthesise an MPC scheme, we distinguish three types of hybrid dynamics. In the most general case, the resulting system dynamics are hybrid of the form

$$\dot{x}(t) = \sum_{i=1}^{N_b} s_i(t) f_i\big(x(t), v(t), w(t)\big), \tag{11.3a}$$

$$s_i(t) = \begin{cases} 1 & \text{when } h_i(x(t), u(t), v(t)) \ge 0, \\ 0 & \text{otherwise,} \end{cases} \qquad i \in \{1, \dots, N_b\}, \tag{11.3b}$$

where s_i are binary variables necessary to describe the hybrid dynamics and N_b is the number of binaries. The binary variables may be associated to physical switches, but it is in general not required. $u(t)$ is the vector of binary manipulated inputs that allows one to force commutations through binary switch control signals. $v(t)$ and $w(t)$ are vectors of exogenous inputs, generally supply or load voltages or currents. Equations (11.3a), (11.3b) may for instance represent the dynamics of a drive system such as shown in Fig. 11.1a.

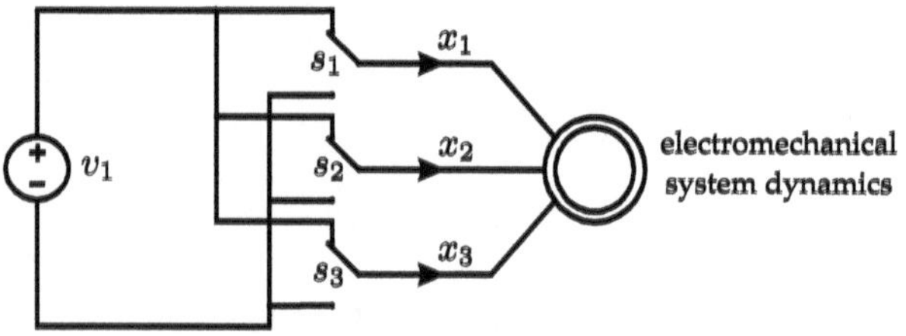

(a) Three-phase drive inverter, dynamics belong to class (11.3)

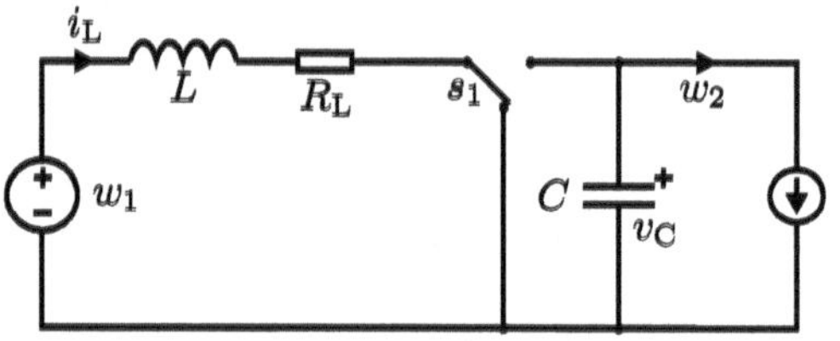

(b) DC–DC boost converter, dynamics belong to class (11.4)

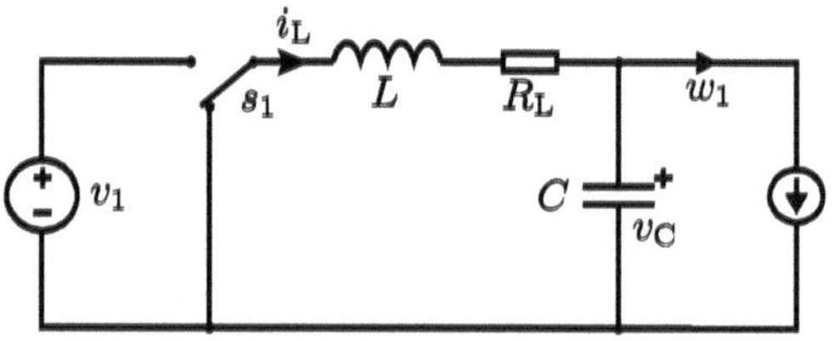

(c) DC–DC buck converter, dynamics belong to class (11.5)

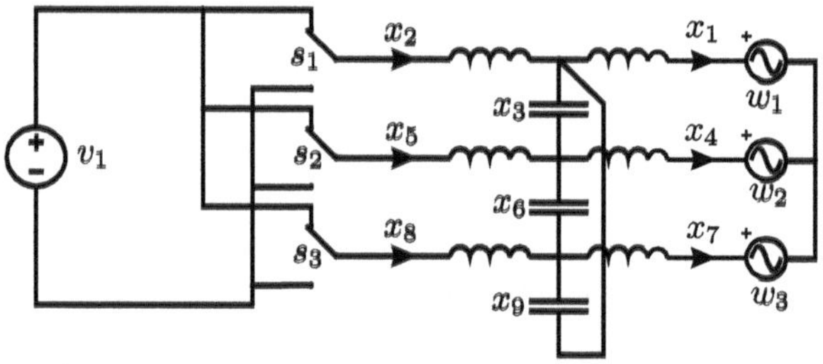

(d) Three-phase grid inverter, dynamics belong to class (11.5)

Fig. 11.1 Converter topologies and corresponding hybrid system classes

The switched dynamics (11.3a) and inequality constraints (11.3b) are often linear, which allows us to rewrite more conveniently the system dynamics (11.3a), (11.3b):

$$\dot{x}(t) = \sum_{i=1}^{N_b} \big(s_i(t)F_i x(t) + s_i(t)G_i v(t)\big) + G_w w(t), \tag{11.4a}$$

$$s_l(t) = \begin{cases} 1 & \text{when } H_{x,l}x(t) + H_{u,l}u(t) + H_{v,l}v(t) \geq 0, \\ 0 & \text{otherwise,} \end{cases} \quad l \in \{1, \ldots, N_b\}. \tag{11.4b}$$

The model (11.4a), (11.4b) may represent the dynamics of a boost converter such as shown in Fig. 11.1b.

The dynamics are often hybrid only with respect to the system inputs, and the autonomous system dynamics are often linear, which allows a further simplification of the hybrid system:

$$\dot{x}(t) = F_0 x(t) + \sum_{l=1}^{N_b} s_l(t) G_l v(t) + G_w w(t), \tag{11.5a}$$

$$s_l(t) = \begin{cases} 1 & \text{when } H_{x,l}x(t) + H_{u,l}u(t) + H_{v,l}v(t) \geq 0, \\ 0 & \text{otherwise,} \end{cases} \quad l \in \{1, \ldots, N_b\}. \tag{11.5b}$$

The model (11.5a), (11.5b) may represent the dynamics of a buck converter or a grid inverter such as shown in Fig. 11.1c and Fig. 11.1d, respectively.

In general, it is difficult to deal directly with the types of hybrid dynamics described above in the MPC framework. Several discrete-time approaches are considered to formulate a tractable control problem

1. *MPC formulation based on PWM system averaged model:* the semiconductor devices are switched indirectly through a PWM at continuous time instants. The controller manipulates the PWM duty cycles that are bounded continuous variables. A discrete-time averaged model accounting for the evolution of the system state as a function of the duty cycles is then employed to formulate the MPC problem. This approach will be discussed in Sect. 11.4.
2. *Enumerative integer MPC:* the semiconductor devices are switched directly at discrete-time instants. The controller manipulates boolean variables that correspond to the ON or OFF state of the switches during the sampling period. A discrete-time model is employed to predict the evolution of the system state as a function of these boolean variables. This approach will be discussed in Sect. 11.5.
3. *MPC formulation based on PWM system intersampling behaviour:* The semiconductor devices are switched through a PWM at continuous time instants. The approach employs a hybrid model accounting for the switching, which allows the formulation of a more complex control objectives as will be discussed in Sect. 11.6.

11.4 MPC Formulations Based on the PWM System Averaged Behaviour

In this section, pulse-width modulation is employed to generate the binary switch functions. Each binary switch function is then characterised by a switching period T_p, which is generally constant, and by a duty cycle $d_i(t)$, which is the manipulated variable. The simplest way to obtain a control model of a PWM system suitable for MPC synthesis is to employ the so-called state space averaged model [14] to approximate the switched converter dynamics with a continuous system describing the average value of the rippling state over a switching period.

In the averaging procedure, the switch functions s_i are replaced by continuous variables $d_i(t)$ satisfying

$$d_{i,k} = \frac{1}{T_s} \int_{k\,T_s}^{(k+1)\,T_s} s_i(\tau)\,\mathrm{d}\tau \quad \forall i \in \{1, \ldots, N_b\}, \tag{11.6a}$$

$$d_i(t) = d_{i,k} \quad \forall t \in \big[k\,T_s, (k+1)\,T_s\big[\;\forall i \in \{1, \ldots, N_b\}, \tag{11.6b}$$

$$T_p = N_s T_s, \quad N_s \in \{1, 2\}, \tag{11.6c}$$

where the switching period T_p has been assumed here to be the same for all switches for simplicity.

It has to be noted that most systems with several switching frequencies, such as for instance converters featuring switches of different voltage ratings, can be modelled using the highest switching frequency and applying suitable input equality constraints on the switches that are slower. The constraints (11.6b) express the discrete time nature of the duty cycles. As the switching and the sampling generally need to be operated in frequency and phase synchronism to obtain a high dynamic performance, (11.6c) states the ratio between the switching and sampling periods.

Using (11.6a)–(11.6c), the dynamics (11.4a), (11.4b) are approximated by

$$\dot{x}(t) = \sum_{i=1}^{N_b} \big(d_i(t) F_i x(t) + d_i(t) G_i v(t)\big) + G_w w(t), \quad d_i(t) \in [0, 1]. \tag{11.7}$$

The duty cycle constraints follows from the averaging procedure and from the binary constraints.

In many applications, the averaged model is sufficiently accurate to design high-performance controllers. Furthermore, it is often relatively simple and allows for the application of standard tools for control synthesis. In the sections below, we formulate and analyse several MPC formulations for switched-mode power converters employing averaged models.

11.4.1 Linearly Constrained MPC

We consider the class of systems (11.5a), (11.5b) that feature dynamics linear in the state. As PWM is employed to generate the switch functions, the model averaging procedure results in continuous dynamics that are also linear in the controlled inputs $d_i(t)$. The resulting averaged model still presents some difficulties as the duty cycles associated to the controlled input are constrained by nature and as the averaged dynamics may still be hybrid due to the natural commutation constraints that follow from (11.5b). The duty cycle constraints can readily be incorporated in the constrained optimal control problem.

The discontinuous conduction can be dealt with in the hybrid control framework presented in Sects. 11.4.2 and 11.6; however it is neglected in this section, which focuses on linearly constrained MPC problems. Under this assumption, i.e., $H_{x,i} = 0$, $H_{v,i} = 0$, $i \in \{1, \dots, N_b\}$, (11.7) reduces to

$$\dot{x}(t) = F_0 x(t) + \sum_{i=1}^{N_b} d_i(t) G_i v(t) + G_w w(t), \quad d_i(t) \in [0, 1]. \tag{11.8}$$

The averaged model is in this case equivalent to relaxing the binary constraints on the switch functions. Assuming that the vectors of exogenous inputs $v(t) = v_l$ and $w(t) = w_l$ are constant over the switching period, the dynamics can be represented exactly between successive sampling instants as

$$x_{l+1} = A_0 x_l + \sum_{i=1}^{N_b} d_{i,l} B_i v_l + B_w w_l, \quad d_{i,l} \in [0, 1], \tag{11.9a}$$

$$A_0 = e^{F_0 T_s}, \qquad B_i = \int_0^{T_s} e^{F_0(\tau)} \, \mathrm{d}\tau G_i, \qquad B_w = \int_0^{T_s} e^{F_0(\tau)} \, \mathrm{d}\tau G_w. \tag{11.9b}$$

Due to the bilinear relation between the exogenous input v_l and the manipulated variables $d_{i,l}$, the averaged model (11.9a), (11.9b) is a linear parameter-varying system with constrained control input. In order to solve parametrically this problem as will be discussed later in Sect. 11.7, it would be preferable to derive a linear time-invariant (LTI) control model. The input constraint can be rewritten as

$$H_d d_l + h_d \leq 0. \tag{11.10}$$

This constraint expresses that the exogenous input can be modulated and that the corresponding effect on the system depends on the magnitude of the exogenous input. In the averaged model sense, it does not make any difference to increase the exogenous input or to increase by the same factor the duty cycle. Exploiting this, the exogenous inputs v_l and the duty cycles d_l can often be combined by defining a new auxiliary input variable, which allows rewriting the dynamics and the input constraints as an LTI dynamic system

$$x_{l+1} = A_0 x_l + B_0 u_l + B_w w_l, \tag{11.11a}$$

$$H_u u_l + H_v v_l \leq 0. \tag{11.11b}$$

The exogenous input v_l does not appear anymore in the dynamics, instead it appears in the constraint (11.11b). A restriction is that v_l must be unipolar. This restriction is weak since most converter stages operate only in one or two quadrants with respect to the modulated exogenous inputs. Enforcing the inequality constraints (11.11b) therefore inherently enforces the inequality constraints (11.10). See Sect. 11.8.1 for an example of the application of this principle.

We consider the reference tracking problem applied to the linear constrained control model (11.11a), (11.11b); the dynamic system output and the tracking error are defined as

$$y_l = C x_l, \tag{11.12a}$$

$$e_l = r_l - y_l. \tag{11.12b}$$

At the equilibrium, the reference r_l is reached if the following condition is satisfied:

$$\begin{pmatrix} I - A_0 & -B_0 \\ C & 0 \end{pmatrix} \begin{pmatrix} x_{t,l} \\ u_{t,l} \end{pmatrix} = \begin{pmatrix} B_w & 0 \\ 0 & I \end{pmatrix} \begin{pmatrix} w_l \\ r_l \end{pmatrix}. \tag{11.12c}$$

The MPC problem is then formulated as minimising the deviation to the target (x_t, u_t):

$$\min \sum_{l=k}^{k+N-1} \left|Q(x_{l+1} - x_t)\right|_p + \left|R(u_l - u_t)\right|_p \tag{11.13}$$

subject to: the model equality and inequality constraints (11.11a), (11.11b) and the reference tracking constraints (11.12a)–(11.12c). The employed norm $p \in \{1, 2, \infty\}$ is selected both according to the objective and to the desired properties of the resulting optimisation problem. State constraints, such as maximum admissible value for the state, can be incorporated in the optimisation problem.

The MPC problem is then stated as

$$\min \sum_{l=k}^{k+N-1} \left|Q(x_{l+1} - x_{t,l})\right|_p + \left|R(u_l - u_{t,l})\right|_p \tag{11.14a}$$

subject to: state model equality constraints

$$x_{l+1} = A_0 x_l + B_0 u_l + B_w w_l, \tag{11.14b}$$

parameter equality constraints

$$v_l = v_{l-1}, \qquad w_l = w_{l-1}, \qquad r_l = r_{l-1}$$
$$\forall l \in [k+1, \dots, k+N], \tag{11.14c}$$

tracking equality constraints

$$\begin{pmatrix} I - A_0 & -B_0 \\ C & 0 \end{pmatrix} \begin{pmatrix} x_{t,l} \\ u_{t,l} \end{pmatrix} = \begin{pmatrix} B_w & 0 \\ 0 & I \end{pmatrix} \begin{pmatrix} w_l \\ r_l \end{pmatrix}, \tag{11.14d}$$

inequality constraints

$$H_x x_l + H_u u_l + H_v v_l \le 0. \tag{11.14e}$$

When the cost is quadratic, problem (11.14a)–(11.14e) is formulated as a *quadratic program* (QP); when it employs the one- or infinity-norm, it is formulated as a linear program (LP). Section 11.7 will discuss how to exploit the fact that these problems can be solved parametrically off-line for fast real-time implementation of the control law, Sect. 11.8.1 will develop an example of linearly constrained MPC.

11.4.2 MPC of Systems with Nonlinear Averaged Models

When the averaged approximation of (11.3a), (11.3b) is not linear, there is in general no systematic way of solving the corresponding control problem, and thus further simplification is needed. The two main approaches for obtaining an approximate solution are:

1. Approximating the nonlinear dynamics with a *piecewise-affine* (PWA) *model*.
2. Solving the non-convex optimisation problem for a finite number of initial conditions and interpolating to obtain a control law.

In the present section, we consider the averaged model (11.7), which is bilinear since the control input multiplies the state vector. This type of system does not fit directly into the standard MPC control framework.

Using PWA approximations [17], it is possible to approximate the averaged model (11.7) with arbitrarily high accuracy. We seek for a model of the form

$$x_{k+1} = A_i x_k + B_i d_k + f_i$$
$$\text{if } \begin{pmatrix} x_k \\ d_k \end{pmatrix} \in \mathscr{P}_i, \tag{11.15}$$

where $\mathscr{P}_i$ is a polyhedral partition of the state and control input space.

There are various ways of deriving a PWA approximation (11.15) of a nonlinear dynamic system. For example, so-called hinging hyperplanes can be applied to approximate any nonlinear vector field. In the present section however, we focus on bilinear systems where the nonlinearities are all of the form dx_i where the control input multiplies a state variable. Given this special structure of the nonlinearities, we can also impose a special structure on the approximation.

We seek for PWA approximations of the terms dx_i of the form

$$dx_i \approx a_{i,j}^\top \begin{pmatrix} x_i \\ d \end{pmatrix} + b_{i,j} \quad \text{if } \begin{pmatrix} x_i \\ d \end{pmatrix} \in \mathscr{P}_{i,j},$$

where $\mathscr{P}_{i,j}$ is a polyhedral partition of the state and control input space. We want to find the PWA function $a_{i,j}$, $b_{i,j}$, $\mathscr{P}_{i,j}$ which minimises some error measure while satisfying two properties: Firstly, we want the PWA approximation to be continuous in state and input so that

$$a_{i,j}^\top \begin{pmatrix} x_i \\ d \end{pmatrix} + b_{i,j} = a_{i,k}^\top \begin{pmatrix} x_i \\ d \end{pmatrix} + b_{i,k} \quad \forall \begin{pmatrix} x_i \\ d \end{pmatrix} \in \mathscr{P}_{i,j} \cap \mathscr{P}_{i,k}.$$

Continuity is important in order to have satisfactory performance of the closed-loop system as was discussed in [1]. Secondly, we want to impose boundary conditions on the PWA approximation. When the duty cycle saturates at either zero or one, certain states (typically inductor currents) may grow very rapidly, and it is therefore important that the control model gives an accurate description of this situation. This can be ensured by imposing that the PWA approximation is exact for saturated values of the duty cycle. For example, we may impose that

$$a_{i,j}^\top \begin{pmatrix} x_i \\ 1 \end{pmatrix} + b_{i,j} = 1 \quad \forall \begin{pmatrix} x_i \\ 1 \end{pmatrix} \in \Omega_{i,j} \ \forall j$$

to make the approximation exact for $d = 1$.

To obtain a PWA approximation satisfying the conditions on continuity and boundary values, one can choose the polyhedra $\mathscr{P}_{i,j}$ as triangular regions as shown in Fig. 11.2. Here, the scalar function dx is approximated with a four-piece PWA function defined on disjoint triangular regions with a common tangent point. We note that we can impose the PWA approximation to be exact at the tangent point, and that this condition, together with boundary conditions stating what values the

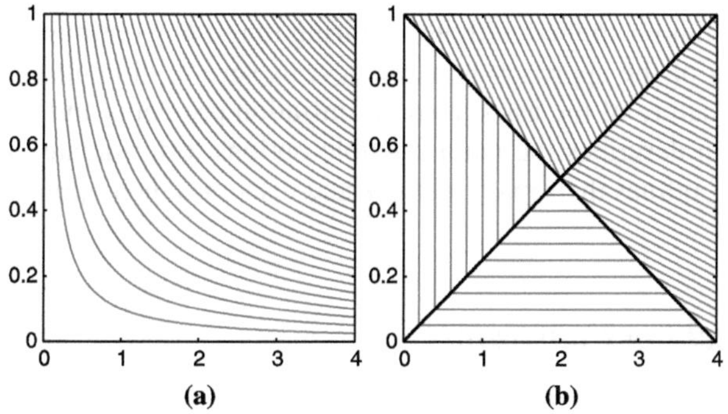

Fig. 11.2 (**a**) Level sets of the product $d\,x$. (**b**) Level sets of the PWA function approximating the product $d\,x$. The *black lines* show the partition of a rectangular region into four triangular regions

PWA approximation should take at the boundary of the box, determines the terms $a_{i,j}$, $b_{i,j}$ of the PWA function uniquely.

Formulating the condition that the state and control input lie in a certain polytope as linear inequalities

$$\begin{pmatrix} x_k \\ d_k \end{pmatrix} \in \mathscr{P}_i \Leftrightarrow H_{x,i} x_k + H_{d,i} d_k \leq h_i,$$

the optimisation problem is written

$$\min \quad \sum_{l=k}^{k+N-1} \left|Q(x_{l+1} - x_t)\right|_p + \left|R(d_l - d_t)\right|_p \tag{11.16a}$$

$$\text{subject to:} \quad x_{l+1} = A_i x_l + B_i d_l + f_i \quad \text{when } H_{x,i} x_l + H_{d,i} d_l \leq h_i. \tag{11.16b}$$

The optimisation problem will be a *mixed-integer linear problem* (MILP) if the one- or infinity- norm is used. It will be a *mixed-integer quadratic problem* (MIQP) if the two-norm is used. Section 11.7 will discuss how to exploit the fact that MILP can be solved parametrically off-line for fast real-time implementation of the control law, Sect. 11.8.2 will develop an example of hybrid MPC for a boost converter.

11.5 Enumerative Integer MPC

Enumerative integer MPC, simply often referred to as predictive control in the literature about the control of power electronics systems [7, 15], is a relatively intuitive approach that can be applied without requiring to use any optimisation software. A closer analysis of the approach however reveals several difficulties that are linked to the integer nature of the associated optimisation problems and to their tractability. The section describes the main possible options to formulate the optimisation problems and outlines their main features and discusses their applicability.

11.5.1 Switching Constraints

An intuitive way of formulating the optimal control problem associated to the dynamics of the switched systems (11.3a)–(11.5b) consists in directly manipulating the switches as boolean variables. This seems simple because the number of control possibilities is finite; however difficulties arise because the switches can in principle be manipulated at any time due to the continuous nature of the switched dynamics. Indeed, determining optimally the continuous time instants at which the system should switch between continuous dynamics is generally an intractable problem.

By allowing transitions of the switches only at discrete-time instants a tractable problem can be obtained. The corresponding constraints are written as

$$s_i(kT_q + \tau) - s_i(kT_q) = 0 \quad \forall \tau \in [0, T_q],\ i \in \{1, \dots, N_b\}, \tag{11.17}$$

where N_b is the number of switches, and T_q is the switching time quantization step.

The selection of a large value for T_q limits the accuracy of the switching pattern that can be generated, while a small value will require a high computational effort.

11.5.2 Discrete-Time Model of the Dynamics

Applying the switching constraints (11.17), the discrete-time dynamics can be written in a relatively straightforward way as depending only on the selected switching combination.

If the switched dynamics are linear between the switching instants as described in (11.4a), (11.4b), then the discrete-time dynamics can be computed exactly for each of the dynamics associated to a switching combination by applying (11.9b)

$$x_{l+1} = A_j x_l + B_j v_l + B_{w,j} w_l, \qquad \Phi_j = \int_0^{T_q} e^{\sum_{i=1}^{N_b} F_i s_i(\tau)} \mathrm{d}\tau, \qquad j = \sum_{i=1}^{N_b} 2^i s_i,$$

$$A_j = e^{\sum_{i=1}^{N_b} F_i s_i T_q}, \; B_l = \Phi_j \sum_{i=1}^{N_b} G_i s_i, \; B_{w,l} = \Phi_j G_w, \tag{11.18}$$

where v_l and w_l are assumed to be constant over the time step T_q. Different dynamics are obtained for each possible binary combination.

If the continuous-time dynamics are nonlinear as in (11.3a), (11.3b), the dynamics of the system can still be obtained relatively accurately using forward Euler approximation, provided that the time step T_q is sufficiently small,

$$x_{k+1} = x_k + \sum_{i=1}^{N_b} f_i(x_l, v_l, w_l) T_q s_i. \tag{11.19}$$

11.5.3 Constrained Optimal Control Problem formulation

The control objective is written as

$$\min \quad \sum_{l=k}^{k+N-1} J_l(x_{l+1}, s_l), \tag{11.20a}$$

$$\text{subject to:} \quad \text{dynamics (11.18) or (11.19)}, \tag{11.20b}$$

$$h_l(x_l) \geq 0 \quad \forall l \in \{k, \ldots, k+N\}, \tag{11.20c}$$

where x_k is the initial state, and $s_l = [s_{1,l}, \ldots, s_{N_b,l}]^\top$ is the vector of manipulated binary input variables at step l. The stage cost $J_l(x_{l+1}, s_l)$ does not need to be a convex function, and therefore the dynamics do not need to be linear.

At each time step, there are 2^{N_b} possible switched dynamics. The number of possible binary input values for the vector of manipulated variables $S_k = [s_k^\top, \ldots, s_{k+N-1}^\top]^\top$ grows exponentially with the number of switch variables and the horizon. The key for computing the solution to (11.20a) in real time is to keep the horizon sufficiently small such that the cost can be evaluated and compared for the $2^{N_b N}$ possible input sequences. As a consequence, the time step T_q can only be very coarse to predict the evolution of the state over a sufficiently long horizon to stabilise the system. For complex converters with many switch combinations, the complexity of the optimisation problem can be reduced by constraining the manipulated binary input to belong to a subset of the possible switching combinations based on the a priori knowledge of the steady-state behaviour of the plant. During transient, this constraint may obviously eliminate switching sequences that would allow reducing the cost criterion.

Due to the binary nature of the manipulated input, the feasibility of the hard constraints (11.20c) may not be guaranteed. It is therefore preferable to replace these hard constraints by soft constraints in the cost to minimise constraint violation.

A further difficulty in the approach is to keep the converter switching frequency bounded. The simple approach that is often employed to this end consists in selecting a time step T_q sufficiently large to avoid an unacceptably high switching frequency. The drawback of this approach is that the output cannot be controlled accurately due to the coarse definition of the converter switching pattern. An alternative approach consists in penalising the transitions of the switches in the cost function. The optimisation problem is reformulated as

$$\min \quad \sum_{l=k}^{k+N-1} J_l(x_{l+1}, s_l) + \left|Q_p(s_l - s_{l-1})\right|_1, \tag{11.21a}$$

$$\text{subject to:} \quad \text{dynamics (11.18) or (11.19).} \tag{11.21b}$$

The first difficulty is related to the tuning of the matrix Q_p that penalises the transitions of the input. The steady-state behaviour is typically highly sensitive to changes in Q_p, and the change of behaviour is often unpredictable. The second difficulty is computational since a sufficiently large horizon N is required in order to capture the trade-off between the primary control objective $J_l(x_{l+1}, s_l)$ and the switching constraints.

11.5.4 Sliding-Mode Enumerative Approach

An MPC approach derived from sliding mode can be employed to circumvent the switching frequency issue [10]. Instead of defining a cost function to be minimised, a target set in which the system must be confined while doing as few transitions as possible is defined as a convex set. The resulting optimisation problem is written as

$$\min \quad \sum_{l=k}^{k+N-1} \left|Q_p(s_l - s_{l-1})\right|_1, \tag{11.22a}$$

$$\text{subject to:} \quad H(x_{k+l}, s_{k+l-1}) \le 0. \tag{11.22b}$$

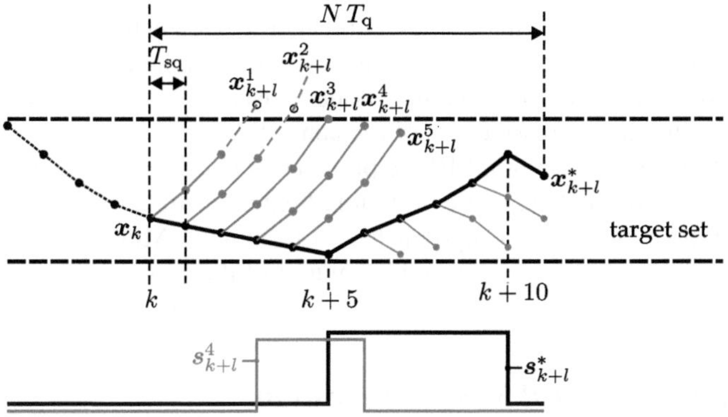

Fig. 11.3 Sliding-mode enumerative MPC approach example

In this optimisation problem, the objective is to minimise the transitions of the switches subject to: some tracking quality constraint. The approach is illustrated by a simple example in Fig. 11.3. Starting from the state $x_k \in \mathbb{R}^1$, the trajectory that is generated by the sequence s^j_{k+l} is denoted x^j_{k+l}. The thick horizontal *dash lines* represent the limits of the set that fulfils (11.22b). Many trajectories such as x^1_{k+l} and x^2_{k+l} are eliminated because they exit the target window and thus do not fulfil (11.22b). Several trajectories that are within the target window such as x^3_{k+l} and x^4_{k+l} exhibit the same number of transitions. It is however likely that selecting x^3_{k+l} will achieve a higher number of transitions over time since the switch makes a transition early. To avoid such issue and to obtain higher costs for, the penalty of the transitions decreases over the horizon, and therefore $Q_{p,l-k}$ depends on the horizon index $l-k$. The optimal solution for the shown example is the trajectory x^*_{k+l}.

11.6 MPC Formulations Based on PWM System Intersampling Behaviour

Due to the integer nature of the manipulated variables and the time quantization, the enumerative integer MPC approaches of Sect. 11.5 generate pulse patterns of limited accuracy. The quantization is severe as short prediction horizons are imposed by tractability issues.

The PWM-based and averaged-model-based MPC approaches of Sect. 11.4 allow us to synthesise control laws that provide satisfactory dynamic performance and constraint satisfaction in most cases. The switching ripple that is inherent in the state of switched mode power converters however has an adverse effect on the closed-loop performance. This ripple can be eliminated from the feedback measurements by filtering, but as this dramatically cuts the achievable closed-loop bandwidth, synchronous sampling is preferred when a high dynamic performance is sought. The ripple signal present in the continuous-time measurements is mostly present at the switching frequency and its harmonics, but it also appears in the baseband due to the sampling aliasing. This causes low-frequency distortion and may even destabilise the system in some cases. The stability issue is usually dealt with satisfactorily by slightly reducing the aggressiveness of the controller.

This section derives a hybrid MPC approach that allows us to circumvent the limitations of averaged models, while providing higher accuracy and better tractability than enumerative integer MPC approaches. It formulates MPC problems with control objectives that shape the switched behaviour of the system.

11.6.1 Hybrid Control Model Derivation

In the first step, a *hybrid model* that predicts the intersampling behaviour of the state is derived. Consider the class of systems (11.5a), (11.5b). We partition the sampling interval in N_q segments indexed by $m \in \{0, \dots, N_q - 1\}$ and model the evolution of the state over each of these time intervals. Several cases depending on the binary decision variables must be considered to establish the model. When all binaries are set to zero, the system is autonomous, and the evolution of the state during one time interval $T_q = T_s / N_q$ is

$$x_{l,m+1}(s = 0) = A_q x_{l,m}, \tag{11.23a}$$

where the double index notation has been introduced to distinguish the state predicted at the sampling instants x_l from the state predicted at and within the sampling instants $x_{l,m}$. The relations $x_l = x_{l,0} = x_{l-1,N_q}$ hold. The argument $s = 0$ states the situation where all the switches are set to 0. The transition matrix A_q determines the evolution of the state over the time interval T_q. If a binary variable value is set to one during the whole time interval T_q, the effect on the dynamics is determined by the transition matrix $B_{q,l}$:

$$x_{l,m+1}(s|s_i = 1) = x_{l,m+1}(s|s_i = 0) + B_{q,i} v_l, \tag{11.23b}$$

where A_q and $B_{q,i}$ are obtained applying (11.8) with T_q as time interval. If a binary variable makes a transition during the time interval T_q, the effect on the dynamics can be approximated using an averaged model as

$$x_{l,m+1}(s|d_{i,l,m}) = x_{l,m+1}(s|s_i = 0) + d_{i,l,m} B_{q,i} v_l, \tag{11.23c}$$

where $d_{i,l,m}$ is the proportion of the time—the duty cycle—during which the binary s_i is set to one over the time interval T_q. The duty cycles $d_{i,l,m}$ over an interval of duration T_q are auxiliary variables that are only used to approximate the PWM signals. They must be distinguished from the duty cycles $d_{i,l}$ that are really applied over the sampling period T_s. The relation between these two duty cycles is

$$d_{i,l} = \frac{1}{N_q} \sum_{m=0}^{N_q-1} d_{i,l,m}. \tag{11.24}$$

Based on (11.23a)–(11.23c), the hybrid dynamics can be written compactly:

$$x_{l,m+1} = A_q x_{l,m} + \sum_{i=1}^{N_b} d_{i,l,m} B_{q,i} v_l, \quad m \in \{0, \dots, N_q - 1\}, \tag{11.25a}$$

$$d_{i,l,m} = \begin{cases} 0 & \text{when } n_l\, m \le n_l\, m_{l,0}, \\ 1 & \text{when } n_l\, m \ge n_l\,(m_{l,0}+1), \quad n_l \in \{0,1\}, \\ N_q\, d_{l,i} - m_{l,0} & \text{otherwise}, \end{cases} \tag{11.25b}$$

$$m_{l,0} = \inf\{d : d \le N_b\big(n_l\, d_{i,l} + (1-n_l)(1-d_{i,l})\big),\ d \in \mathbb{N}\}, \tag{11.25c}$$

$$x_{l,0} = x_l, \qquad x_{l+1} = x_{l,N_q}. \tag{11.25d}$$

Equation (11.25a) is linear in the state and in the auxiliary variables $d_{i,l,m}$. The difficulty comes from the switching logic described in (11.25b) and (11.25c). In (11.25b), the integer variable $m_{l,0}$ indicates in which time interval the associated switch signal s_l makes a transition. $m_{l,0}$ depends on the partition of the duty cycles defined in (11.25c). The binary variable n_l specifies if the associated pulses are leading or trailing or, in other words, if the transition is from 1 to 0 or vice versa. This allows reproducing many PWM scheme behaviours. The condition (11.25c) can be translated into linear inequalities involving binary variables depending on the dynamics of n_l.

In direct current–direct current (DC–DC) converters, leading edge PWM is simply translated by $n_l \equiv 0$. In two-level direct current–alternate current (DC–AC) converters, the most commonly employed PWM scheme translates as $n_l = 1$ when l is even and $n_l = 0$ when l is odd. The vast majority of PWM schemes can be described employing (11.25a)–(11.25d), with a suitable definition of n_l.

Figure 11.4 compares the evolution of the state as predicted by an averaged model (*left*) with the state as predicted by a model of the form (11.25a)–(11.25d) (*right*). The *solid thin lines* represent the output of the real PWM system. On both sides, the disks show the discrete-time system output predicted at the sampling instants. On the *left*, the *solid thick black lines* show the large prediction error of the averaged model, while the *circles* on the *right* show the accuracy of the additional intersamples predicted by the hybrid model (11.25a)–(11.25d).

11.6.2 MPC Problem Formulation

The hybrid model (11.25a)–(11.25d) can been employed to formulate MPC problems. Since the state is only measured or estimated at the sampling instant designated by $(l, 0)$, the control objective can be formulated only based on the predicted

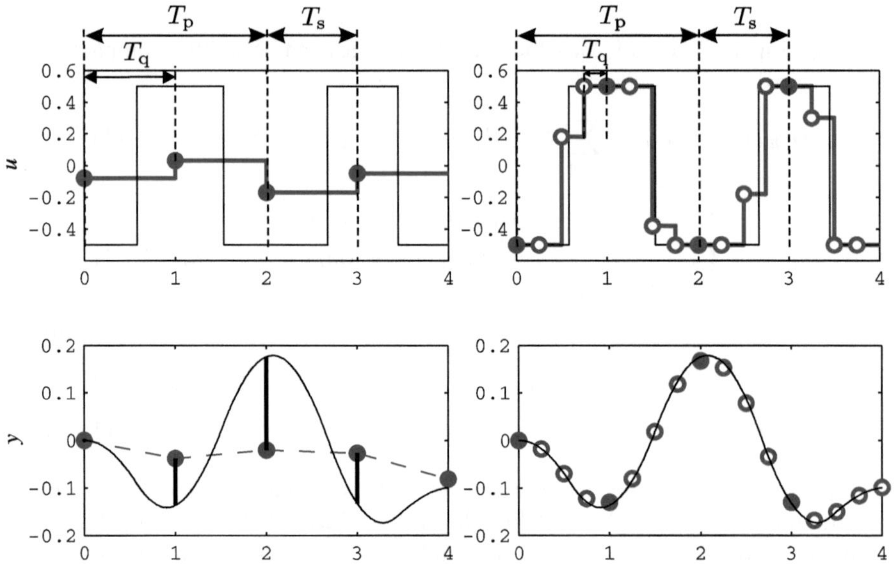

Fig. 11.4 Comparison of PWA model with four partitions and real hybrid system. *Solid thin lines*: real input and output of the hybrid system; *solid thick lines*: input of the PWA model; *solid disks*: output of the PWA model at sampling instants; *circles*: output of the PWA model between sampling instants

value of the state at these instants:

$$\min \quad \sum_{l=k}^{N} \left|Q(x_{t,l,0} - x_{l,0})\right|_p + \left|R(d_{t,l} - d_l)\right|_p + |S\Delta d_l|_p \tag{11.26a}$$

$$\text{subject to:} \quad (11.25\text{a})\text{–}(11.25\text{d}) \tag{11.26b}$$

$$\Delta d_l = d_l - d_{l-1} \quad \forall l \in \{k+1, \dots, k+N-1\}, \qquad \Delta d_k = d_k - d_{k-1}^*. \tag{11.26c}$$

The difficulty is to produce the target $x_{\mathrm{t},l,0}$ that must feature the ripple component $x_{l,0}$ should have. Employing an averaged reference model, the predicted signal $x_{l,0}$ will be attracted to the wrong target, since the correct target should feature a ripple. The produced output will be distorted. The delta formulation, which consists in reformulating the problem in terms of the input variations $\Delta d_{l,0}$ and in penalising these variations in the cost function, may be employed when it is difficult to produce a target for the control input or to avoid chattering of the control input when employing the one- or infinity-norm.

An alternative approach consists in predicting a rippleless reference and in minimising the distance of the state to this reference at the sampling instants and between the sampling instants:

$$\min \quad \sum_{l=k}^{N}\sum_{m=1}^{N_q} \left|Q(r_{l,m} - x_{l,m})\right|_p + \left|R(d_{t,l} - d_l)\right|_p + |S\Delta d_l|_p \tag{11.27a}$$

$$\text{subject to:} \quad (11.25a)\text{–}(11.25d). \tag{11.27b}$$

In this case, an averaged reference can reasonably be employed since the real control objective is indeed to minimise the distance between the continuous-time state, which features a ripple approximated by our signal comprising intersamples and an averaged model reference which does not comprise any ripple. If the ripple is sufficiently accurately approximated, this powerful approach even allows us to minimise the distortion caused by the switching. The complexity of solving (11.27a), (11.27b) may be too high, and an alternative approach may be preferable.

A possible solution consists of minimising the distance between the averaged model reference and the averaged state computed from the predicted intersamples:

$$\min \quad \sum_{l=k}^{N} \left|Q(r_l - \bar{x}_l)\right|_p + \left|R(d_{t,l} - d_l)\right|_p + |S\Delta d_l|_p \tag{11.28a}$$

$$\text{subject to:} \quad (11.25a)\text{–}(11.25d) \text{ and} \tag{11.28b}$$

$$\bar{x}_l = \frac{1}{N_q}\sum_{i=0}^{N_q} x_{l,i}. \tag{11.28c}$$

Section 11.8.3 will present an application of these MPC formulations.

11.7 Explicit MPC

In the application of MPC, an optimisation problem is solved at every sampling instant to obtain the control input. Because solving the optimisation problem is often time consuming, MPC has usually been reserved for relatively slow and/or small systems. For certain classes of MPC problems, however, the optimal solution can be written as an explicit function of the initial state. For these problems, a considerable part of the computational effort can be moved off-line. In a first stage, the aforementioned explicit function is computed and stored in some suitable data structure. Computing the function may be computationally expensive, but since the computations are done off-line, computational complexity is less restrictive. The controller can then be applied on-line simply by evaluating the precomputed function. The evaluation is often much less computationally expensive than solving the optimisation problem. The approach of computing the optimal solution off-line and representing it as a function is referred to as explicit MPC, and it has opened up new areas of application for MPC. In the section below we outline the main ideas and principles of explicit MPC.

11.7.1 Parametric Programming: Problem Classes and Solution Structure

The MPC problem (11.14a)–(11.14e) depends on the initial condition $\hat{x}_k$. The initial condition is referred to as the parameter and is assumed to lie in some polytope $\mathscr{X}$ called the parameter set. The optimisation problem is called multi-parametric where the word "multi" refers to the fact that the parameter $\hat{x}_k$ may have dimension larger than one.

If in problem (11.14a)–(11.14e) the norm is defined using the one- or infinity-norm, i.e., $p \in \{1, \infty\}$, the problem can be reformulated as a *multi-parametric linear problem* (mp-LP). If the problem is defined using the two-norm, i.e., $p = 2$, the problem can be reformulated as a *multi-parametric quadratic problem* (mp-QP). As mentioned above, for certain classes of multi-parametric optimisation problems, the optimal solution can be written as an explicit function of the parameter. Two problem classes for which this holds are mp-LP and mp-QP.

It has been shown [5, 6] that the optimal solution of the mp-LP and mp-QP stemming from (11.14a)–(11.14e) is a continuous PWA function of the parameter $\hat{x}_k$. In other words, the optimal solution can be written as

$$u^*_{N,k}(\hat{x}_k) = K_i \hat{x}_k + k_i \quad \text{if } \hat{x}_k \in \mathscr{P}_i,\ i = 1, \ldots, N_r, \tag{11.29}$$

where

$$\mathscr{P}_i := \{x \in \mathbb{R}^n \mid n_i^\top x \le o_i\}, \quad i = 1, \ldots, N_r,$$

is a polyhedral partition of the parameter set $\mathscr{X}$, and N_r is the number of regions in the partition. From (11.29) it immediately follows that the first control input

$$u^*_k(\hat{x}_k) = \begin{pmatrix} I^m & 0^{m\times m} & \ldots & 0^{m\times m} \end{pmatrix} u^*_{N,k}(\hat{x}_k)$$

is also a continuous PWA function of the initial condition.

In [5, 6], the authors also suggest methods for computing the affine gains (K_i, k_i) and hyperplanes defined by their normal vector n_i and offset o_i defining the explicit solution of the mp-LP and mp-QP, respectively. For the solution of mp-QP, a more efficient algorithm was introduced in [18].

11.7.2 Implementation of the Explicit Solution

Because of the PWA structure, implementing the control law $u_k^*(\hat{x}_k)$ corresponds to executing the following three operations:

1. Measure or estimate the current state $\hat{x}_k$.
2. Determine which polytope $\mathscr{P}_l$ contains the state $\hat{x}_k$. (This is referred to as the point location problem.)
3. Multiply the state with the corresponding gains K_l, k_l.

Operations 1 and 3 above are easily implemented also for very large problems. However operation 2 (the *point location problem*) may be prohibitive. The most straightforward way of solving the point localisation problem is to go through the polytopes sequentially until the polytope containing the current state has been identified. For each polytope, the set of hyperplanes which define the polytope would have to be checked. In the worst case, all polytopes and all hyperplanes would have to be evaluated. In many applications, the number of polytopes is very large, and the "straightforward" approach to the point location problem becomes too computationally expensive. Having to store all polytopes also puts high requirements on memory capacity.

In [2] the authors suggest a method for evaluating PWA functions which does not require storage of the hyperplanes and thus reduces the memory requirements. By using the structure of the cost functions in LPs and QPs, the polyhedra can be represented by integer arrays rather than hyperplanes, and the point location problem can thus be solved by comparing integers. However, computation time can still be large since the method requires that the integer arrays representing the polyhedra be checked sequentially.

To reduce the computational burden, the structure of the polyhedral partition can be utilised. The fact that each hyperplane partitions the parameter set into two halves can be exploited to build a binary search tree where each node corresponds to one of the hyperplanes (see Fig. 11.5). At each node in the tree, a linear inequality is evaluated to determine if the current point lies above or below the corresponding hyperplane. The outcome of the evaluation decides if the traversal proceeds to the left or the right child-node, and the search will end in a leaf which corresponds to the polytope containing the current point. If the hyperplanes are checked in such an order to give a balanced search tree, the point location problem can be solved efficiently.

The algorithm in [6] for solving the mp-QP could also be used to generate the binary search tree for dealing with the point location problem. However, it is not

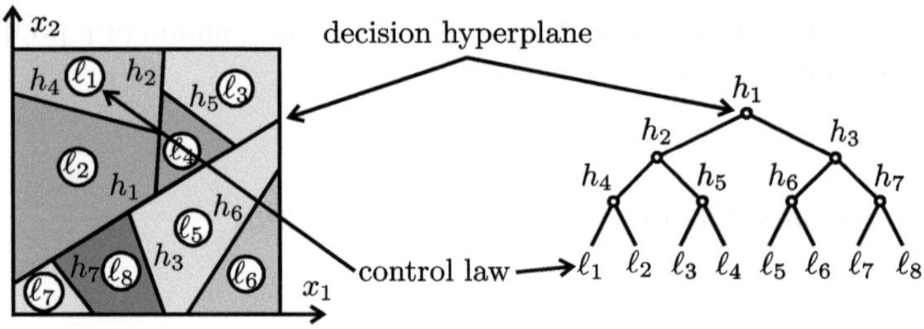

Fig. 11.5 Explicit solution and its binary search tree implementation

obvious that the resulting tree would be balanced. The problem of generating a well-balanced search tree with small depth was considered in [19]. With the resulting tree, computation time for the point location problem is logarithmic in the number of regions of the PWA function.

11.7.3 Explicit Solutions to Mixed-Integer Problems

When MPC problems are formulated using hybrid system models, the resulting optimisation problem will not belong to the mp-LP or mp-QP problem classes discussed above. Indeed, problem (11.16a), (11.16b) obtained by using PWA approximations of the system dynamics will generate *multi-parametric mixed-integer linear problems* (mp-MILP) or *multi-parametric mixed-integer quadratic problems* (mp-MIQP) depending on which norm is used to define the cost function.

The optimal solution of mp-MILP [3, 4, 8] has the same PWA structure as the mp-LP discussed above. This implies that the explicit solution can be computed and implemented using the same tool as for mp-LP. It should be noted however, that for mp-MILP, the complexity of the explicit solution (the number of regions of the partition of the parameter set) increases very rapidly with the number of integer variables. If the complexity is too high, it is no longer practical to implement the solution.

The optimal solution of mp-MIQP does *not* have the PWA structure of the problems discussed so far, and this makes the explicit solution more difficult to represent. Nevertheless, for mp-MIQP with a relatively small number of integer variables, the solution can be represented using the explicit solutions of mp-QPs. This holds because when the integer variables of the mp-MIQP are fixed, the problem structure reduces to an mp–QP. Thus, to represent the solution of an mp-MIQP, one could enumerate all combinations of integer variables and solve the corresponding mp-QPs explicitly. In the on-line application, all explicit mp-QP solutions would then be evaluated, and the cost would be compared in order to pick the optimiser of the mp-MIQP.

11.8 Application Examples

This section will illustrate the presented MPC concepts through several applications. Section 11.8.1 will investigate the application of linearly constrained explicit MPC, using a linear averaged model, and the application of enumerative integer MPC, using discretized switching instants, to the same buck converter system. Section 11.8.2 will investigate the application of hybrid MPC, using a nonlinear averaged model, to a boost converter. Section 11.8.3 will investigate how to shape the switched behaviour of a DC–AC converter using hybrid MPC.

11.8.1 Buck Converter Control

11.8.1.1 Linearly Constrained Explicit MPC

The synchronous step-down (buck) DC–DC converter depicted in Fig. 11.1c has dynamics that belong to the class of systems (11.5a), (11.5b). The averaged dynamics belong to the class of systems (11.8). They are linear in the state and bilinear in the manipulated and exogenous inputs.

Defining the state as $x := (v_C, i_L)^\top$, where v_C is the capacitor voltage, and i_L is the inductor current, the averaged dynamics are written as

$$\dot{x}(t) = F_0 x(t) + d_1(t) G_1 v_1(t) + G_w w_1(t), \tag{11.30}$$

where v_1 is the source voltage, w_1 is the load current, and

$$F_0 = \begin{pmatrix} 0 & \frac{1}{C} \\ -\frac{1}{L} & -\frac{R_L}{L} \end{pmatrix}, \qquad G_1 = \begin{pmatrix} -\frac{1}{C} \\ 0 \end{pmatrix}, \qquad G_w = \begin{pmatrix} 0 \\ \frac{1}{L} \end{pmatrix}.$$

This model, which in this form is linear parameter varying in the exogenous input, can be reformulated more conveniently as an LTI model

$$\begin{aligned} x_{l+1} &= A_0 x_l + B_1 u_{1,l} + B_{w,1} w_{1,l}, \quad 0 \le u_{1,l} \le v_{1,l}, \\ d_{1,l} &= \frac{u_{1,l}}{v_{1,l}}, \quad v_{1,l} > 0, \end{aligned} \tag{11.31}$$

where A_0, B_1 and $B_{w,1}$ are readily obtained from (11.9a), (11.9b).

The bilinearity has been eliminated by substituting the product of the manipulated duty cycle with the exogenous input by a new auxiliary input. At each sampling period, this substitution requires to compute the inverse of the supply voltage $v_{1,l}$ to obtain the duty cycle to be applied to the plant from the auxiliary input. The

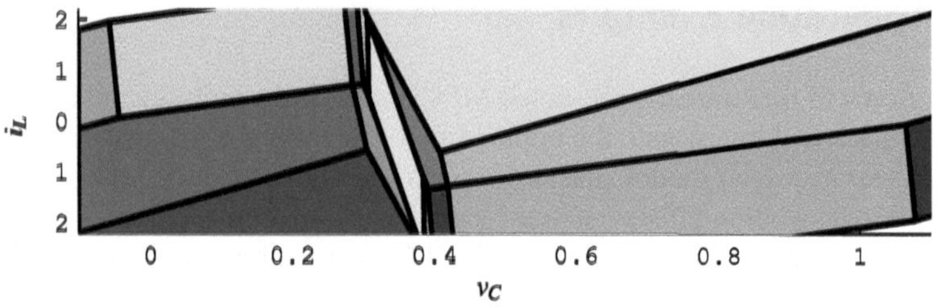

Fig. 11.6 Explicit MPC law for the buck converter system

supply voltage then only appears in the inequality constraints and not anymore in the equality constraints.

The primary control objective is that the converter output voltage v_C tracks a reference as quickly as possible despite disturbances, while respecting the maximum allowed coil current $i_{L,\max}$. The maximum admissible current constraint is a state constraint formulated as

$$-i_{L,\max} \le i_{L,l} \le i_{L,\max}. \tag{11.32}$$

The tracking MPC problem (11.14a)–(11.14e) is formulated with a quadratic cost. Compared to a formulation using one- or infinity-norms, the quadratic cost allows us to obtain a better closed-loop performance and a simpler parametric solution. The penalty matrices are designed iteratively based on plant simulation by tuning the weight on the plant output while keeping the other weights to 1. A few iterations quickly lead to $Q = \operatorname{diag}\left(8000\ 1\right)$, $R = 1$ and $N = 3$ for the targeted scenario.

The resulting constrained finite-horizon optimal control problem is solved parametrically, yielding an explicit solution with 19 regions in a six-dimensional space. A two-dimensional cut of the associated polyhedral partition of the state is represented in Fig. 11.6 for constant reference $v_{\text{ref}} = 0.5\,\text{V}$, maximum admissible current $i_{L,\max} = 2\,\text{A}$, supply voltage $v_1 = 1\,\text{V}$ and load current $w_1 = 0.45\,\text{A}$. Each polyhedral region is associated to an affine control law, the overall control law being PWA. All constraints are inactive in the region in light grey in the middle, where is the equilibrium $(0.33, 0.45)^\top$. At least one constraint is active in the other regions which are darker.

Simulation results are shown for this explicit MPC solution in Fig. 11.7 for a buck converter operating at 40 kHz. In the considered scenario a reference profile is tracked starting from the initial state $(0, 0)^\top$. The analysis of the results illustrates the activation of constraints. At start-up, the maximum input constraint is the first activated. It then becomes inactive as the maximum current constraint is activated. In the next applied abrupt transients at 1 ms and 1.5 ms, the maximum and minimum current constraints become active to ensure the fastest convergence to the reference. This constraint mechanism ensures a very fast dynamic operation of the system close to its limits. The time-varying maximum voltage constraint follows the supply voltage (see step at 2 ms), which allows one to fulfil the maximum duty cycle

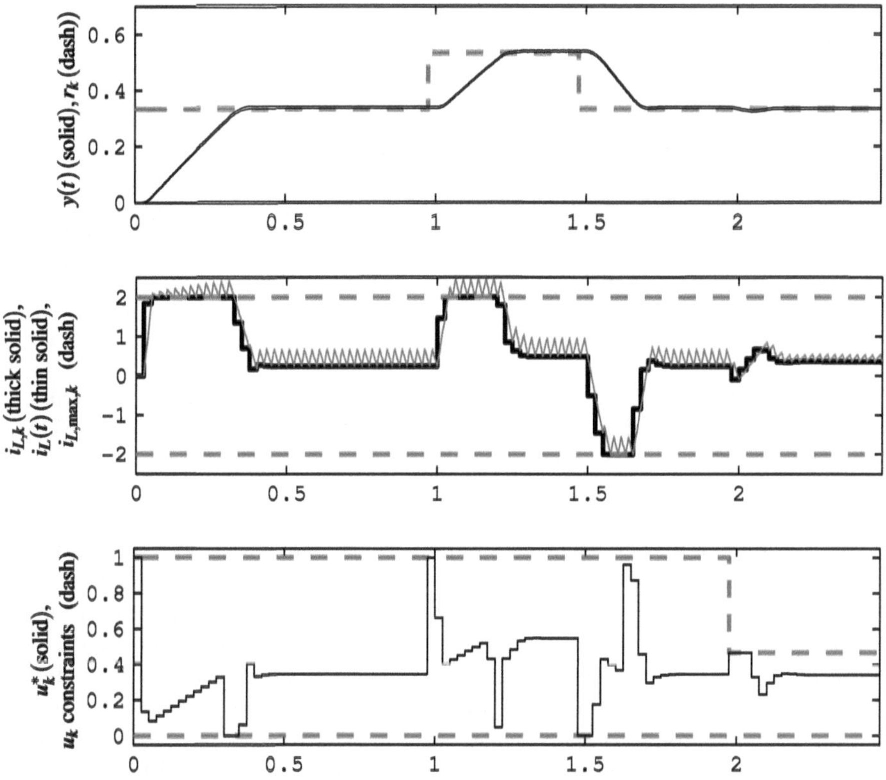

Fig. 11.7 Buck converter controlled by explicit MPC with $f_{sw} = 40\,\text{kHz}$ (time axis t in ms)

constraint all the time. The accuracy of the averaged control model is nearly independent of the operating conditions, resulting in an MPC effective for all operating points. Comparing the sampled current plot (*thick black solid curve*) and the continuous time current plot (*thin grey solid curve*), it however appears that although the sampled current does not violate the maximum admissible constraint, the actual continuous-time current does slightly violate the current constraint. This behaviour was expected as the averaged model does not predict the peak current. If the peak current is to be exactly limited, the MPC formulation can be slightly adjusted to eliminate this problem, or alternately a hybrid model as described in Sect. 11.6 can be employed.

11.8.1.2 Enumerative Integer MPC

We will now synthesise a control law using enumerative MPC, keeping the same control objectives for the same converter rated 40 kHz.

The control objective can be formulated by using (11.21a), (11.21b)

$$s_k^* = \min \sum_{l=k}^{k+N-1} |r_k - v_{C,l}|_2 + Q_p |s_{1,l} - s_{1,l-1}|_1$$
$$+ Q_{sc}\big[|i_{\mathrm{L,max}} - i_{L,l} + z_{\max,l}|_1 + |-i_{\mathrm{L,max}} - i_{L,l} - z_{\min,l}|_1\big] \tag{11.33a}$$

$$\text{subject to:} \quad x_{l+1} = A_0 x_l + s_{1,l} B_1 v_{1,k} + B_w w_{1,k}, \tag{11.33b}$$

$$s_{1,l} \in \{0, 1\}, \qquad z_{\max,l} \le 0, \qquad z_{\min,l} \le 0. \tag{11.33c}$$

The current hard constraints have been reformulated as soft constraints highly penalised by Q_{sc} and where the imposed switching frequency has been reformulated as a penalty Q_p on the switch control signal transitions.

The second step consists in determining the quantization time, the prediction horizon and the penalty weights. The quantization time should be a fraction of the targeted switching period in order to have a sufficiently good definition of the applied switching pattern. For simplicity, the same fast rate is employed for the sampling $T_s = T_q$. For the considered scenario, a few tests show that an input quantization time five times smaller than the rated switching period yields a good trade-off between computational complexity and performance: $T_q = T_p/5 = 5\ \mu\text{s}$. The matrices describing the system dynamics, A_0, B_1 and B_w, are computed for a sampling period of $T_s = T_q = 5\ \mu\text{s}$.

In the first simulation example illustrated in Fig. 11.8, the maximum admissible current violation is highly penalised with $Q_{sc} = 100$, and the switching transitions are not penalised, $Q_p = 0$. A satisfactory performance is obtained with a horizon of 5. With shorter horizons, the performance deteriorates, while longer horizons quickly lead to intractable problems due to the number of combinations to evaluate that grows exponentially with the horizon for a performance improving only very slowly.

The first analysis that focuses on the two upper plots in Fig. 11.8 reveals results similar to those obtained by the explicit MPC in Fig. 11.7: the current constraints are always satisfied, which ensures that the system is operated close to its dynamic performance limits. The slight overshoot is due to the limited length of the horizon. The peak current constraints are better enforced than for the explicit MPC since the switched nature of the system is taken into account.

The second part of the analysis focuses on the two bottom plots in Fig. 11.8. The fourth plot represents the switching frequency computed by measuring the time between two consecutive transitions of the switching signal from low to high. The third plot is equivalent to the optimal control input in Fig. 11.7: it is computed by measuring the averaged voltage applied to the converter filter. These two plots reveal two very important limitations of this control approach. The switching period varies, being constrained to a multiple of the quantization time. As a consequence,

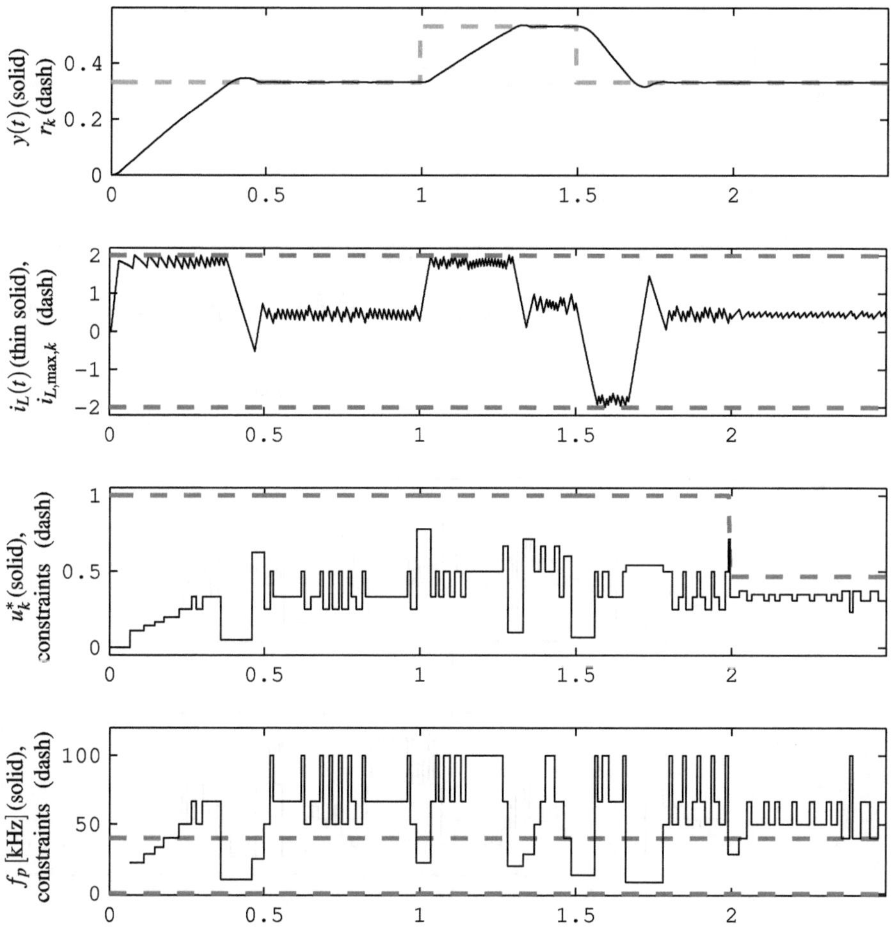

Fig. 11.8 Buck converter controlled by enumerative MPC without penalty on transitions (t in ms)

the averaged voltage applied to the converter filter is also quantized. This limits the accuracy of the control and causes the switching frequency to exceed the converter ratings.

In the second simulation example illustrated in Fig. 11.9, the switching transitions are penalised by setting $Q_p = 3 \times 10^{-6}$ in order to try keeping the converter switching losses below the maximum admissible value. The switching frequency is reduced in most places; however it is not possible to obtain a switching frequency respecting the constraints while keeping an acceptable ripple on the current and output voltage. The reason is that the switching frequency appears to be very sensitive to the operating point and to the penalty coefficient Q_p. For this reason, it is preferable to employ the sliding-mode enumerative integer MPC formulation of (11.22a), (11.22b) to minimise the switching frequency, while satisfying some performance constraints.

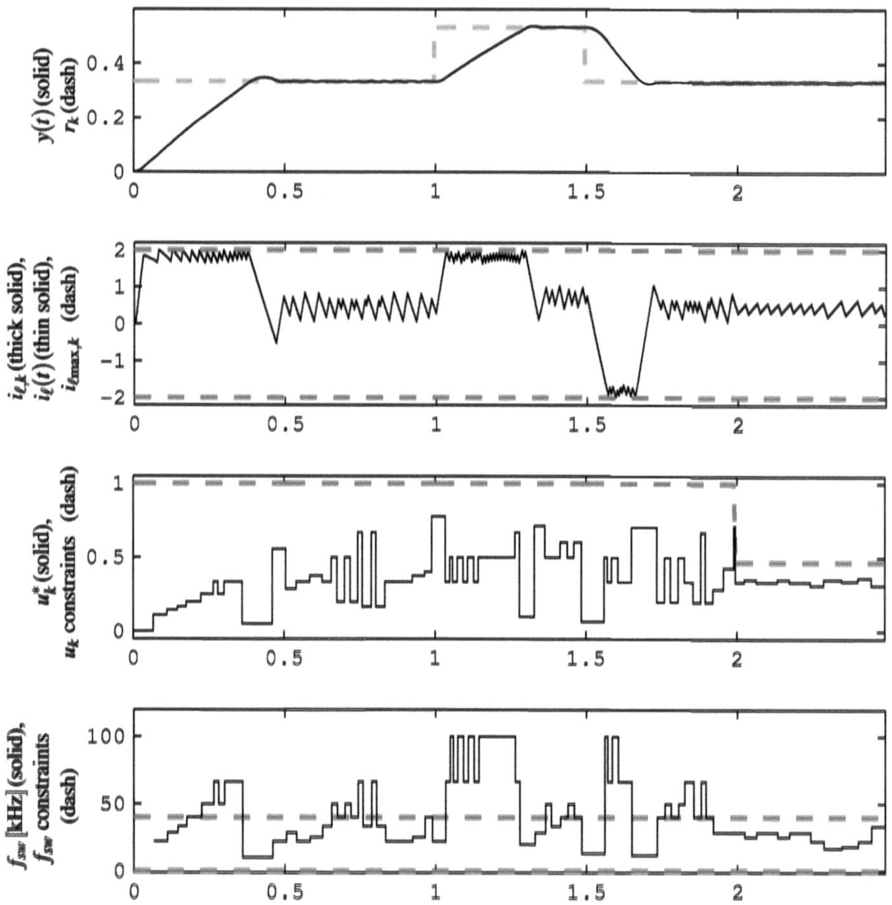

Fig. 11.9 Buck converter controlled by enumerative MPC with a penalty on transitions (t in ms)

11.8.2 Boost Converter Control

Consider the synchronous step-up (boost) DC–DC converter depicted in Fig. 11.1b. The averaged model of the system dynamics belongs to the class of systems (11.7).

Defining the state as $x := (v_C, i_L)^\top$, the averaged model can be written as

$$\dot{x}(t) = F_0 x(t) + d_1(t) F_1 x(t) + G_w w(t), \tag{11.34}$$

where the vector of exogenous inputs $w = (w_1\ w_2)^{\mathrm{T}}$ contains the source voltage w_1 and load current w_2, and where the system matrices are

$$F_0 = \begin{pmatrix} 0 & 0 \\ 0 & -\frac{R_L}{L} \end{pmatrix}, \qquad F_1 = \begin{pmatrix} 0 & \frac{1}{C} \\ -\frac{1}{L} & 0 \end{pmatrix}, \qquad G_w = \begin{pmatrix} 0 & \frac{1}{C} \\ \frac{1}{L} & 0 \end{pmatrix}. \quad (11.35)$$

We consider control of the boost converter based on a PWA approximation of (11.34). The model has two nonlinear terms since the duty cycle multiplies both states. Ideally, both products should be approximated by PWA functions, but in order to reduce complexity, we only approximate the product between the duty cycle and inductor current with a PWA function while the remaining product is approximated by a linear function. The PWA approximation is defined over a four-piece partition as illustrated in Fig. 11.2.

The corresponding optimisation problem (11.16a), (11.16b) was solved using the one-norm and cost matrices $Q = \text{diag}(4\ 1)$ and $R = 0$. The prediction horizon was $N = 2$, and the inductor current was imposed to satisfy the constraint $-i_{L,\max} \leq i_L \leq i_{L,\max}$ with $i_{L,\max} = 3\ \text{A}$. The resulting explicit solution had 1279 regions.

Simulation results are shown in Fig. 11.10. The simulation shows an initial start-up transient followed by a 50% increase in the load current w_2. The load then returns to the nominal value, and after the steady state is reached again, the load current decreases 50% and then returns to the nominal value. In all scenarios, the controller returns the capacitor voltage to the reference value. We note that the solution does not display any chattering even though there is no penalty on the control input, and the steady-state solution moves between regions of the PWA approximation (see Fig. 11.2). This is a consequence of the continuity of the PWA approximation.

11.8.3 DC–AC Converter Control

We now examine the hybrid MPC of one phase of the grid inverter depicted in Fig. 11.1d, applying some of the hybrid MPC formulations described in Sect. 11.6. This type of converters is typically employed to inject power from renewable energy sources into the grid. A resonant inductor–capacitor–inductor filter is employed to mitigate high-order harmonics, while the low-order harmonics must be taken care by the control. In order to minimise the switching losses, leading and trailing edge PWM are alternated. The PWM system is controlled and sampled twice per switching period at the instants depicted by disks in Fig. 11.11. The binary variable n_1 introduced in (11.25a)–(11.25d) is required to describe that leading edge and trailing edge pulses alternate; therefore $n_1 = 1$ when l is even and $n_1 = -1$ otherwise.

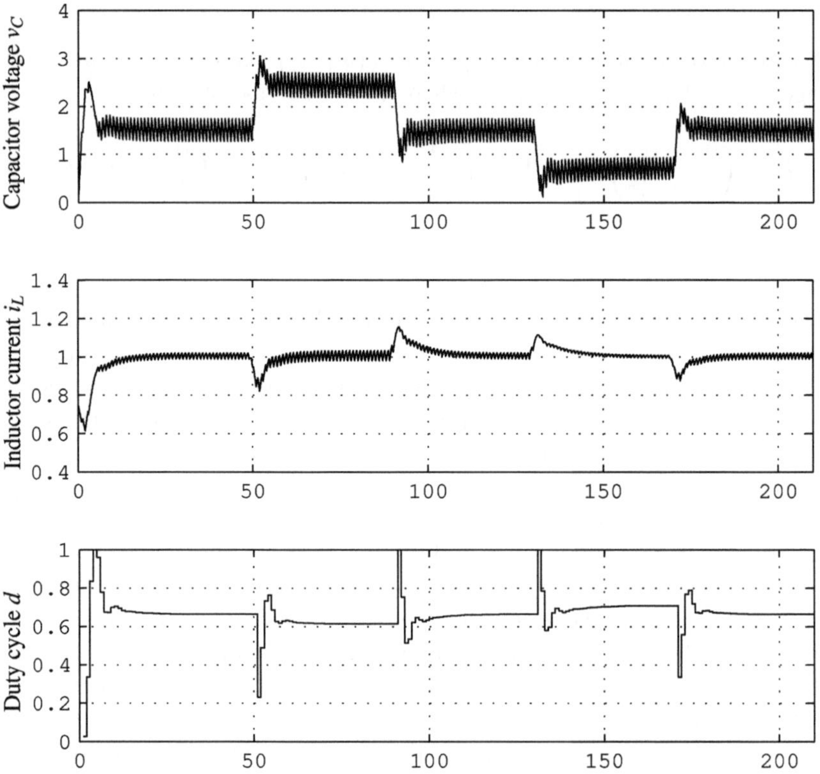

Fig. 11.10 Boost converter controlled by MPC using a PWA approximation of the bilinear averaged dynamics

For each phase j, the objective is that the output current y_j tracks its reference value r_j. As the system is switched, the objective is reformulated as generating a limit cycle as close as possible to the reference, and the objective is written as

$$\min \quad \sum_{l=1}^{N}\sum_{m=1}^{N_q}\left|Q(x_{l,m}-x_{t,l,m})\right|_p+\left|R(d_{1,l}-d_{t,1,l})\right|_p+|S\Delta d_{1,l}|_p \tag{11.36a}$$

$$\text{subject to:}\quad d_{1,m,l}=\begin{cases}0: & n_1 m\le n_1 m_{1,0},\\ 1: & n_1 m\ge n_1 (m_{1,0}+1),\\ N_q d_{1,l}-m_{1,0}: & \text{otherwise},\end{cases}$$

$$n_1=\begin{cases}1: & \forall l \text{ odd},\\ -1: & \forall l \text{ even},\end{cases} \tag{11.36b}$$

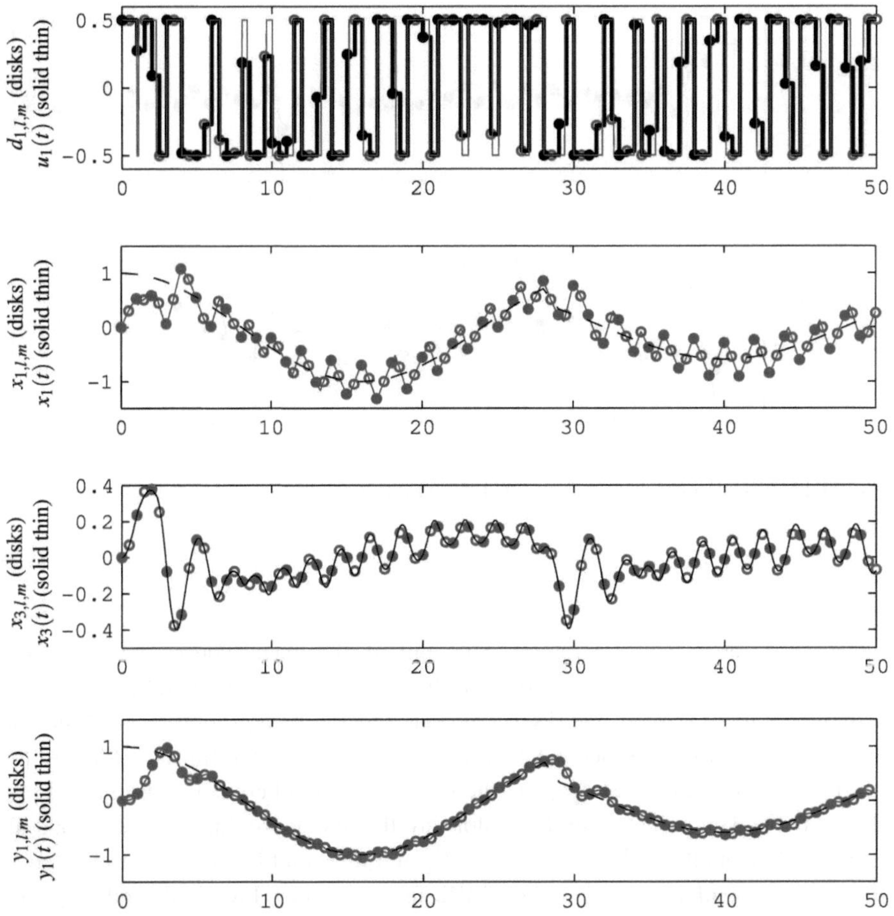

Fig. 11.11 Waveforms of single-phase two-level inverter with resonant filter. Hybrid model employs a discretization time such that $N_q = 2$

$$m_0 = \inf\{d : d \le N_q\big(n_1\, d_{1,l} + (1-n_1)\,(1-d_{1,l})\big),\ d \in \mathbb{N}\}, \tag{11.36c}$$

$$d_{i,l} = \frac{1}{N_q}\sum_{m=0}^{N_q-1} d_{i,l,m}, \tag{11.36d}$$

where the target $(x_{t,l,m}, d_{t,1,l})$ has been obtained using relation (11.12c).

The hybrid MPC scheme is analysed through simulation in Fig. 11.11. The norm $p = 1$ is employed with penalty matrices $Q = \operatorname{diag}\big(0.1\ 200\ 1\big)$, $R = 0$ and $S = 1$ and with a prediction horizon $N = 2$. The sampling period is divided into $N_q = 2$

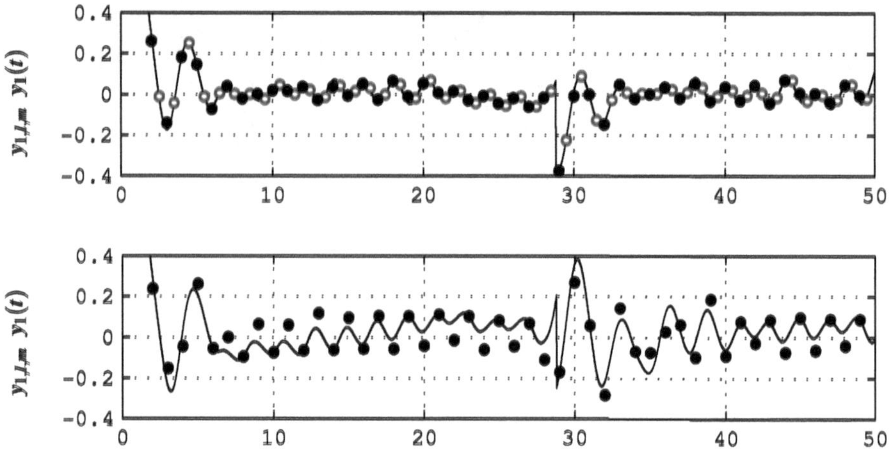

Fig. 11.12 Tracking errors on output for different quantization time $T_q = T_s/N_q$. *Top* $N_q = 2$, *bottom* $N_q = 1$

intervals. A piecewise sine reference is applied, the magnitude of the sine being given by the desired power injection or absorption. The top plot compares the approximated input (*circles* and *disks*, connected by the *bold plain lines*) and the real applied PWM input. The second plot shows the switched current to be filtered, the third plot the capacitor voltage, and the fourth plot the filtered current injected into the grid. The key points from these plots are that the approximate is very good already with a small number of intervals N_q and that the filtered current injected in the grid that results from this hybrid MPC features a very low ripple and very well damped transients.

In Fig. 11.12, we further investigate the performance of the hybrid MPC scheme by comparing the tracking error on the output current for two MPC schemes, the hybrid MPC illustrated in Fig. 11.11 where $N_q = 2$ and another MPC with $N_q = 1$. The second MPC is equivalent to a scheme based on an averaged model. The *thin plains curves* are the continuous-time variables, while the *disks* feature the predicted optimal sequence over one sampling period. The hybrid MPC accurately predicts the samples, while there is a significant error for the averaged MPC. The error present low-frequency components, which results in a low-frequency nonlinear distortion between the reference and the generated waveform that is difficult to eliminate. The clear interest of the hybrid MPC formulations for this kind of problem is to both allow increasing dynamic performance and reducing harmonic distortion.

Both schemes have an explicit solution of the same structure, mp-MILP and mp-LP, respectively, as discussed in Sect. 11.7.1. The mp-MILP associated to the hybrid MPC, i.e., when $N_q \geq 2$, however quickly becomes intractable as N_q and N increase.

11.9 Summary

MPC allows us to formulate a wide range of control objectives for systems where an explicit model of the plant dynamics is available. It is in general required to discretise the system dynamics. The hybrid dynamics of switched electronics systems result in several MPC classes. The enumerative integer MPC approach applicability is very limited due to accuracy and tractability issues. PWM systems with linear averaged models result in linearly constrained MPC that yield a very good performance. These MPC problems can be very effectively presolved parametrically off-line. The resulting piecewise affine control law can be computed very quickly on-line using a binary search tree algorithm, which makes the approach the best suited for real-time applications.

The potential of hybrid MPC to formulate complex objective is high. When used with continuous PWA models with hinging hyperplanes, hybrid MPC can tackle PWM systems with nonlinear averaged models relatively systematically. When used with models that represent the intersampling behaviour of the system, hybrid MPC can shape the switched behaviour of PWM systems.

The main limitation is however the complexity of the resulting mixed-integer programs. The future challenge is solving in real time these problems for more than a few simple examples. Hybrid MPC schemes that result in mixed-integer linear programs can be computed in the same way as linearly constrained MPC problems, however the optimisation problem quickly becomes intractable as the number of constraints grows very quickly. Hybrid MPC schemes that result in mixed-integer quadratic programs can be computed as separate quadratic programs when the number of binary variables remains small.

References

1. Almér, S., Mariéthoz, S., Morari, M.: Piecewise affine modeling and control of a step-up DC–DC converter. In: Proc. of the American Control Conference, Baltimore, Maryland, USA, pp. 3299–3304 (2010)
2. Baotic, M., Borrelli, F., Bemporad, A., Morari, M.: Efficient on-line computation of constrained optimal control. SIAM J. Control Optim. **47**(5), 2470–2489 (2008)
3. Bemporad, A., Borrelli, F., Morari, M.: Optimal controllers for hybrid systems: stability and piecewise linear explicit form. In: Proc. of the IEEE Conference on Decision and Control, Sydney, Australia, pp. 1810–1815 (2000)
4. Bemporad, A., Borrelli, F., Morari, M.: Piecewise linear optimal controllers for hybrid systems. In: Proc. of the American Control Conference, Chicago, Illinois, USA, pp. 1190–1194 (2000)
5. Bemporad, A., Borrelli, F., Morari, M.: Model predictive control based on linear programming—the explicit solution. IEEE Trans. Autom. Control **47**(12), 1974–1985 (2002)
6. Bemporad, A., Morari, M., Dua, V., Pistikopoulos, E.N.: The explicit linear quadratic regulator for constrained systems. Automatica **38**(1), 3–20 (2002)
7. Cortes, P., Kazmierkowski, M.P., Kennel, R.M., Quevedo, D.E., Rodriguez, J.: Predictive control in power electronics and drives. IEEE Trans. Ind. Electron. **55**(12), 4312–4324 (2008)
8. Dua, V., Pistikopoulos, E.N.: An algorithm for the solution of multiparametric mixed integer linear programming problems. Ann. Oper. Res. **99**(1–4), 123–139 (2000)

9. García, C.E., Prett, D.M., Morari, M.: Model predictive control: Theory and practice—A survey. Automatica **25**(3), 335–348 (1989)
10. Geyer, T., Papafotiou, G., Morari, M.: Model predictive direct torque control—Part I: Concept, algorithm, and analysis. IEEE Trans. Ind. Electron. **56**(6), 1894–1905 (2009)
11. Mariéthoz, S., Morari, M.: Explicit model-predictive control of a PWM inverter with an LCL filter. IEEE Trans. Ind. Electron. **56**(2), 389–399 (2009)
12. Mariéthoz, S., Almér, S., Bâja, M., Beccuti, A.G., Patino, D., Wernrud, A., Buisson, J., Cormerais, H., Fujioka, H., Geyer, T., Jönsson, U.T., Kao, C.Y., Morari, M., Papafotiou, G., Rantzer, A., Riedinger, P.: Comparison of hybrid control techniques for buck and boost DC–DC converters. IEEE Trans. Control Syst. Technol. **18**(5), 1126–1145 (2010)
13. Mayne, D.Q., Rawlings, J.B., Rao, C.V., Scokaert, P.O.M.: Constrained model predictive control: Stability and optimality. Automatica **36**(6), 789–814 (2000)
14. Middlebrook, R.D., Ćuk, S.: A general unified approach to modeling switching-converter power stages. In: Proc. of the IEEE Power Electronics Specialist Conference, Cleveland, Ohio, USA, pp. 18–34 (1976)
15. Naouar, M.W., Naassani, A.A., Monmasson, E., Belkhodja, I.S.: FPGA-based predictive current controller for synchronous machine speed drive. IEEE Trans. Power Electron. **23**(4), 2115–2126 (2008)
16. Qin, S.J., Badgwell, T.A.: A survey of industrial model predictive control technology. Control Eng. Pract. **11**(7), 733–764 (2003)
17. Sontag, E.D.: Nonlinear regulation: The piecewise linear approach. IEEE Trans. Autom. Control **26**(2), 346–358 (1981)
18. Tøndel, P., Johansen, T.A., Bemporad, A.: An algorithm for multi-parametric quadratic programming and explicit MPC solutions. Automatica **39**(3), 489–497 (2003)
19. Tøndel, P., Johansen, T.A., Bemporad, A.: Evaluation of piecewise affine control via binary search tree. Automatica **39**(5), 945–950 (2003)

Chapter 12
Adaptive PI Stabilisation of Switched Power Converters Described by Port-Hamiltonian Models

Michael Hernandez-Gomez, Romeo Ortega, Françoise Lamnabhi-Lagarrigue, and Gerardo Escobar

12.1 Introduction and Problem Formulation

In [9] it was shown that input-affine nonlinear systems which are passifiable via a constant control action can be stabilised with a linear PI controller that does not require the knowledge of the constant term. This simple fact was then used to propose a methodology to design PI controllers for a class of switched power converters that satisfy a technical condition needed for passifiability via constant control.

Switched power converters, with linear capacitors and inductors, are described in the so-called port-Hamiltonian form [3]

$$\dot{x} = \left(J_0 + \sum_{i=1}^{m} J_i u_i - R \right) \nabla H(x) + E, \tag{12.1}$$

M. Hernandez-Gomez (✉) · R. Ortega · F. Lamnabhi-Lagarrigue
Laboratory of Signals and Systems, CNRS, Plateau de Moulon, 91192 Gif-sur-Yvette, France
e-mail: hernandez@lss.supelec.fr

R. Ortega
e-mail: ortega@lss.supelec.fr

F. Lamnabhi-Lagarrigue
e-mail: lamnabhi@lss.supelec.fr

G. Escobar
ABB Switzerland Ltd, Segelhofstrasse 1K, 5405 Baden-Dattwil, Switzerland
e-mail: gerardo.escobar@ch.abb.com

F. Vasca, L. Iannelli (eds.), *Dynamics and Control of Switched Electronic Systems*, Advances in Industrial Control,
DOI 10.1007/978-1-4471-2885-4_12,

where $x \in \mathbb{R}^n$ is the converter state, and $u \in \mathbb{R}^m$ denotes the duty ratio of the switches. The total energy stored in inductors and capacitors is

$$H(x) = \frac{1}{2} x^\top Q x, \quad Q = Q^\top > 0,$$

$\nabla = \frac{\partial}{\partial x}$, $J_i = -J_i^\top$, $i \in \bar{m} := \{0, \dots, m\}$, are the interconnection matrices, $R = R^\top \geq 0$ represents the dissipation matrix, and the vector $E \in \mathbb{R}^n$ contains the (fixed) external voltage and current sources.

The reader is referred to [3] for further details of this model, which is obtained by applying Kirchhoff's laws to the switching topologies.

In [9] it was shown that, if a technical condition (involving Q, J_i and E), i.e., condition (11) of Proposition 2, is satisfied, then there exists a matrix $C_\rho \in \mathbb{R}^{m \times n}$ such that the map $u \to y_\rho := C_\rho x$ is passive, where C_ρ depends linearly on a free vector $\rho \in \mathbb{R}^n$. Consequently, system (12.1) with the inclusion of switching external sources can be stabilised with a PI around y_ρ. Furthermore, if y_ρ is a detectable output for the closed loop for a given desired equilibrium, then the state of the system will converge to this equilibrium. Although the condition is a linear matrix inequality that can be easily verified numerically, unfortunately it is highly technical and has no obvious physical interpretation. Moreover the requirement of detectability, which should be satisfied with a suitable selection of ρ, is a non-trivial task—as witnessed by the involved derivations of Sect. 12.3 for a three-phase rectifier example in [9].

In this chapter we use the stronger property of passivity of the *nonlinear incremental* model [5], instead of passivity of the standard nonlinear model, to extend the result of [9]. For ease of reference, we recall now the following definition of nonlinear incremental model from [5].

Definition 12.1.1 Given a nonlinear system of the form

$$\begin{aligned} \dot{x} &= f(x) + g(x)u, \\ y &= h(x), \end{aligned} \tag{12.2}$$

where $x \in \mathbb{R}^n$, $u, y \in \mathbb{R}^m$, with $m \leq n$, and the matrix $g(x)$ has full rank. Fix an assignable equilibrium state[1] $x^\star \in \mathbb{R}^n$ and define the constant input and

[1] See [5] for the definition of the assignable equilibrium set.

output vectors associated to $x^\star$ as

$$\begin{aligned} u^\star &:= \left(g^\top(x^\star)g(x^\star)\right)^{-1}g^\top(x^\star)f(x^\star), \\ y^\star &:= h(x^\star). \end{aligned} \tag{12.3}$$

The nonlinear incremental model of (12.2) is given by

$$\begin{aligned} \dot{x} &= f(x) + g(x)u^\star + g\tilde{u}, \\ \tilde{y} &= h(x) - h(x^\star), \end{aligned} \tag{12.4}$$

where $\tilde{(\cdot)} := (\cdot) - (\cdot)^\star$ are the incremental variables.

The fact that the nonlinear incremental model of power converters is passive was first established in the foundational paper [10], where a stabilising switching control for the duty ratio is proposed. As indicated in [10], passivity of the nonlinear incremental model is a particular case of incremental passivity [2], and it is called in that paper "relative passivity". In this chapter, instead of the latter, we follow [5] and refer to this property with the more explicit name of "passivity of the nonlinear incremental model". The property has been used in [13] to design a stabilising switching control for a power converter connected to the load through a transmission line. It has also been reported in essentially the same form used here in [6], highlighting the interesting fact that the new passive output has units of power. It should be underscored that incremental passivity has been recently revisited in various other contexts by the control community, see e.g. [5] and references therein.

In [10] a direct adaptive controller is proposed to estimate the desired operating equilibrium that depends on uncertain parameters, in particular, the load resistance. Even though the estimator of [10] is not implementable with the usual measurable data, i.e. the converter state, an ingenious scheme to indirectly measure a signal that is required by the algorithm is suggested—this signal is a linear combination of voltages and currents, which depends on the specific converter topology; the reader is referred to [10] for further details, see also Sect. 12.4. In this chapter an indirect adaptive scheme is proposed to estimate the uncertain load resistance that, besides the system state, does not require any additional measurements.

A large class of converters that can be stabilised via adaptive PI control. More specifically:

1. The class consists of power converters that include switching external sources and are described by the averaged model

$$\dot{x} = \left(J_0 + \sum_{i=1}^{m} J_i u_i - R\right)\nabla H(x) + \left(G_0 + \sum_{i=1}^{m} G_i u_i\right)E, \tag{12.5}$$

where G_i, $i \in \bar{m}$, are $n \times n$ matrices. Comparing (12.1) with (12.5), we notice that, besides the fixed sources $G_0 E$, the model contains switching sources $\sum_{i=1}^{m} G_i u_i E$.

2. No additional assumption on the system, except the obviously necessary condition of existence of an equilibrium point, is imposed.
3. As indicated above, we define a new output y—that is a *linear* combination of the states—such that the nonlinear incremental model of (12.5) with this new output is passive.[2]
4. For unknown (but constant) load resistance, three adaptive versions of the PI that preserve stability of the closed loop are proposed. The first scheme is a standard model reference estimator, while the other two are based on the Immersion and Invariance (I&I) methodology recently developed in [1]. For all three schemes, conditions to ensure parameter convergence under normal operating conditions—an essential feature for the stability proof—are given.

12.2 Passivity of the Nonlinear Incremental Model

The following result relates the concepts of passivity and nonlinear incremental model by considering the model (12.5).

Proposition 12.2.1 *Consider switched power converters described by* (12.5). *Let* $x^\star \in \mathbb{R}^n$ *be an admissible equilibrium point, that is,* $x^\star$ *satisfies*

$$0 = \left(J_0 + \sum_{i=1}^{m} J_i u_i^* - R \right) \nabla H(x^\star) + \left(G_0 + \sum_{i=1}^{m} G_i u_i^\star \right) E, \qquad (12.6)$$

for some $u^\star \in \mathbb{R}^m$. *The nonlinear incremental model of the system for the output* $y = \mathscr{C} x$, *where*

$$\mathscr{C} := \begin{pmatrix} E^\top G_1^\top - (x^\star)^\top Q J_1 \\ \vdots \\ E^\top G_m^\top - (x^\star)^\top Q J_m \end{pmatrix} Q \in \mathbb{R}^{m \times n}, \qquad (12.7)$$

---

[2]As shown later, the fact that we are dealing with incremental outputs makes the detectability condition easier to satisfy.

is passive. More precisely, the system verifies the dissipation inequality

$$\dot{V} \leq \tilde{y}^\top \tilde{u},$$

where $\tilde{x} := x - x^\star$, $y^\star = \mathscr{C} x^\star$, *and the (positive definite) storage function* $V : \mathbb{R}^n \to \mathbb{R}_+$ *is given by*

$$V(x) = \frac{1}{2} \tilde{x}^\top Q \tilde{x}. \tag{12.8}$$

Proof To simplify the notation, define the matrix function $G_N : \mathbb{R}^n \to \mathbb{R}^{n \times m}$ by

$$G_N(x) := [J_1 Q x + G_1 E \mid \cdots \mid J_m Q x + G_m E], \tag{12.9}$$

which allows us to write (12.5) in the compact form

$$\dot{x} = (J_0 - R) \nabla H(x) + G_0 E + G_N(x) u.$$

Now,

$$\begin{aligned} \dot{x} &= (J_0 - R) \nabla V(x) + G_N(x)(\tilde{u} + u^\star) - G_N(x^\star) u^\star \\ &= \left(J_0 + \sum_{i=1}^{m} J_i u_i^* - R \right) \nabla V(x) + G_N(x) \tilde{u}, \end{aligned}$$

where (12.6) and (12.8) have been used to obtain the first identity, and

$$\left[G_N(x) - G_N(x^\star) \right] u^\star = \sum_{i=1}^{m} J_i u_i^* Q \tilde{x},$$

which follows from (12.9), is used to get the second one.

Now, using the skew-symmetry of the matrices J_i, we see that $\tilde{y}$ may be written as

$$\tilde{y} = G_N^\top(x^\star) Q \tilde{x}.$$

On the other hand, it is easy to verify that

$$G_N^\top(x^\star) Q \tilde{x} = G_N^\top(x) Q \tilde{x}.$$

The proof is completed evaluating the derivative of (12.8) along the trajectories of the system, replacing the expression of $V(x)$ and using the non-negativity of R to get

$$\begin{aligned}\dot{V} &= -\nabla^\top V(x) R \nabla V(x) + \tilde{y}^\top \tilde{u} \\ &= -\tilde{x}^\top Q R Q \tilde{x} + \tilde{y}^\top \tilde{u} \\ &\leq \tilde{y}^\top \tilde{u}.\end{aligned} \tag{12.10}$$

□

Similarly to [9], where the new passive output y_ρ is a linear function of x, y above is also a linear function of x. However, the passivity property established in Proposition 12.2.1 corresponds to the mapping $\tilde{u} \to \tilde{y}$. This fact makes a fundamental difference when it comes to satisfy the detectability condition necessary for convergence of the state to $x^\star$. Another interesting observation stemming from the skew-symmetry of J_l is that

$$y^\star = \mathscr{C} x^\star = \begin{pmatrix} E^\top G_1^\top \\ \vdots \\ E^\top G_m^\top \end{pmatrix} Q x^\star, \tag{12.11}$$

which has two important consequences. First, $y^\star$ is linear in $x^\star$—a property that is essential for the adaptive design; second, $y^\star = 0$ for converters without switched external sources.

12.3 PI Stabilisation

An immediate corollary of passivity of the incremental model is the following result, whose proof is established with standard arguments, see [5], and is given here only to streamline the proof of Proposition 12.4.2 in Sect. 12.4 that refers to the adaptive PI.

Proposition 12.3.1 *Consider a switched power converter described by* (12.5) *in closed loop with the PI controller*

$$\begin{aligned}\dot{z} &= -\tilde{y}, \\ u &= -K_p \tilde{y} + K_i z,\end{aligned} \tag{12.12}$$

with $\tilde{y} = \mathscr{C}\tilde{x}$ *and* $\mathscr{C}$ *given by* (12.7), *and* $K_p = K_p^\top > 0$, $K_i = K_i^\top > 0$ *and* $x^\star, u^\star$ *satisfying* (12.6). *For all initial conditions* $(x(0), z(0)) \in \mathbb{R}^{n+m}$, *the trajectories of the closed-loop system are bounded and such that*

$$\lim_{t\to\infty} \tilde{y}_a(t) = 0,$$

where y_a is an augmented "output signal" defined as

$$\tilde{y}_a := \begin{pmatrix} \mathscr{C} \\ RQ \end{pmatrix} \tilde{x}. \tag{12.13}$$

Moreover,

$$\lim_{t\to\infty} x(t) = x^\star$$

if $\tilde{y}_a$ is detectable, that is, if for any solution $x(t)$ of the closed-loop system, the following implication is true:

$$\tilde{y}_a(t) \equiv 0 \quad \Rightarrow \quad \lim_{t\to\infty} x(t) = x^\star. \tag{12.14}$$

Proof Consider the Lyapunov function candidate

$$W(x,z) := V(x) + \frac{1}{2}\tilde{z}^\top K_I \tilde{z},$$

where $z^\star := K_I^{-1} u^\star$. Taking the time derivative of $W(x,z)$ and using (12.10) yield

$$\begin{aligned} \dot{W} &= -\tilde{x}^\top QRQ\tilde{x} + \tilde{y}^\top \tilde{u} - \tilde{z}^\top K_I \tilde{y} \\ &= -\tilde{x}^\top QRQ\tilde{x} - \tilde{y}^\top K_p \tilde{y}. \end{aligned}$$

The previous equation proves that the trajectories are bounded and, via La Salle's invariance principle [8], that $\tilde{y}_a \to 0$. The convergence of the state, under the detectability assumption (12.14), follows from [12]. □

The detectability condition (12.14) of Proposition 12.3.1 may be verified by computing the zero dynamics (associated to the output $\tilde{y}_a$) and then proving that it is asymptotically stable. However, this analysis may be quite involved. The proposition below gives a verifiable condition on the matrices J_i and R that ensures detectability for power dissipating converters[3] without external switching sources.

Proposition 12.3.2 *Assume that the power converter does not have external switching sources, i.e. $G_i \equiv 0$, $i = 1, \dots, m$, and that it dissipates power at the given equilibrium $x^\star$, that is,*

$$(x^\star)^\top QRQx^\star > 0. \tag{12.15}$$

[3]The "power dissipating" qualifier is needed for the technical proof. It is always satisfied in applications, where power has to be furnished to an external load.

- *The detectability condition* (12.14) *of Proposition* 12.3.1 *can be replaced by*

$$y_a(t) \equiv 0 \quad \Rightarrow \quad x^\top(t)x^\star = |x(t)|\,|x^\star|, \tag{12.16}$$

where $|\cdot|$ *is the Euclidean norm.*

- *The implication* (12.16) *is true if either* $n = 2$ *or* $m = n - 1$ *and*

$$\operatorname{rank}\begin{pmatrix} (x^\star)^\top Q J_1 \\ \vdots \\ (x^\star)^\top Q J_{n-1} \end{pmatrix} = n - 1.$$

Proof The proof of the first claim is established showing that if (12.16) holds, then $\lim_{t\to\infty} x(t) = x^\star$. From the Schwarz inequality we have

$$x^\top(t)x^\star = |x(t)||x^\star| \quad \Leftrightarrow \quad \exists \alpha : \mathbb{R}_+ \to \mathbb{R}$$

such that $x(t) = \alpha(t)x^\star$, that is, if x and $x^\star$ are collinear. Hence,

$$\begin{aligned} x^\star \dot{\alpha} &= \dot{x} \\ &= \left(J_0 + \sum_{i=1}^m J_i u_i - R\right) Q x^\star \alpha + G_0 E \\ &= \left(J_0 + \sum_{i=1}^m J_i u_i - R\right) Q x^\star \alpha - \left(J_0 + \sum_{i=1}^m J_i u_i^\star - R\right) Q x^\star, \end{aligned}$$

where the equilibrium equation (12.6) has been used to get the last equation. Multiplying the equation above on the right by $(x^\star)^\top Q$ and using the skew-symmetry of J_i, one gets

$$(x^\star)^\top Q x^\star \dot{\alpha} = -(x^\star)^\top Q R Q x^\star \alpha + (x^\star)^\top Q R Q x^\star.$$

From (12.15) one also has that $|x^\star| \neq 0$. Therefore, the equation above takes the from

$$\dot{\alpha} = -\mu(\alpha - 1),$$

where

$$\mu := \frac{(x^\star)^\top Q R Q x^\star}{(x^\star)^\top Q x^\star} > 0,$$

with positivity of μ following from (12.15). Clearly, $\lim_{t\to\infty} \alpha(t) = 1$, which implies that $\lim_{t\to\infty} x(t) = x^\star$. □

The second part of the proposition is established invoking the following linear algebra lemma.

Lemma 12.3.3 *Given a vector $b \in \mathbb{R}^n$ and the matrices $J_i \in \mathbb{R}^{n\times n}$, $J_i + J_i^\top = 0$, $i = 1, \dots, n-1$. Define the matrix*

$$W := \begin{pmatrix} b^\top J_1 \\ \vdots \\ b^\top J_{n-1} \end{pmatrix} \in \mathbb{R}^{(n-1)\times n}.$$

Assume rank $W = n-1$. *For all $a \in \mathbb{R}^n$ such that $Wa = 0$, we have $a^\top b = |a|\,|b|$.*

Proof For $n = 2$, the proof follows immediately from the fact that, in that case,

$$W = db^\top \begin{pmatrix} 0 & 1 \\ -1 & 0 \end{pmatrix}$$

for some $d \in \mathbb{R}$. To avoid cluttering, the proof is given for the case $n = 3$. As will become clear below, the same derivations apply for the general case. An arbitrary 3×3 skew symmetric matrix may be written in the form

$$J_i = \begin{pmatrix} 0 & g_i & r_i \\ -g_i & 0 & m_i \\ -r_i & -m_i & 0 \end{pmatrix}$$

for some $g_i, r_i, m_i \in \mathbb{R}$. Then, for any vector $b = \begin{pmatrix} b_1 & b_2 & b_3 \end{pmatrix}^\top$, one gets

$$W = \begin{pmatrix} -b_2g_1 - b_3r_1 & b_1g_1 - b_3m_1 & b_1r_1 + b_2m_1 \\ -b_2g_2 - b_3r_2 & b_1g_2 - b_3m_2 & b_1r_2 + b_2m_2 \end{pmatrix}.$$

Given that rank $W = 2$, there are two linearly independent columns in W. Assume that they are the first two—otherwise, swap the elements of a. Then, partition the matrix as

$$W = \begin{bmatrix} W_1 & W_2 \end{bmatrix}, \quad W_1 \in \mathbb{R}^{2\times 2},\ W_2 \in \mathbb{R}^2.$$

Using this notation, one has

$$Wa = 0 \quad \Leftrightarrow \quad W_1 a_{12} = -W_2 a_3, \tag{12.17}$$

where a has been partitioned as $a = \begin{pmatrix} a_{12}^\top & a_3^\top \end{pmatrix}^\top$.

Partition $b = \begin{pmatrix} b_{12}^\top & b_3 \end{pmatrix}^\top$ and note that

$$W_1 b_{12} = -b_3 \begin{pmatrix} b_1r_1 + b_2m_1 \\ -b_1r_2 + b_2m_2 \end{pmatrix} = -b_3 W_2.$$

Further, since $|W_1| \neq 0$ if and only if $b_3 \neq 0$, one can write

$$W_2 = \frac{-1}{b_3} W_1 b_{12}.$$

The proof is completed replacing the last identity in (12.17) and noting that, since W_1 is full rank, $a = a_3 b / b_3$. □

Notice that, in contrast to (12.14), the new condition (12.16) is not asymptotic, therefore it cannot be claimed to be weaker than detectability. However, it is easily verifiable from the converter data. Indeed, for systems without external switching sources, the augmented output y_a takes the form

$$y_a = \begin{pmatrix} x^\top Q J_1 Q x^\star \\ \vdots \\ x^\top Q J_m Q x^\star \\ RQ(x - x^\star) \end{pmatrix}.$$

Clearly, the implication (12.16) is true, provided that some rank condition on J_l and R is satisfied. This is illustrated by the second claim of Proposition 12.3.2. Unfortunately, in the latter the fact that $\tilde{y}_a(t) = 0 \Rightarrow RQ(x - x^\star) = 0$ is not exploited, adding conservativeness to the result.

12.4 Adaptive PI Stabilisation

In this section the case of unknown load resistance is considered. From (12.7) and (12.11) it is clear that the passive output $\tilde{y}$—only signal needed for the controller implementation—depends linearly on x and the desired equilibrium $x^\star$. However, $x^\star$ will, in general, depend nonlinearly on the resistors. Indeed, the power balance equation of the converter model (12.5) is

$$\dot{H} = -x^\top QRQx + x^\top Q\left(G_0 + \sum_{l=1}^{m} G_l u_l\right)E.$$

Hence, the desired equilibrium $x^\star$ and the corresponding control $u^\star$ must satisfy the equation

$$0 = -(x^\star)^\top QRQx^\star + (x^\star)^\top Q\left(G_0 + \sum_{l=1}^{m} G_l u_l^\star\right)E, \tag{12.18}$$

which is a nonlinear smooth function of the elements of R. Consequently, denoting by $\theta \in \mathbb{R}_+$ the unknown load resistance (or conductance), $\tilde{y}$ may be expressed in the form

$$\tilde{y} = \left[A_0 + \psi_0(\theta)\right]x + b_0 + \psi_1(\theta), \tag{12.19}$$

where $A_0 \in \mathbb{R}^{m \times m}$ and $b_0 \in \mathbb{R}^m$ are independent of θ, while $\psi_0, \psi_1 : \mathbb{R}_+ \to \mathbb{R}^m$ are known smooth functions. The nonlinear dependence on θ stymies the application of direct adaptive control. On the other hand, the linear dependence of $\tilde{y}$ on $x^\star$ suggests to consider direct estimation of the latter, as done in [10].

This task seems unattainable in the present scenario of restricted complexity controller, i.e. a PI, where the only property that is exploited is the dissipation inequality (12.10). Indeed, replacing $x^\star$ in the control by its estimate and computing the derivative of the Lyapunov function candidate $W(x,z)$, one sees that the term multiplying the parameter error is $\tilde{y}$, which is not measurable.[4] Therefore, it is not clear how the design can be completed in this case without assuming additional measurements—as done in [10].

Since direct adaptive control is ruled out, an indirect formulation is then adopted. As is well known [1, 11], in this approach consistent estimates of θ are required to carry out the stability analysis. This, in its turn, translates into some form of persistency of excitation condition, which is imposed in our developments.

For ease of presentation, it is assumed that the load resistance appears in the kk term of the matrix R, that is

$$R = R_0 + \theta e_k e_k^\top, \tag{12.20}$$

where $R_0 \in \mathbb{R}^{n\times n}$ contains the known resistances of the converter, and $e_k \in \mathbb{R}^n$ is the kth vector of the standard n-dimensional Euclidean basis. Under this assumption, the converter model (12.5) can be written as

$$\dot{x} = m(x,u) - \theta\chi e_k, \tag{12.21}$$

where $m : \mathbb{R}^n \times \mathbb{R}^m \to \mathbb{R}^n$ is the known function

$$m(x,u) := \left(J_0 + \sum_{i=1}^m J_i u_i - R_0\right) Qx + \left(G_0 + \sum_{i=1}^m G_i u_i\right) E, \tag{12.22}$$

and

$$\chi := e_k^\top Qx. \tag{12.23}$$

Three important observations are in order.

- χ represents, either, the current flowing through the load resistance or the voltage across its conductance. As explained above, in normal operation the converter furnishes power to the load; therefore, the desired value for χ is non-zero.
- As clearly seen from (12.21), the parameter θ is not identifiable if $\chi \equiv 0$. In view of these considerations, it is reasonable to assume that

$$\chi^\star := e_k^\top Qx^\star \neq 0. \tag{12.24}$$

[4] This situation happens in [10], where $\tilde{y}$ is denoted $Q(Bx+b)$.

- It is clear from the previous derivations that assumption (12.20) is done without loss of generality. Indeed, as long as it appears linearly, the unknown resistance may be in any other element of the matrix R (or even repeatedly), which would lead to a new linear combination of the elements of x for the definition of χ.

Three estimators that ensure parameter convergence, that is,

$$\lim_{t\to\infty} \tilde{\theta}(t) = 0, \tag{12.25}$$

where $\tilde{\theta} := \hat{\theta} - \theta$, are presented. The first one is a classical model reference scheme, while the other two estimators are obtained applying the Immersion and Invariance (I&I) methodology recently developed in [1]. The convergence of the estimators will be proved under different assumptions, which in all three cases are satisfied under normal operating conditions, i.e., when χ converges to a constant value that, as indicated in (12.24), is non-zero.

Proposition 12.4.1 *Consider a switched power converter described by* (12.5), *with dissipation matrix of the form* (12.20) *and definitions* (12.22) *and* (12.23).

- *The model reference estimator*

$$\begin{aligned}\dot{\hat{\chi}} &= -\lambda(\hat{\chi} - \chi) - Q_{kk}\chi\hat{\theta} + e_k^\top Qm(x,u),\\ \dot{\hat{\theta}} &= \gamma\chi(\hat{\chi} - \chi),\end{aligned} \tag{12.26}$$

with $Q_{kk} = e_k^\top Q e_k$ *and* $\lambda, \gamma > 0$, *ensures* (12.25) *if* $\chi(t)$ *is persistently exciting, that is, if*

$$\frac{1}{T}\int_t^{t+T} \chi^2(\tau)\,\mathrm{d}\tau \geq \delta$$

for some $T, \delta > 0$ *and for all* $t \geq 0$.

- *The first I&I estimator*

$$\begin{aligned}\dot{\xi} &= \frac{\lambda}{\gamma}\chi\left\{e_k^\top Qm(x,u) - \left(\gamma\xi - \frac{\lambda}{2}\chi^2\right)Q_{kk}\chi\right\},\\ \hat{\theta} &= \gamma\xi - \frac{\lambda}{2}\chi^2,\end{aligned} \tag{12.27}$$

ensures (12.25) *if* $\chi(t)$ *is not square integrable.*

- *The second I&I estimator*

$$\dot{\xi} = \frac{\lambda}{\gamma\chi}\{e_k^\top Qm(x,u) - (\gamma\xi - \lambda\ln\chi)Q_{kk}\chi\},$$
$$\hat{\theta} = \gamma\xi - \lambda\ln\chi, \qquad (12.28)$$

ensures (12.25) *if* $\chi(t)$ *is bounded away from zero.*

Proof For the model reference estimator (12.26), define the error $\tilde{\chi} := \hat{\chi} - \chi$, which satisfies the equation

$$\dot{\tilde{\chi}} = -\lambda\tilde{\chi} - Q_{kk}\chi\tilde{\theta}.$$

The latter, together with the second equation in (12.26), yields the classical model reference error model of adaptive systems [11] with regressor χ and the strictly positive real transfer function $Q_{kk}(s+\lambda)^{-1}$ (recall that $Q_{kk} > 0$ because $Q > 0$). Direct application of the well-known Theorem 2.6.5 of [11] yields the exponential stability of the error system if χ is persistently exciting.

To make the presentation of the I&I estimators more didactic and illustrate the flexibility of the approach, (12.27) and (12.28) are derived following step by step the I&I methodology described in [1]. Define the parameter error[5]

$$\tilde{\theta} := \gamma\xi + \lambda\beta(\chi) - \theta, \qquad (12.29)$$

where the function $\beta : \mathbb{R} \to \mathbb{R}$ is to be designed. The objective in I&I is to select $\dot{\xi}$ and $\beta(\chi)$ to ensure an asymptotically stable dynamics for $\tilde{\theta}$. In this way, an estimate of θ is given by $\hat{\theta} = \gamma\xi + \lambda\beta(\chi)$.

Differentiating the error, one gets

$$\begin{aligned}\dot{\tilde{\theta}} &= \gamma\dot{\xi} + \lambda\beta'(\chi)\dot{\chi}\\ &= \gamma\dot{\xi} + \lambda\beta'(\chi)e_k^\top Q\dot{x}\\ &= \gamma\dot{\xi} + \lambda\beta'(\chi)\big[e_k^\top Qm(x,u) - Q_{kk}\chi\theta\big]\\ &= \gamma\dot{\xi} + \lambda\beta'(\chi)\big\{e_k^\top Qm(x,u) - Q_{kk}\chi\big[\gamma\xi + \lambda\beta(\chi) - \tilde{\theta}\big]\big\},\end{aligned}$$

where $(\cdot)'$ denotes differentiation with respect to its argument, (12.23) has been used in the second identity, (12.21) in the third one, and (12.29) in the last one. Selecting

$$\dot{\xi} = -\frac{\lambda}{\gamma}\beta'(\chi)\big\{e_k^\top Qm(x,u) - \big[\gamma\xi + \lambda\beta(\chi)\big]Q_{kk}\chi\big\} \qquad (12.30)$$

[5] The coefficient γ, although is not needed for the analysis, has been added to improve numerical conditioning in the simulations.

yields the error dynamics

$$\dot{\tilde{\theta}} = \lambda\beta'(\chi)Q_{kk}\chi\tilde{\theta}. \tag{12.31}$$

The design is completed by selecting a suitable function $\beta(\chi)$ that ensures stability of the $\tilde{\theta}$ dynamics. They are, respectively, taken as

$$\beta(\chi) = -\frac{1}{2}\chi^2 \quad \Rightarrow \quad \dot{\tilde{\theta}} = -\lambda Q_{kk}\chi^2\tilde{\theta},$$

$$\beta(\chi) = -\ln\chi \quad \Rightarrow \quad \dot{\tilde{\theta}} = -\lambda Q_{kk}\tilde{\theta}.$$

For the first error system, the standing assumption $\chi \notin \mathscr{L}_2$ ensures (12.25). For the second design, the assumption $|\chi(t)| \geq \epsilon > 0$ is needed to avoid a singularity in the computation of $\beta'(\chi)$.

The estimators are obtained by replacing $\beta(\chi)$ in (12.30). □

We are now ready to prove the stability result with adaptive PI controller.

Proposition 12.4.2 *Consider a switched power converter described by* (12.5), *with dissipation matrix of the form* (12.20), *in closed loop with the adaptive PI controller*

$$\begin{aligned} \dot{z} &= -\left[A_0 x + b_0 + \psi_0(\hat{\theta})x + \psi_1(\hat{\theta})\right], \\ u &= -K_p\left[A_0 x + b_0 + \psi_0(\hat{\theta})x + \psi_1(\hat{\theta})\right] + K_i z, \end{aligned} \tag{12.32}$$

together with (12.19), *with* $\tilde{y} = \mathscr{C}\tilde{x}$, $\mathscr{C}$ *given by* (12.7), *and* $K_p = K_p^\top > 0$, $K_i = K_i^\top > 0$. *The estimated parameter* $\hat{\theta}$ *is generated by either one of the estimators of Proposition* 12.4.1.

For all initial conditions such that the trajectories of the closed-loop system are bounded and satisfy the corresponding standing assumption on χ, *one has*

$$\lim_{t\to\infty} y_a(t) = 0, \qquad \lim_{t\to\infty} \tilde{\theta}(t) = 0,$$

with y_a *defined in* (12.13). *Moreover,* $\lim_{t\to\infty} x(t) = x^\star$ *if* $\tilde{y}_a$ *is detectable.*

Proof To study the effect of the parameter uncertainty in the closed-loop, write the control (12.32) in the perturbed form

$$\begin{aligned} \dot{z} &= -\tilde{y} - \tilde{\psi}, \\ u &= -K_p\tilde{y} + K_i z - K_p\tilde{\psi}, \end{aligned}$$

where, to avoid cluttering, the signal

$$\tilde{\psi} := \left[\psi_0(\hat{\theta}) - \psi_0(\theta)\right]x + \psi_1(\hat{\theta}) - \psi_1(\theta)$$

is defined. Notice that, if the trajectories are bounded, $\lim_{t\to\infty} \tilde{\psi}(t) = 0$. Considering the same Lyapunov function candidate as in Proposition 12.3.1, after some simple calculations, one gets

$$\dot{W} \leq -\tilde{y}^\top K_p \tilde{y} - (\tilde{y}^\top K_p - \tilde{z}^\top K_I)\tilde{\psi}.$$

The term multiplying $\tilde{\psi}$ is bounded, and, since $\tilde{\psi}$ is (monotonically) converging to zero, after some time this disturbance term is "dominated" by the first negative right-hand term, and stability is achieved. □

The following remarks are in order.

Remark 12.4.3 Persistency of excitation of χ is necessary to ensure parameter convergence of the model reference estimator. Using I&I estimators [1], it is possible to replace this assumption by either the condition that χ is non-square integrable or that it is bounded away from zero. The transient performance of the PI with the various estimators will, of course, be different. Some simulation comparisons are presented in Sect. 12.5 for a three-phase rectifier, and experimental results in Sect. 12.6.

Remark 12.4.4 The assumption of boundedness of trajectories is necessary to complete the proof. This is, clearly, a restrictive assumption but seems unavoidable in the present scenario of restricted complexity controller, i.e. a PI. In the spirit of Corollary 3.1 of [1], it might be possible to relax this condition with an I&I controller, but this would considerably complicate the design.

Remark 12.4.5 The linear dependence of $\tilde{y}$ on $x^\star$ suggests to consider direct estimation of the latter, as done in [10]. Unfortunately, as it happens in [10], the term multiplying the parameter error in the derivative of the Lyapunov function candidates is $\tilde{y}$ (denoted $Q(Bx + b)$ in that paper), which is not measurable. Therefore, it is not clear how the design can be completed in this case.

12.5 Application to Three-Phase Rectifier

In this section the controller presented above will be applied to a three-phase (boost-type) rectifier feeding a load that is modelled as a pure constant current source. An electrical scheme of the converter is depicted in Fig. 12.1.

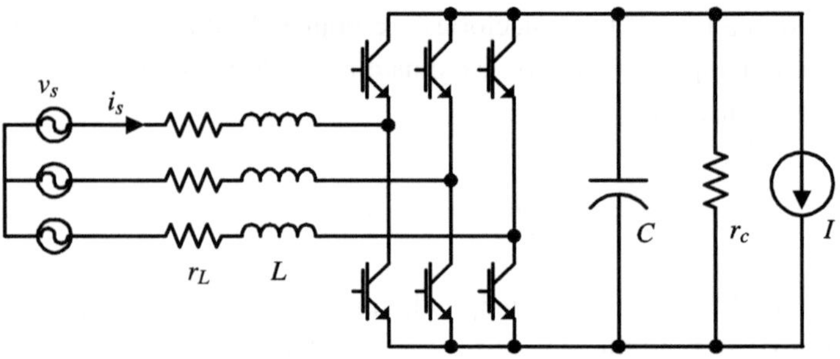

Fig. 12.1 Schematic of the three-phase rectifier

12.5.1 Mathematical Model

The mathematical model of the converter, after the standard 3 → 2 transformation of the average model and rotation to the dq reference frame [7], is:

$$\frac{\mathrm{d}}{\mathrm{d}t}\phi_d = -\frac{r_L}{L}\phi_d + \omega\phi_q - \frac{\mu_0}{C}u_1 q_C + V,$$

$$\frac{\mathrm{d}}{\mathrm{d}t}\phi_q = -\frac{r_L}{L}\phi_q - \omega\phi_d - \frac{\mu_0}{C}u_2 q_C,$$

$$\frac{\mathrm{d}}{\mathrm{d}t}q_C = \frac{\mu_0}{L}u_1\phi_d + \frac{\mu_0}{L}u_2\phi_q - \frac{1}{Cr_c}q_C - I,$$

where ϕ_d and ϕ_q are the direct and quadrature fluxes in the inductors, q_C is the charge in the capacitor, u_1 and u_2 are the direct and quadrature modulation indices, V is the constant direct supply voltage, I is the constant output current, L, C, r_L, r_c, μ_0 are the positive model parameters, and ω represents the frequency of the voltage source in radians.

See [3] for further details of the model, in particular, the motivation to select fluxes and charges as state variables instead of the more standard voltages and currents. It should be noted that the latter choice is used in [9].

12.5.2 Controller

The system can be expressed in the port-Hamiltonian form (12.5) with the definition $x = (\phi_d\ \phi_q\ q_C)^\top$ and the following matrices:

$$G_0E = \begin{pmatrix} V \\ 0 \\ -I \end{pmatrix}, \qquad R = \begin{pmatrix} r_L & 0 & 0 \\ 0 & r_L & 0 \\ 0 & 0 & \frac{1}{r_c} \end{pmatrix}, \qquad Q = \begin{pmatrix} \frac{1}{L} & 0 & 0 \\ 0 & \frac{1}{L} & 0 \\ 0 & 0 & \frac{1}{C} \end{pmatrix},$$

$$J_0 = \omega L \begin{pmatrix} 0 & 1 & 0 \\ -1 & 0 & 0 \\ 0 & 0 & 0 \end{pmatrix}, \qquad J_1 = \mu_0 \begin{pmatrix} 0 & 0 & -1 \\ 0 & 0 & 0 \\ 1 & 0 & 0 \end{pmatrix},$$

$$J_2 = \mu_0 \begin{pmatrix} 0 & 0 & 0 \\ 0 & 0 & -1 \\ 0 & 1 & 0 \end{pmatrix}.$$

The control objective is the regulation of the capacitor voltage to a given constant value v_d, i.e. the regulation of the charge x_3 to a desired reference $x_3^\star > 0$ where $Cv_d = x_3^\star$. Moreover, it is desirable that (in steady state) the power factor be as close as possible to one, an objective that in the present formulation is achieved driving the flux x_2 to zero. The set of admissible values for $x_1^\star$ is obtained from the power balance equation (12.18) evaluated at $x^\star = (x_1^\star, 0, x_3^\star)$:

$$\frac{r_L}{L^2}\left(x_1^\star\right)^2 + \frac{1}{C^2 r_c}\left(x_3^\star\right)^2 = \frac{V}{L}x_1^\star - \frac{I}{C}x_3^\star,$$

which defines a circle in the (x_1, x_3) plane. For a given $x_3^\star$, the solution of this equilibrium equation is

$$x_1^\star = \frac{L}{2r_L}\left(V - \sqrt{V^2 - \frac{4r_L}{C^2 r_c}x_3^{*2} - \frac{4r_L}{C}Ix_3^\star}\right),$$

where the negative value of the square root has been chosen to reduce the value of $x_1^\star$. Since the circuit does not have switched external sources $y^\star = 0$, we have

$$\tilde{y} = y = \frac{x_3^\star \mu_0}{LC}\begin{pmatrix} \frac{x_1^\star}{x_3^\star}x_3 - x_1 \\ -x_2 \end{pmatrix}, \tag{12.33}$$

which is obtained from (12.7).

The rectifier verifies the conditions of Proposition 12.3.2, namely, $m = n - 1$, and the matrix

$$\begin{pmatrix} (x^\star)^\top QJ_1 \\ (x^\star)^\top QJ_2 \end{pmatrix} = \mu_0 \begin{pmatrix} \frac{x_3^\star}{C} & 0 & -\frac{x_1^\star}{L} \\ 0 & \frac{x_3^\star}{C} & 0 \end{pmatrix}$$

is of rank two. Therefore, the detectability condition (12.14) is satisfied, and the equilibrium can be rendered *globally asymptotically stable* with the PI controller (12.12).

For the adaptive version of the PI, it is assumed that the resistance r_c, which captures the losses of the leaky capacitor, is known. Furthermore, it is assumed that the load current source I can be represented as a variable conductance θ, that is, $I = \theta x_3/C$. The passive output $\tilde{y}$ (12.33) can be written in the form (12.19) with

$$A_0 = -\frac{\mu_0 x_3^\star}{LC}\begin{pmatrix} -1 & 0 & 0 \\ 0 & -1 & 0 \end{pmatrix},$$

$$\psi_0(\theta) = \eta_1 \begin{pmatrix} 0 & 0 & 1 \\ 0 & 0 & 0 \end{pmatrix},$$

where

$$\eta_1 = \frac{\mu_0}{2r_L C}\left(V - \sqrt{V^2 - \frac{4r_L}{C^2}\left(\frac{1}{r_c} + \theta\right)(x_3^\star)^2}\right)$$

and $b_0 = \psi_1(\theta) = 0$.

Noticing that $\chi = \frac{1}{C}x_3$ yields the model reference estimator (12.26)

$$\dot{\hat{\chi}} = -\lambda\left(\hat{\chi} - \frac{x_3}{C}\right) - \frac{x_3}{C^2}\hat{\theta} + e_3^\top Qm(x,u),$$

$$\dot{\hat{\theta}} = \gamma\frac{x_3}{C}\left(\hat{\chi} - \frac{x_3}{C}\right),$$

while the first I&I estimator (12.27) is given by

$$\dot{\xi} = \frac{\lambda}{\gamma}\frac{x_3}{C}\left\{e_3^\top Qm(x,u) - \left(\gamma\xi - \lambda\frac{x_3^2}{2C^2}\right)\frac{x_3}{C^2}\right\},$$

$$\hat{\theta} = \gamma\xi - \lambda\frac{x_3^2}{2C^2},$$

and the second I&I estimator (12.28) is

$$\dot{\xi} = \frac{\lambda}{\gamma}\frac{C}{x_3}\left\{e_3^\top Qm(x,u) - \left(\gamma\xi - \lambda\ln\left(\frac{x_3}{C}\right)\right)\frac{x_3}{C^2}\right\},$$

$$\hat{\theta} = \gamma\xi - \lambda\ln\left(\frac{x_3}{C}\right),$$

where

$$e_3^\top Qm(x,u) = \frac{1}{C}\left(\frac{\mu_0}{L}u_1 x_1 + \frac{\mu_0}{L}u_2 x_2 - \frac{1}{Cr_c}x_3\right)$$

is obtained from (12.22).

Table 12.1 Tuning gains of the estimators for the rectifier example

	λ	γ
Model reference estimator	10^3	10^{-3}
First I&I estimator	8×10^{-6}	7×10^5
Second I&I estimator	2.515×10^{-1}	10^3

12.5.3 Simulations

Simulations were realised to compare the performance of the various controllers. The parameters were chosen as $L = 1\,\text{mH}$, $C = 4700\,\mu\text{F}$, $r_L = 1\,\text{m}\Omega$ and $r_c = 22\,000\,\Omega$. The gains of the PI controller were set to $K_p = 10^{-2}$ and $K_i = 10^{-4}$ for all control schemes. The values of the tuning parameters γ and λ of the estimators are given in Table 12.1. A voltage reference step change from $v_d = 700\,\text{V}$ to $v_d = 1000\,\text{V}$, and a ramp change in the load current from $I = 25\,\text{A}$ to $I = 50\,\text{A}$, as depicted in Fig. 12.2, were simulated. The initial conditions were taken as

$$x(0) = \begin{pmatrix} 7.85 \times 10^{-3} & 7.85 \times 10^{-3} & 1.034 \end{pmatrix}^\top,$$

where the units are Teslas for the inductances and Coulombs for the capacitances, and Table 12.2 shows the initial condition for the parameter $\tilde{\theta}$ in the different con-

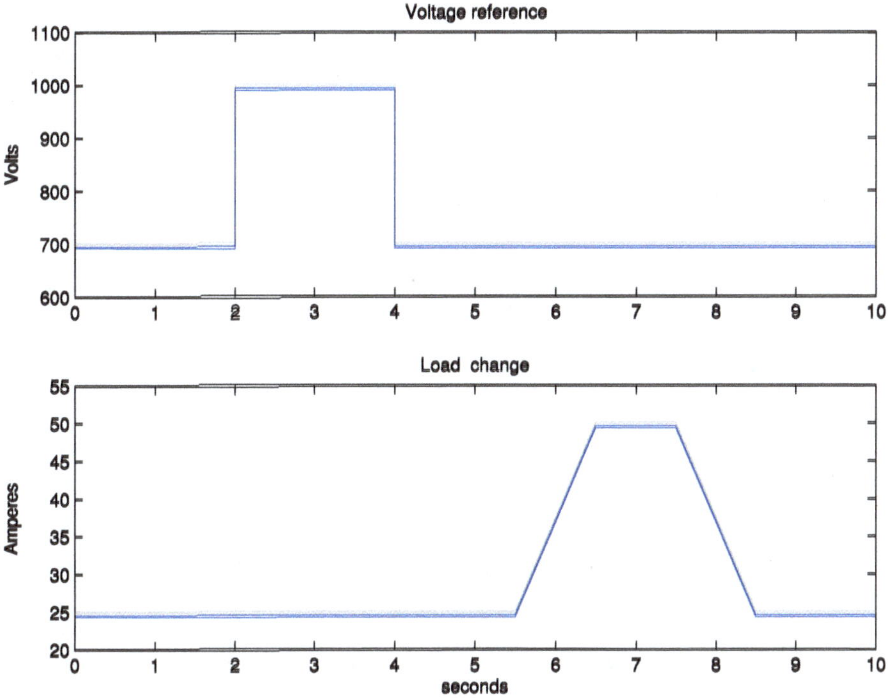

Fig. 12.2 Reference and load disturbance

Table 12.2 Initial condition of the parameter estimation $\tilde{\theta}$

	$\tilde{\theta}(0)$
Model reference estimator	−0.3025
First I&I estimator	−0.3661
Second I&I estimator	−0.269

trollers. In Figs. 12.3, 12.4, 12.5, 12.6 we show the behaviour of the state variables and the estimation errors for the known parameter PI and the various adaptive versions. As shown in the plots, the required parameter convergence is achieved in all cases.

12.6 Application to Quadratic Boost Converter

In this section we apply the proposed PI controllers to the quadratic converter shown in Fig. 12.7.

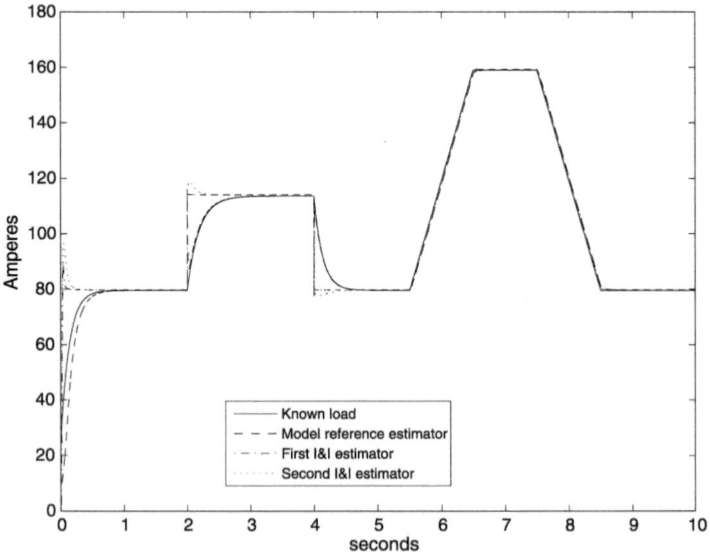

Fig. 12.3 Real current component of the system (x_1/L)

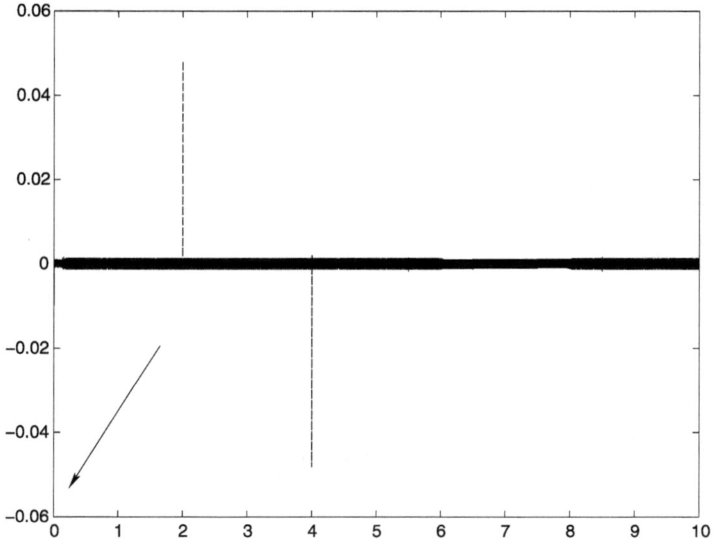

Fig. 12.4 Reactive current component of the system (x_2/L)

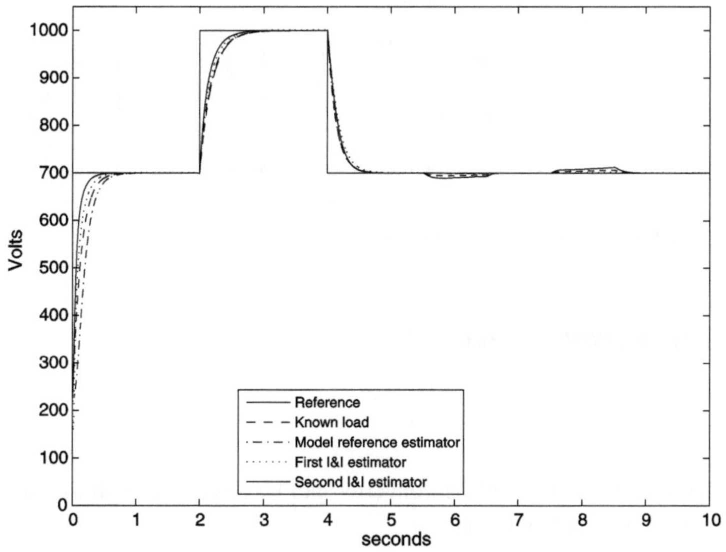

Fig. 12.5 Capacitor voltage compared and its reference

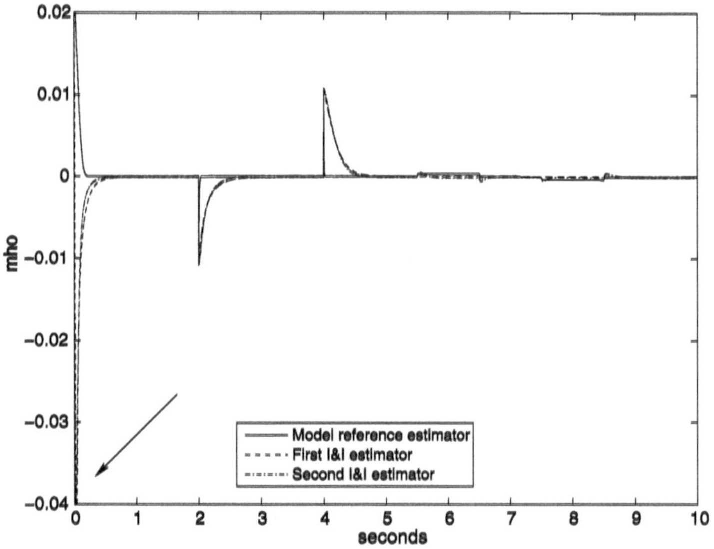

Fig. 12.6 Evolution of the parameter estimation errors $\tilde{\theta}$ for the different estimators

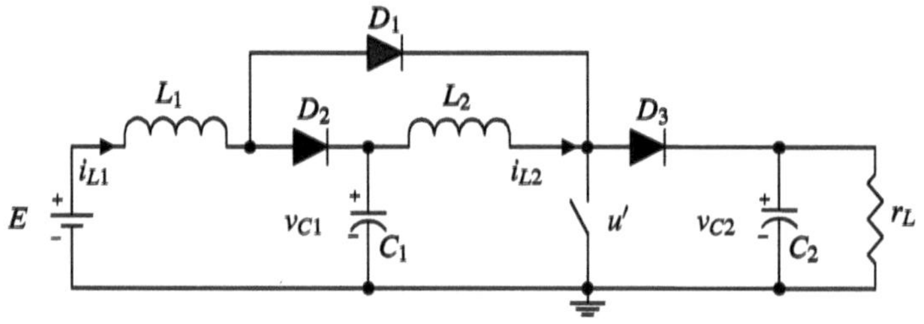

Fig. 12.7 Schematic of the quadratic converter

12.6.1 Mathematical Model

The dynamic behaviour of the converter can be described with the averaged model given by the equations

$$\frac{\mathrm{d}}{\mathrm{d}t} i_{L1} = \frac{1}{L_1}(E - v_{C1}u), \tag{12.34a}$$

$$\frac{\mathrm{d}}{\mathrm{d}t} i_{L2} = \frac{1}{L_2}(v_{C1} - v_{C2}u), \tag{12.34b}$$

$$\frac{d}{dt}v_{C1} = \frac{1}{C_1}(i_{L1}u - i_{L2}), \tag{12.34c}$$

$$\frac{d}{dt}v_{C2} = \frac{1}{C_2}\left(i_{L2}u - \frac{1}{r_L}v_{C2}\right), \tag{12.34d}$$

where i_{L1}, i_{L2} are the currents in the inductances, v_{C1}, v_{C2} are the voltages in the capacitors, L_1, L_2, C_1 and C_2 are the values of inductances and capacitances, respectively, r_L is the load, E the input voltage, and $u = 1 - u'$ is a continuous control equal to the slew rate of the pulse width modulator (PWM).

To simplify the analysis, the inductors are assumed lossless.

12.6.2 Robust PI Control

For the controller implementation, the full state is assumed measurable, and the input voltage known. On the other hand, none of the parameters needs to be known, and, in particular, an adaptation scheme will be added to estimate the load resistance.

To derive a PI controller for the quadratic converter, we will apply the technique described in [4].

The model (12.34a)–(12.34d) can be expressed in port-Hamiltonian form [3]

$$\dot{x} = (J_0 + J_1 u - R)\frac{\partial H}{\partial x} + B$$

with the definitions

$$x = \begin{pmatrix} i_{L1} & i_{L2} & v_{C1} & v_{C2} \end{pmatrix}^\top, \qquad B = \begin{pmatrix} \frac{E}{L_1} & 0 & 0 & 0 \end{pmatrix}^\top,$$

$$R = \begin{pmatrix} 0 & 0 & 0 & 0 \\ 0 & 0 & 0 & 0 \\ 0 & 0 & 0 & 0 \\ 0 & 0 & 0 & \frac{1}{r_L C_2^2} \end{pmatrix}, \qquad Q = \begin{pmatrix} L_1 & 0 & 0 & 0 \\ 0 & L_2 & 0 & 0 \\ 0 & 0 & C_1 & 0 \\ 0 & 0 & 0 & C_2 \end{pmatrix},$$

$$J_0 = \frac{1}{C_1 L_2}\begin{pmatrix} 0 & 0 & 0 & 0 \\ 0 & 0 & 1 & 0 \\ 0 & -1 & 0 & 0 \\ 0 & 0 & 0 & 0 \end{pmatrix}, \qquad J_1 = \begin{pmatrix} 0 & 0 & -\frac{1}{L_1 C_1} & 0 \\ 0 & 0 & 0 & -\frac{1}{L_2 C_2} \\ \frac{1}{L_1 C_1} & 0 & 0 & 0 \\ 0 & \frac{1}{L_2 C_2} & 0 & 0 \end{pmatrix},$$

where

$$H(x) = \frac{1}{2}x^\top Qx$$

is the energy stored in the circuit.

The goal of the control is to regulate the voltage v_{C2} across the load around a constant value v_d, which is equivalent to regulation of the capacitor voltage x_4 to the constant value $x_4^\star = v_d$. The admissible equilibria of system (12.34a)–(12.34d) can be parameterised by the reference $x_4^\star$ as follows:

$$x^\star := \begin{pmatrix} \frac{1}{r_L(u^\star)^2} & \frac{1}{r_L u^\star} & u^\star & 1 \end{pmatrix}^\top x_4^\star,$$

where $u^\star = \sqrt{E/x_4^\star}$ is the corresponding constant control.

12.6.3 Identifying a Suitable Output

It is shown in [4] that the output signal

$$\tilde{y} = -\sqrt{Ev_d}x_1 - v_d x_2 + \frac{v_d^2}{Er_L}x_3 + \frac{v_d}{r_L}\sqrt{\frac{v_d}{E}}x_4 \tag{12.35}$$

satisfies the following properties.[6]

- The mapping from the incremental input $\tilde{u} = u - u^\star$ to the output $\tilde{y}$ is passive. That is, the system (12.34a)–(12.34d), (12.35), satisfies the power balance inequality

  $$\dot{V} \leq \tilde{u}^\top \tilde{y}$$

  with

  $$V(\tilde{x}) = \frac{1}{2}\tilde{x}^\top Q\tilde{x}$$

  the storage function.
- The output $\tilde{y}$ is zero-state detectable. That is, the following implication is true:

  $$\tilde{y}(t) = 0 \quad \Rightarrow \quad \lim_{t\to\infty} x(t) = x^\star.$$

[6] See Sect. IV of [4] for more details.

12.6.4 Proposed PI Control

As discussed in [4], the two aforementioned properties allow us to prove that the equilibrium $x^\star$ can be rendered *globally asymptotically stable* with the PI controller:

$$\dot{z} = \tilde{y},$$
$$u = -K_p\tilde{y} - K_i z,$$

where K_p, K_i are positive tuning gains.

It is important to underscore that the only parameters required for the implementation of the controller are r_L and E, and that the tuning gains can take arbitrary positive values—hence the "robust" qualifier claimed for the controller.

Even though it is reasonable to assume that the voltage E is known, the load resistance may be highly uncertain. To further robustify the proposed controller, an adaptive version of the PI controller is constructed as follows. First, notice that $x^\star$ is linear in the unknown conductance $\theta := 1/r_L$. Hence, the passive output $\tilde{y}$ can be written in the linear regression form

$$\tilde{y} = \psi_0(x)\theta + A_0(x)$$

with

$$A_0(x) = -\sqrt{Ev_d}x_1 - v_d x_2,$$
$$\psi_0(x) = \frac{v_d^2}{E}x_3 + v_d\sqrt{\frac{v_d}{E}}x_4.$$

In [4] three estimators for θ, with different stability properties, are proposed. All estimators are parameterised by two tuning gains, λ and γ, which can take arbitrary positive values. First, a classical model reference estimator is a second-order system of the form

$$\dot{\hat{\chi}} = -\lambda(\hat{\chi} - x_4) + \frac{1}{C_2}(ux_2 - \hat{\theta}x_4),$$
$$\dot{\hat{\theta}} = \gamma x_4(\hat{\chi} - x_4).$$

Second, an immersion and invariance (I&I) estimator, derived using the techniques studied in the book [1], is given by the first-order equation

$$\hat{\theta} = \gamma\xi - \lambda {x_4}^2,$$
$$\dot{\xi} = \frac{\lambda}{\gamma C_2}(ux_2 - \hat{\theta}x_4)x_4.$$

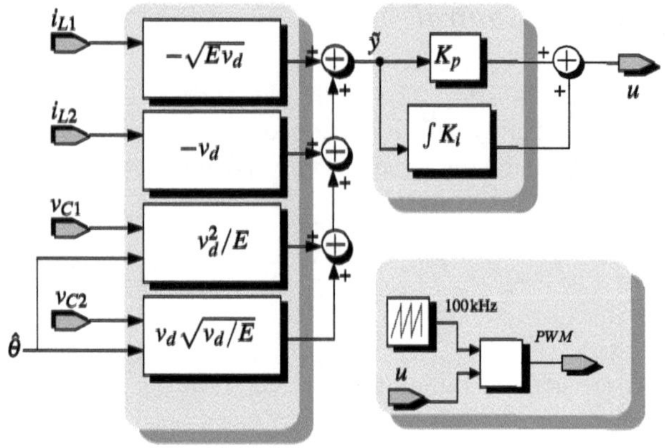

Fig. 12.8 Diagram of the proposed control with unknown load

Finally, a second I&I estimator, which assumes that x_4 is bounded away from zero, is described by

$$\hat{\theta} = \gamma\xi - \lambda \ln(x_4),$$
$$\dot{\xi} = \frac{\lambda}{\gamma C_2}(ux_2 - \hat{\theta}x_4)\frac{1}{x_4}.$$

To incorporate the estimator in the PI controller, we define the new output

$$\tilde{y} = -\sqrt{Ev_d}x_1 - v_d x_2 + \hat{\theta}\left(\frac{v_d^2}{E}x_3 + v_d\sqrt{\frac{v_d}{E}}x_4\right).$$

It is important to note that in all estimators the only system parameter that is required is C_2. The PI control block diagram is depicted in Fig. 12.8, while in Fig. 12.9 we detail the implementation of every estimator.

12.6.5 Experimental Testbench

The experimental card was assembled using low-cost commercial electronic elements placed on a card designed in the laboratory. In Fig. 12.10 we show the diagram of the experimental set-up, consisting of the quadratic circuit card that receives control signals from the Digital Signal Processor (DSP) DsPIC33F128GP802 from Microchip. inc, which is a low-cost DSP.

In Fig. 12.10 we show the main card which is formed by the quadratic converter, a PWM circuit, and some signal conditioners. The DSP card acquires, using an analog-to-digital converter integrated in the chip, the capacitors voltages and inductors currents signals previously conditioned from the quadratic converter. Two DC power supplies are necessary to operate the whole system, one to provide energy

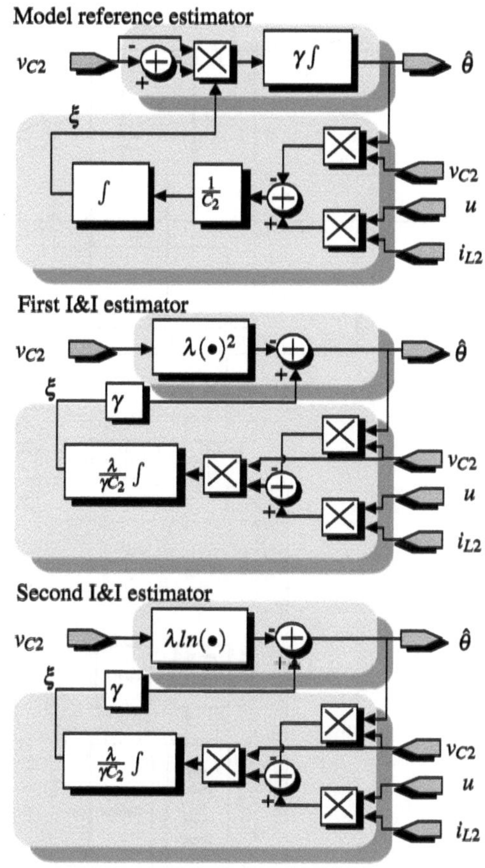

Fig. 12.9 The tree proposed parameter estimators

to the boost system (denoted as power supply E), and the other one to feed the electronic parts of the card.

Two current sensors together with two current-to-voltage converters are introduced to measure the inductors currents i_{L1} and i_{L2} (in this way, in the form of a voltage signal we can feed it into the DSP card to be used in the control law). In the case of the capacitors voltages v_{C1} and v_{C2}, we put two voltage divisors, so we can reduce the level of those signals in such a way that its final value is always in the range 0 V to 3.3 V by rail-to-rail amplifiers.

The following remarks concerning the practical implementation are in order:

- Our objective is to assess the dynamic performance of the proposed PI controller, together with the proposed estimators.
- We have chosen the inductor and the capacitor the experimental circuit close to the values calculated for the selected switched frequency (100 kHz) and the load power of 50 W. With those values, given for a typical commercial application, a significant ripple is present. This ripple, unfortunately, tends to mask the transient performance differences between the various controllers.

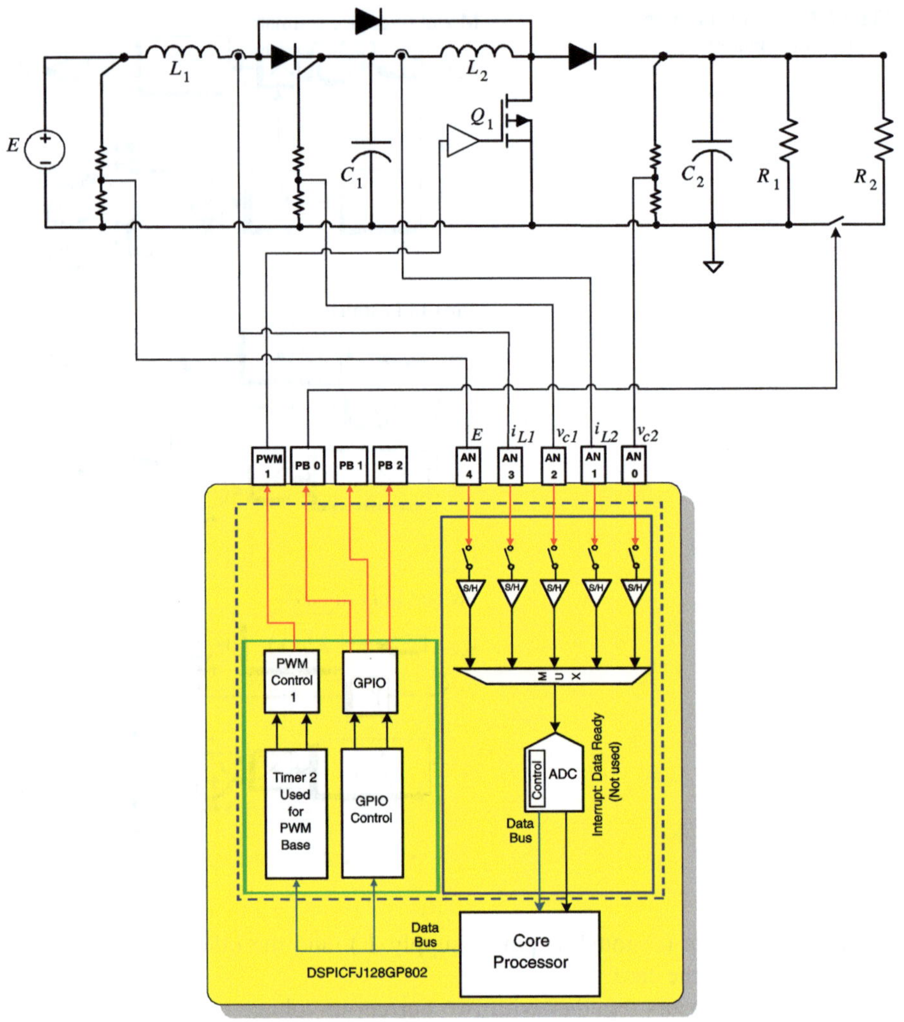

Fig. 12.10 Schematic of the quadratic converter board

Table 12.3 Electronic components used in the experimental setup

Element	Value
Mosfet	IRFS38N20D
Diodes Schottky	MBRB20200CT
Current Sensor	HSX-50
Capacitances	CKG57NX7R2A475M

- Aiming at a practical solution, the controller implementations are carried out with a low-cost DSP. This low-cost DSP runs at 40 MIPS (Millions of instruction per second) at 16 bits per instruction. Due the values of the control parameters, the

Fig. 12.11 Experimental setup quadratic converter board

Table 12.4 Tuning parameters of the different controllers for the quadratic boost

	K_p	K_i	λ	γ
Model reference estimator	1.5×10^{-1}	100	10^5	1×10^{-4}
First I&I estimator	1.5×10^{-1}	100	2×10^{-6}	10^3
Second I&I estimator	1.5×10^{-1}	100	10^{-3}	2×10^{-3}

16-bit resolution is not enough; for that reason, the control algorithm is running in 32 bits. To compensate for the performance degradation, the program is entirely written in assembler code, completes the algorithm in 10 µs, and it is important that a continuous controller is approximated by a discrete-time controller at a fast sampling time.

12.6.6 Experimental Results

The quadratic converter shown in Fig. 12.11 is composed by two inductors, two capacitors, a resistive charge, three diodes and a switch, the latter implemented by interconnecting a MOSFET transistor and two rapid diodes in a suitable manner, all these elements fed by a DC power supply. See Table 12.3. The values of its elements are taken as $E = 12\,\text{V}$, $L_1 = 53\,\mu\text{H}$, $L_2 = 231\,\mu\text{H}$, $C_1 = 4.7\,\mu\text{F}$, $C_2 = 4.7\,\mu\text{F}$ for all control schemes.

The PI control law described in the previous section has been implemented together with the proposed estimators in the quadratic circuit card. Their performance is compared with the following tests:

- A voltage reference step change from $v_d = 80\,\text{V}$ to $v_d = 120\,\text{V}$;
- A change in the load resistance from $r_L = 330\,\Omega$ to $r_L = 198\,\Omega$.

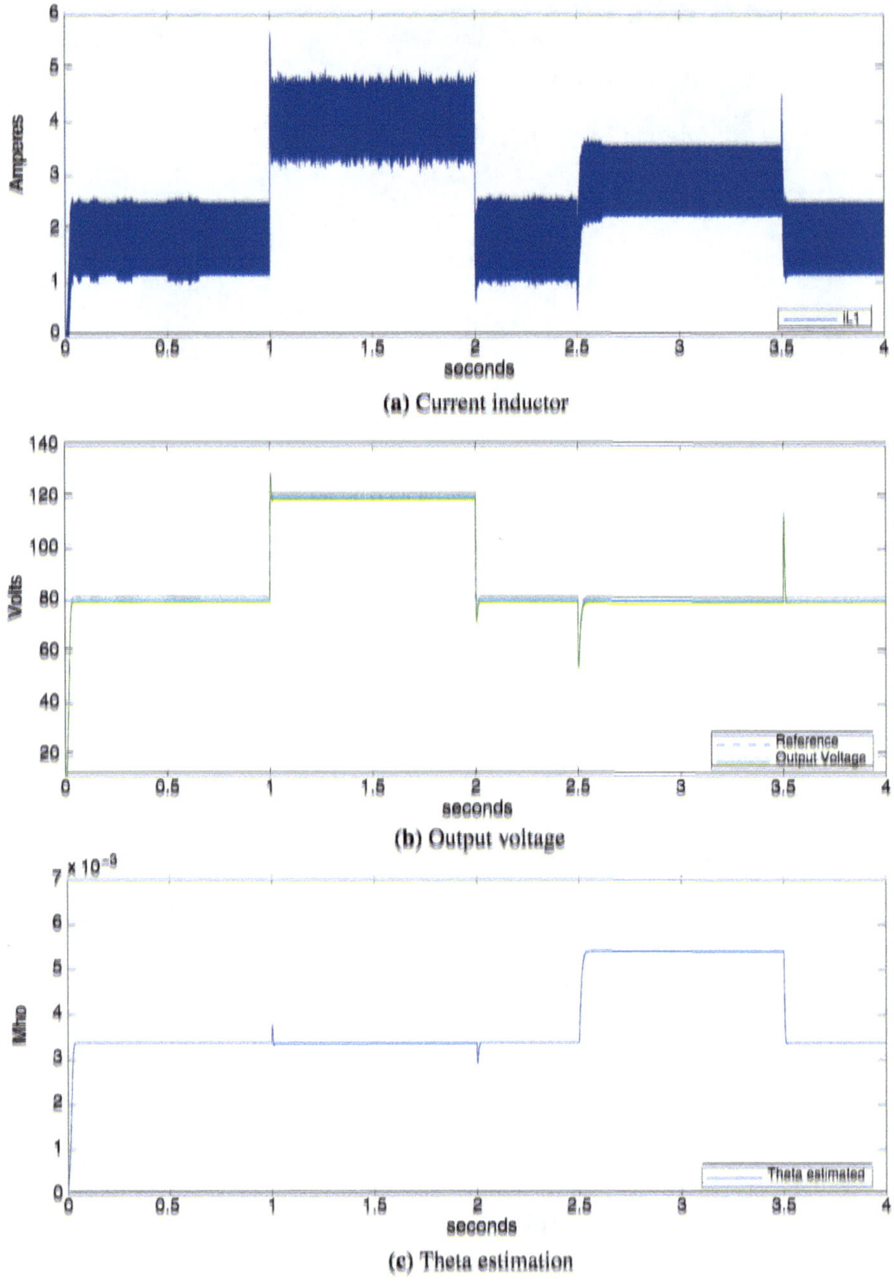

Fig. 12.12 Model reference experimental results

The gains of the PI controller and the tuning parameters γ and λ of the estimators are given in Table 12.4.

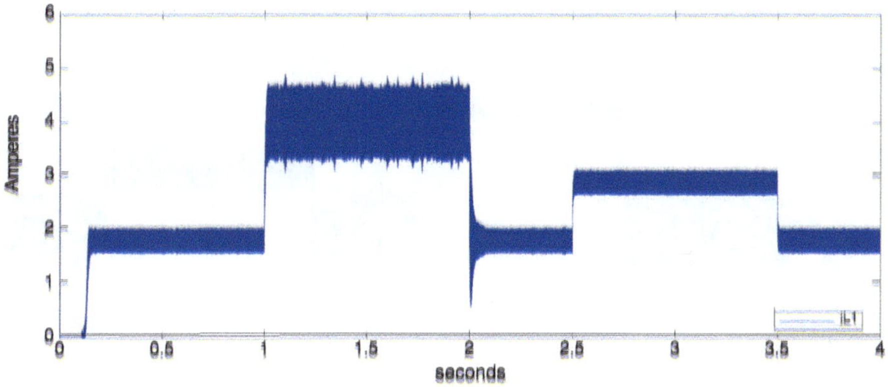

(a) Current inductor

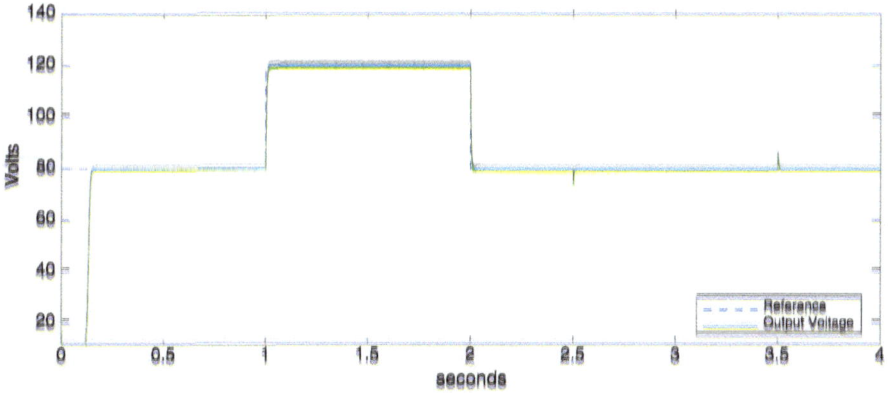

(b) Output voltage

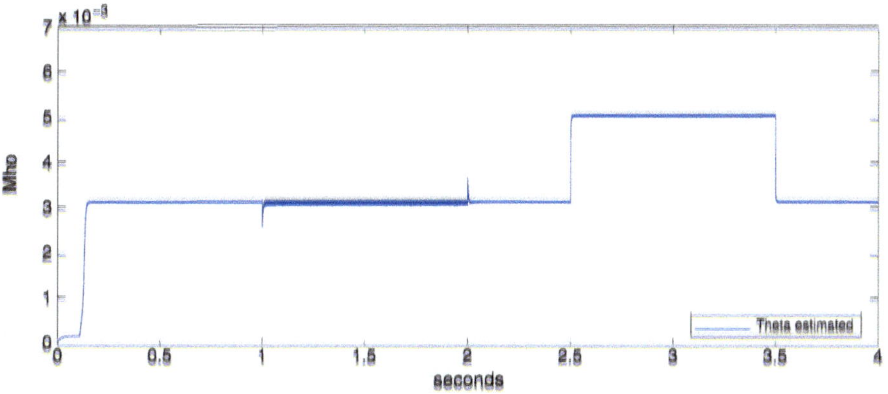

(c) Theta estimation

Fig. 12.13 I&I 1 experimental results

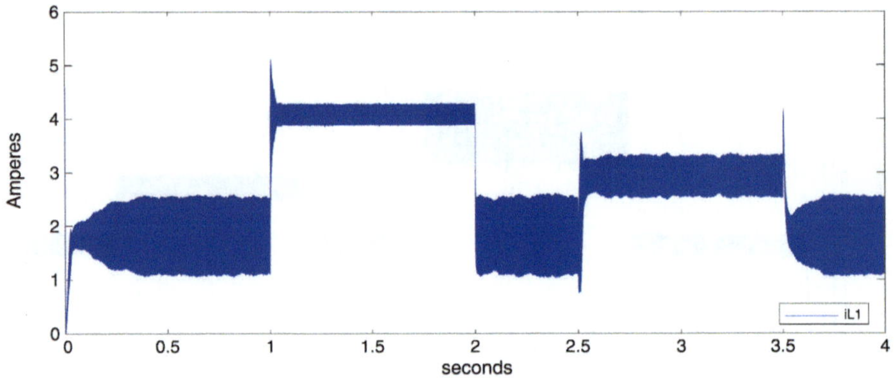

(a) Current inductor

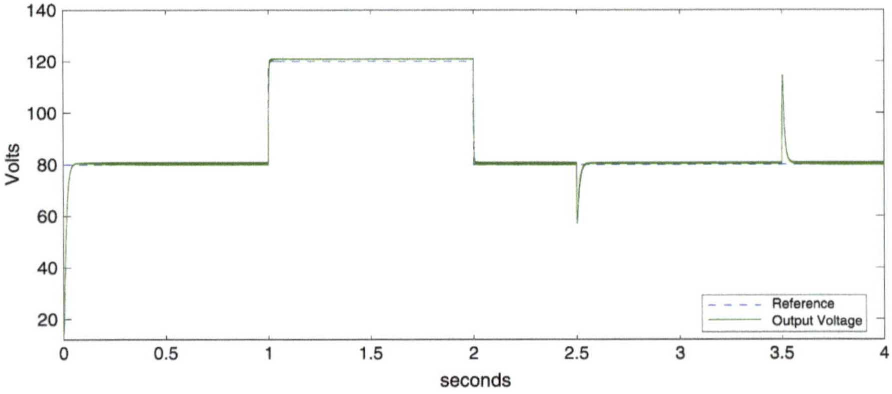

(b) Output voltage

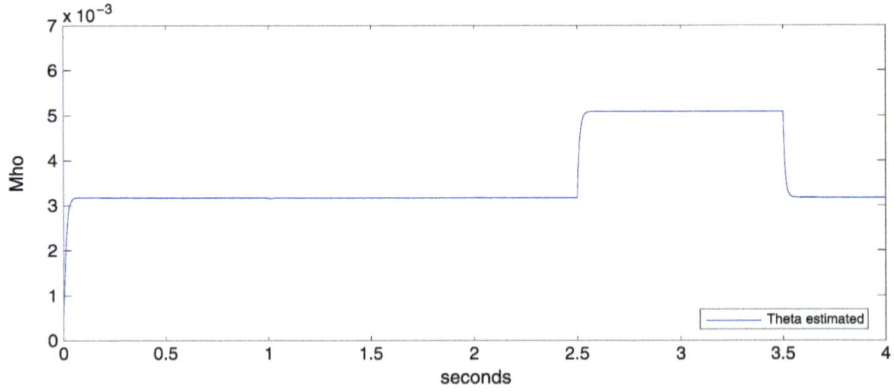

(c) Theta estimation

Fig. 12.14 I&I 2 experimental results

In Figs. 12.12, 12.13, 12.14 we show the behaviour of the state variables and the estimation error for the adaptive PI with the various estimators. As shown in the plots, parameter convergence is achieved in all cases. Due to the reduction of the amplitude in the peaks of v_{C2}, less noisy inductor current i_{L1} and quicker parameter convergence observed in the graphs, we conclude that the I&I 1 controller is more efficient than I&I 2, and this, in its turn, is better than the model reference one.

12.7 Summary

In Sect. 12.1 we proposed three adaptive PI controllers applicable to a large class of switched power converters with unknown load resistance. The work significantly extends the recent results reported in [9]—removing an unnecessary technical condition and adding adaptation. It effectively exploits the key property of passivity of the nonlinear incremental model first reported in [10]. It should be underscored that, as explained in Definition 12.1.1, passivity of the nonlinear incremental model pertains to the original nonlinear system (12.5), not to its linear approximation, and consequently the stability results are global.

It is shown that we can always generate a new output with respect to which passivity can be established. Hence, the incremental output will be driven to zero with a simple linear PI. However, the fact that this output goes to zero does not imply that the control objective is achieved, i.e. that the full state will be driven to the desired equilibrium. Therefore, the additional step of verifying detectability is needed. A simple algebraic (rank) test for the latter is also provided.

It is shown that the I&I approach of [1] generates a family of estimators, parameterised in terms of a free function, $\beta(\chi)$, whose only constraint is to ensure stability of the error dynamics (12.31). Two specific examples, which lead to controllers with different performances, are presented. The estimators add only one nonlinear differential equation to the basic PI and hence are easy to implement in practical applications. It is shown, on the other hand, that a standard model reference estimator is of dimension two with the usual open-loop integration to generate $\hat{\theta}$ that, as is well known, suffers from serious robustness problems in the presence of noise [1, 11].

Detailed calculations are given for a three-phase rectifier and a DC–DC quadratic converter as shown in [4], and the resulting performance is illustrated for the different estimators with simulations and studied in an experimental setup of 50 W.

References

1. Astolfi, A., Karagiannis, D., Ortega, R.: Nonlinear and Adaptive Control with Applications. Springer, Berlin (2007)
2. Desoer, C.A., Vidyasagar, M.: Feedback Systems: Input–Output Properties. Academic Press, New York (1975)

3. Escobar, G., van Der Schaft, A.J., Ortega, R.: A Hamiltonian viewpoint in the modeling of switching power converters. Automatica **35**(3), 445–452 (1999)
4. Hernandez-Gomez, M., Ortega, R., Lamnabhi-Lagarrigue, F., Escobar, G.: Adaptive PI stabilization of switched power converters. IEEE Trans. Control Syst. Technol. **18**(3), 688–698 (2010)
5. Jayawardhana, B., Ortega, R., Garcia-Canseco, E., Castaños, F.C.: Passivity of nonlinear incremental systems: Application to PI stabilization of nonlinear RLC circuits. Syst. Control Lett. **56**(9–10), 618–622 (2007)
6. Jeltsema, D.: Input–output passivity of switched power converters. Technical report (2006)
7. Kassakian, J., Schlecht, M., Verghese, G.: Principles of Power Electronics. Wesley, Reading (1991)
8. Khalil, H.K.: Nonlinear Systems, 3rd edn. Prentice Hall, Upper Saddle River (2002)
9. Perez, M., Ortega, R., Espinoza, J.: Passivity-based PI control of switched power converters. IEEE Trans. Control Syst. Technol. **12**(6), 881–890 (2004)
10. Sanders, S., Verghese, G.C.: Lyapunov-based control for switched power converters. IEEE Trans. Power Electron. **7**(1), 17–24 (1992)
11. Sastry, S., Bodson, M.: Adaptive Control: Stability, Convergence, and Robustness. Prentice Hall, Englewood Cliffs (1989)
12. van der Schaft, A.: L_2-Gain and Passivity Techniques in Nonlinear Control. Springer, London (2000)
13. Zainea, M., van der Schaft, A., Buisson, J.: Stabilizing control for power converters connected to transmission lines. In: Proc. of the American Control Conference, New York, USA, pp. 3476–3481 (2007)

Part IV
Simulating Controlled Converters: A Challenging Issue

Chapter 13
An Overview of Simulation Tools

John Schönberger

13.1 Introduction

Simulation tools are useful for investigating new converter topologies, optimising existing designs and creating virtual prototypes of switched electrical systems. Although the focus of a power electronic simulation tends to be on the converter, the converter's interaction with the controller, source, load and other system modules are also of importance, especially when considering the entire system.

In this section the concept of a modelling hierarchy will be introduced. At each level of the hierarchy, the purpose of the simulation model and the required level of modelled detail are different.

13.1.1 Level of Modelling Detail

A simulation model can be classified as a system, circuit or component-level model as shown in Table 13.1.

> A system-level model can be used to simulate the interaction between different modules in a system such as the converter, controller, source and load. The functionality of slow outer control loops can also be studied. At this level, the low-level switching behaviour of a converter is of no major influence, and a converter can be represented using its transfer function.

J. Schönberger (✉)
Plexim GmbH, Technoparkstrasse 1, Zürich, Switzerland
e-mail: john@plexim.com

F. Vasca, L. Iannelli (eds.), *Dynamics and Control of Switched Electronic Systems*, Advances in Industrial Control,
DOI 10.1007/978-1-4471-2885-4_13,

Table 13.1 Modelling at the system, circuit and component levels

Modelling Level	Purpose	Detail
System	System module interaction Outer control loop functionality	Transfer function representation of converter
Circuit	Converter functionality Inner control loop functionality	Behavioural switch models Instantaneous switching
Component	Switching transients Parasitic effects	Physical/detailed switch models Diode reverse recovery Parasitic circuit elements

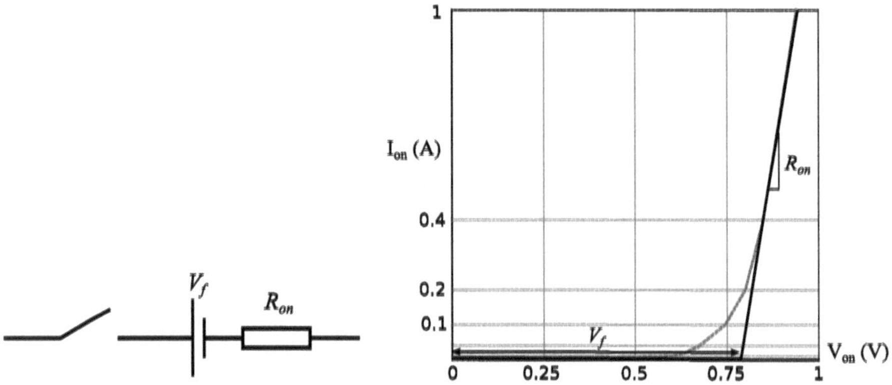

(a) Behavioural diode model with simplified forward characteristic

(b) Forward characteristic of behavioural diode model

Fig. 13.1 Behavioural diode model for use in circuit-level simulations

> The chief purpose of a circuit-level model is to simulate the functionality or large-signal behaviour of a switching converter, and the functionality of the faster inner control loops. At this modelling level, it is sufficient to represent semiconductor devices with simplified behavioural models or ideal models rather than detailed physical models. Switching transitions can also be simplified with the use of instantaneous or idealised switching.

An example behavioural model of a diode that is suitable for use in a circuit-level model is shown in Fig. 13.1. The forward characteristic is modelled in a simplified fashion using two parameters: forward voltage drop, V_f, and ON resistance, R_{on}, as shown in Fig. 13.1a. The reverse blocking characteristic is typically modelled with an infinite or very large resistor.

With appropriately selected parameters, a behavioural model can provide a good approximation of an actual forward characteristic as shown in Fig. 13.1b. The behavioural diode model has no knee at its turn-on point; however, the knee point has no major influence on the large-signal behaviour of the diode when the diode is operated as a switch and conducts currents larger than 0.2 A. To model the exact forward characteristic of the diode, a detailed physical model such as that used in SPICE® would be required. Such a model comprises fourteen electrical parameters and requires a Newton–Raphson-based iteration technique to calculate each time step.

The purpose of a component-level model is to study phenomena such as device voltage and current transients, diode reverse recovery current and device turn-on speed. These phenomena influence the switching-level behaviour of a converter and typically occur within a timeframe in the order of ns.

To create a component-level simulation model, detailed physical device models and circuit stray parameters are needed. The time step must also be very small to simulate these fast phenomena. Due to the increased level of detail and the step size constraints, the simulation speed of component-level models is slower than that of circuit or system-level models.

The three simulation tools described in this chapter are primarily targeted at circuit and system-level modelling. Semiconductor models and switching transition details tend to be simplified because the focus is on modelling the functionality of the electrical circuit and the control system rather than studying the low-level behaviour that occurs during the switching transitions. Appropriate simplification of switch models and other low-level details also improves the simulation speed without adversely affecting the large-signal voltage and current waveforms.

13.2 Solver Types

The simulation tools presented in this chapter use continuous or discrete solvers to simulate the transient response of the system. These solvers, based on different numerical techniques, have their own advantages and drawbacks in terms of accuracy and simulation speed. In this section solver algorithms will also be discussed since this is one key difference between the solvers overviewed in this chapter.

13.2.1 Continuous Solvers

The response of a continuous system can be approximated in a piecewise fashion using numerically constructed Taylor series polynomials as shown in Fig. 13.2. A continuous solver calculates a point, y_{n+1}, that lies on a Taylor series polynomial by evaluating the system dynamics at several intermediate

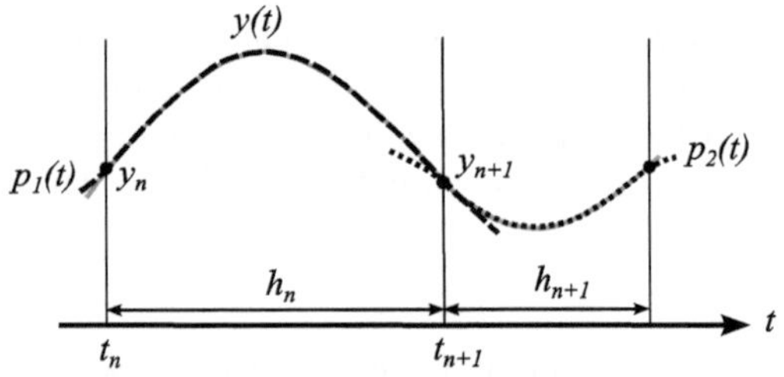

Fig. 13.2 Piecewise approximation of the continuous system state $y(t)$ with the Taylor series polynomials $p_1(t)$ and $p_2(t)$

points between t_n and t_{n+1} and weighting the response. The larger the number of intermediate steps, the higher the order of the solver and the more accurate the result. The result calculated by an nth order continuous solver has the same level of accuracy as an analytical Taylor series polynomial of the same order [1].

Solvers used in switched electrical systems are typically fifth-order solvers, and different families of solvers exist for solving stiff and non-stiff systems.

A continuous solver is often referred to as a variable-step solver due to its adaptive step size capabilities. The solver calculates the next step size based on the relative tolerance, tol_{rel}, and the relative error, ε, as follows:

$$h_{n+1} = \left(\frac{tol_{\text{rel}}}{\varepsilon}\right)^{\frac{1}{5}} h_n. \tag{13.1}$$

The relative tolerance defines the worst-case accuracy limit for the calculated points. The relative error, the relative difference between the fourth- and fifth-order solutions, indicates the actual accuracy of the current point, y_n. The next time step is increased or decreased depending on the accuracy of the current point.

Continuous solvers also include a *zero-crossing detection* algorithm to calculate the exact instant that discontinuities such as switching or comparator events occur. Zero-crossing detection, based on an iteration process, is very precise but adds additional computational overhead since all continuous system states must be updated at each iteration.

Continuous solver algorithms are available for both numerically non-stiff and *stiff* systems. A numerically stiff system is a system with time constants that differ by several orders of magnitude. This causes the system to possess both very small and very large eigenvalues. Power electronic systems often exhibit stiff behaviour since they comprise a mix of fast and slow dynamics.

Non-stiff solvers calculate the next state values in a single step using the system differential equations and the previous state values. On the other hand, stiff solvers require multiple iteration steps to calculate the next state values. The solver uses an estimate of the next state to calculate the next state. It then iterates until the estimate is equal to the calculated result. A stiff solver therefore requires more computations to calculate the next state values and is inherently less efficient than a non-stiff solver.

Although stiff solvers have a larger computational overhead, they are more efficient for solving numerically stiff models. A non-stiff solver typically becomes locally unstable when used to solve a stiff system unless its time step is set to a prohibitively small value. The stability region of a stiff solver is larger than that of a non-stiff solver, and much larger time steps can be taken without encountering instability problems.

13.2.2 Discrete Solvers

Discrete solvers are simple and can provide accurate simulation results with a properly selected step size.

Discrete solvers are typically based on the trapezoidal integration method, which approximates the transient response of a continuous system between two adjacent time points with a line segment. The operation of the trapezoidal integration method is shown in Fig. 13.3. If the system dynamics are described by an equation of the form

$$\frac{\mathrm{d}y}{\mathrm{d}t} = f(t), \tag{13.2}$$

then the trapezoidal rule can be expressed by

$$y_{n+1} = y_n + \frac{h}{2}\big(f(t_n) + f(t_{n+1})\big). \tag{13.3}$$

The accuracy of the trapezoidal integration method is linked to both the integration sample time, h, and the transient frequency of the system. If the sample time is too large in relation to the highest transient frequency in the system, the integrator underestimates the transient response [3]. To ensure the per-step underestimate is less than 3%, the integrator sample time should be ten times smaller than the period of the highest frequency in the system. Thus there is a tradeoff between speed and accuracy.

In continuous systems, it is the highest transient frequency that imposes a limit on the integrator sample time. For switched electrical systems, the converter switching

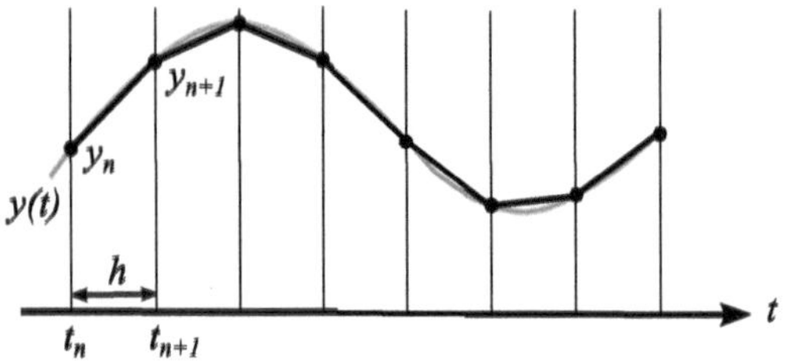

Fig. 13.3 Trapezoidal integration method

frequency normally limits the integrator sample time, since the switching resolution is governed by the sample time. In general, the sample time should be at least 100 times smaller than the switching period to ensure adequate switching resolution. For example, a converter with a switching frequency of 10 kHz should have sample time that is less than 1 μs.

However, when transients occur within a switching cycle, the limiting factor once again becomes the period of the high-frequency transient. Sub-switching cycle transients are generally caused by switch parasitics such as diode reverse recovery charge or switch snubbers. However, such phenomena are often not modelled in system-level simulations, and it is therefore the switching frequency that typically limits the integrator sample time.

Discrete solvers typically do not include a sophisticated zero-crossing algorithm such as that in a continuous solver. Discontinuous events such as switching are generally only processed at each sampling instant. However some discrete solvers implement a simple form of zero-crossing detection based on backward interpolation [10].

When the current through a diode becomes negative for example, the current is linearly interpolated backwards to predict the turn-off instant that would normally occur between sampling instants. This helps improve the simulation accuracy.

13.2.3 Comparison of Solver Types

To highlight the difference between a variable-step and fixed-step solver, a simulation model of the simple resistor–inductor–capacitor (RLC) circuit shown in Fig. 13.4 is used. The example circuit in Fig. 13.4a is moderately stiff since the resistor–capacitor (RC) time constant is 100 times larger than the resistor–inductor (RL) time constant. From the transient response of the current in Fig. 13.4b it can be seen that the circuit is overdamped. The fast rise time is due to the RL time constant, and the slow decay time is dominated by the RC time constant.

The circuit was simulated in PLECS Standalone using a fifth-order variable-step stiff solver and a fixed-step solver. The time steps taken by the solvers are depicted

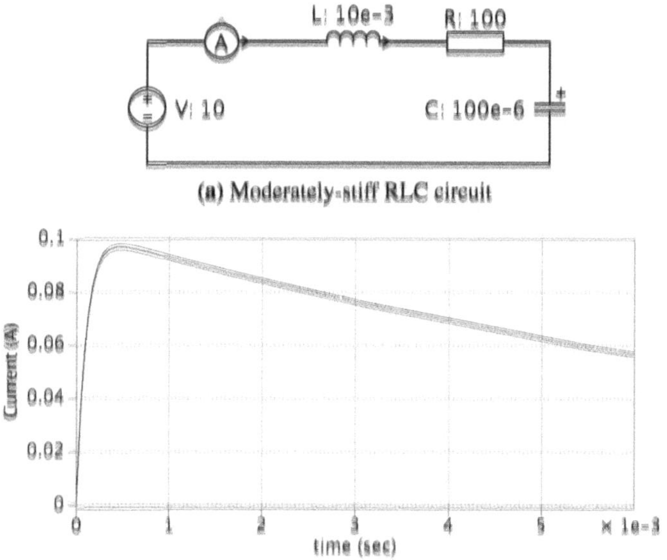

(a) Moderately-stiff RLC circuit

(b) Transient response of current

Fig. 13.4 Transient response of an RLC circuit

in Fig. 13.5. The adaptive time step of the variable-step solver can be clearly seen in Fig. 13.5a. With the relative tolerance set to 10^{-6}, the step size is 9 μs at the beginning of the simulation and 2.38 ms after 3 ms. The time steps taken by the fixed-step solver are depicted in Fig. 13.5b. The sample time was set to 50 μs in order to accurately simulate the fast transient at the beginning of the simulation.

In terms of speed, the variable-step solver is faster, simulating the model 3.5 times faster than the fixed-step solver. Although the variable-step solver must construct an equivalent Taylor series polynomial to calculate the next point, its larger time step more than compensates for this additional computational overhead. Once the initial transient response has decayed, the step size taken by the variable-step solver is approximately 50 times larger than the step size of the fixed-step solver.

Although the variable-step solver is more efficient than the fixed-step solver for the example RLC circuit, this may not always be the case.

The exact performance of a solver is dependent on both the type of circuit and the type of solver. A large circuit with many internal transition events may simulate slower with a variable-step solver due to heavy use of the iterative zero-crossing algorithm.

In terms of accuracy, both solvers can produce accurate simulation results. Accurate results are easier to achieve with a variable-step solver since the solver automatically adjusts its time step to meet the accuracy constraint defined by the relative tolerance. With a fixed-step solver, the accuracy is indi-

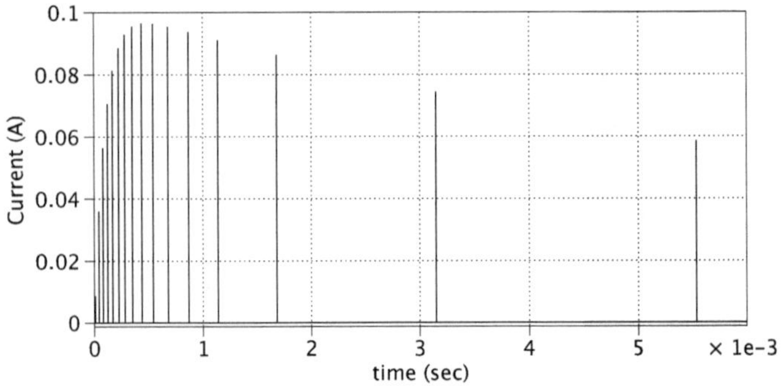

(a) Variable-step stiff-solver with relative tolerance set to 10^{-6}

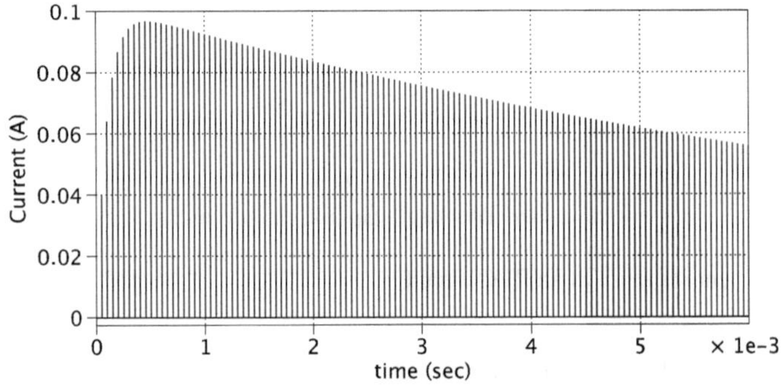

(b) Fixed-step solver with sample time set to 50 μs

Fig. 13.5 Time steps taken with a variable-step and fixed-step solver

rectly controlled using the step size. The accuracy of the simulation can only be verified by comparing the results with a continuous simulation or repeating the simulation for a smaller step size and observing changes in the transient response.

13.3 SimPowerSys

SimPowerSystems (SPS) is an electrical toolbox for Simulink that is produced by the Mathworks. SPS allows circuits to be modelled directly in Simulink using electrical components rather than representing the circuit with differential equations. Although SPS is targeted at simulating power systems, it can also be used for simulating switched power electronic converters. SPS is dependent on both Matlab and Simulink and is available in cross-platform versions.

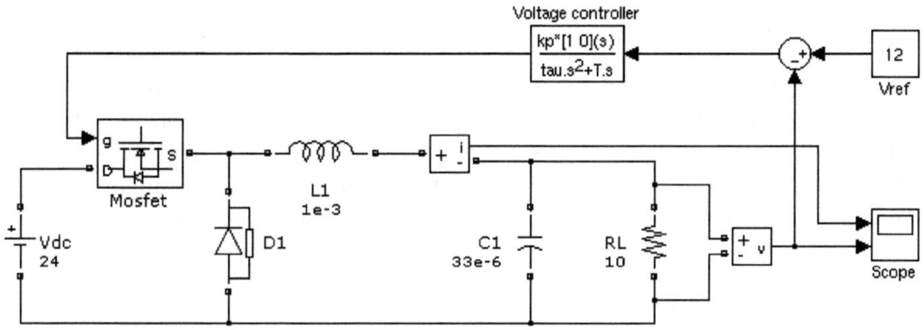

Fig. 13.6 Buck converter modelled in SPS. SPS components are used for the electrical circuit, and Simulink components for the controllers

An example model of a buck converter created in SPS is shown in Fig. 13.6. SPS elements are used to model the electrical circuit and measurements, and Simulink components are used to model the control system. The Simulink library offers a wide range of continuous and discrete control elements, and components and other tools offered by other Simulink toolboxes can also be used in SPS models.

13.3.1 Features

In this section, an overview of some of the important features of SPS will be given. For an exhaustive description of features, one should refer to the SimPowerSystems User's Guide [6].

13.3.1.1 Library

The SPS library includes individual electrical components and switches, and a range of power electronic converter and machine models. Specialised components such as transmission lines, transformers and three-phase loads are included for modelling power systems. Custom components can also be created using the Simulink subsystem block.

SPS and Simulink components are vectorisable. Vectorisation also allows multiple components to be created from a single component by entering the component parameters in array format. It also permits single-line representation of multiple signals. Blocks such as control gains, integrators and electrical elements are dynamically sized when a vectorised signal is detected. Vectorisation helps prevent clutter on the circuit diagram for multiphase systems since elements that are common to each phase do not require duplication.

Another useful component in the Simulink library is the S-function, which allows the user to write a custom function in C or Matlab code. In power electronic system models, the S-function can be used for tasks such as executing C control code at a fixed sample rate.

13.3.1.2 Scope

The Simulink scope is used for displaying the results of an SPS simulation. The scope supports multiple plots and multiple traces on a single plot but otherwise offers rudimentary features. The more-advanced Matlab figure tool, run from the Matlab command line, can also be used for post-viewing saved simulation results. The figure tool offers fine-grained control over many aspects of a plot's appearance.

13.3.1.3 Analysis Tools

Matlab provides a comprehensive scripting environment for SPS models. SPS models can be run from the Matlab command line, and component parameters can be referenced by variables. This scripting functionality is useful for performing advanced parameter sweeps, where multiple component values or operating conditions are changed during each simulation run.

SPS itself includes a number of specific inbuilt analysis tools such as load flow analysis for determining the steady-state operating point of a power system, impedance analysis for determining the frequency-dependent impedance at a specific point in a network and Fast Fourier Transform (FFT) analysis for calculating the Fourier spectrum of a signal.

13.3.1.4 Other Features

A useful simulation mode in Simulink is the rapid accelerator mode. As long as the SPS model contains blocks that support code generation, running a model in this mode converts the model into C-code and compiles it into a standalone executable. This simulation mode is useful for speeding up the simulation time of large models for which the simulation run time is significantly longer than the model compile time.

Additional components and analysis tools from other toolboxes in the Simulink ecosystem can also be used with SPS. One toolbox worth mentioning is the Real-Time Workshop, a toolbox for generating C-code from a selected model or subsystem. The C-code can be used for simulation acceleration, real-time controller prototyping and hardware in the loop applications. Another option is to create C-code that can be directly imported into the development environment of a DSP.

13.3.2 Switch Models

In SPS, switches are classified [6] as either purely resistive switches, inductive switches or ideal switches. The purely resistive switch is modelled as a current source and is part of the linear state-space model of the circuit [6]. The switch must have a finite on resistance. When it is off, its resistance is infinite. At each switching event, the state-space system is updated using a process similar to that described in [2].

The purely resistive switch works with both continuous and discrete solvers, and it is the default switch used by SPS for switches with no internal inductance. However, this switch model has the major drawback that RC snubbers must be connected in parallel with each switch to ensure that a state-space system can be derived when the switch is in series with an inductor or an open switch. When used with a continuous solver, the RC snubbers significantly slow down the simulation since their dynamics must be simulated at each switching event. In addition, the snubbers can impact on the precision of the simulation if not chosen correctly.

A range of forced and naturally commutated switches have been implemented in SPS as purely resistive switches. Internally, the only difference between the device models is the implementation of the switching logic and the ability to add a tail current to some devices.

A switch that does include a series inductance is referred to as an inductive switch [6]. Inductive switches are modelled as current sources that are external inputs to a continuous state-space model of the passive circuit elements. The continuous state-space system is not updated at switching events; instead, the external current sources are a function of the device voltage. More detailed information about the modelling of a converter as continuous state-space system with discontinuous current source inputs can be found in [4].

The inductive switch can only be used with a continuous solver and requires a numerical snubber connected in parallel to ensure numerical convergence during switching. Without the snubber, connecting the switch in series with an inductor, current source or an opened switch is not possible. Care must be taken when choosing the snubber values to ensure the system does not become numerically ill-conditioned. Since this type of current source switch requires the use of a numerical snubber and is restricted to use with a continuous solver, it is rarely used in practice.

In SPS, switches can be configured to switch ideally. Ideal switching is enabled with the Power Graphical User Interface (GUI), an SPS component used for global configuration of the SPS circuit. When using ideal switching, switch snubbers can also be globally disabled.

Ideal switching should not be confused with the ideal switch element, a component that is available in the SPS library. The ideal switch element includes an on-resistance and optional snubber element, and can be switched using the classical method or the ideal switching method.

With ideal switching, the switch is modelled as a resistor that has an infinite resistance when OFF and a zero resistance when ON. An optional ON resistance and a forward voltage drop can be added to the switch. The switches thus appear as resistive elements in the state-space matrix that represents the circuit topology. During a switching transition, the new state-space model of the new circuit topology is recalculated, and the switching occurs instantly.

Ideal switching is only possible with a continuous solver. Snubbers are not needed to ensure numerical convergence, and therefore the snubber dynamics do not need to be simulated. The number of time steps is therefore reduced, and the simulation speed is improved compared to models that employ a continuous solver and purely resistive switches.

13.3.3 Solver Algorithms

13.3.3.1 Continuous

An SPS circuit is solved using the Simulink solver, which includes continuous and discrete solver algorithms. The default continuous solver in Simulink is the ode45 solver, a fifth-order, variable-step, non-stiff solver that is based on the Dormand–Prince method. SPS models that contain switches tend to be stiff, and therefore a stiff solver should be used to achieve the best simulation performance. The recommended stiff solvers for SPS models are the ode23tb and ode15s solvers [6].

13.3.3.2 Fixed-Step Discrete

As mentioned in Sect. 13.3.2, using purely resistive or inductive switches with a continuous solver results in a slow simulation due to the numerical snubbers. Discretising an SPS circuit (in the current version it is possible only with purely resistive switches) can address this performance issue. With an appropriate sample time, the transient response of the numerical snubbers is effectively ignored, yet the important dynamics of the system are captured. For large continuous systems that experience a large number of zero-crossings, discretisation can also improve the simulation speed since the discrete solvers do not perform zero-crossing detection.

An SPS circuit model is discretised by selecting the discrete simulation option in Power GUI interface and entering a sample time. The circuit is discretised using the Tustin method, which is equivalent to trapezoidal integration. A discretised circuit can have a continuous control system, since Simulink supports hybrid systems. However, to achieve a faster simulation speed, a discrete control system should be used [6].

In a discretised circuit, forced-commutation switches are turned on and turned off at the sampling instants. However, naturally commutated switches turn off at the zero-crossing of the current, which typically occurs between sampling instants. In SPS, naturally commutated devices are not turned off at the zero-crossing of the current but rather at the sampling instants. This results in a small negative current which can lead to numerical oscillations [6].

Aside from addressing simulation speed issues, another reason to use a fixed-step discrete algorithm is to allow the SPS model to be ported to the Real-Time Workshop, a Simulink toolbox that converts a model into C-code. The Real-Time Workshop is only compatible with fixed-step solvers.

13.3.4 Summary

Although SPS was developed for simulating power systems, it can be also used for power electronic simulations. The converter and machine modelling capabilities of SPS are complemented by the control modelling capabilities of Simulink and the scripting options of Matlab. The simulation environment can also be further extended by using other toolboxes from the Simulink ecosystem. Both variable- and fixed-step simulations are supported; however, the use of non-ideal switches with a continuous solver should be avoided for performance reasons.

When considering SPS as a tool for simulating power electronics, one shortcoming is the lack of support for modelling semiconductor conducting and switching losses as a function of voltage, current and temperature. This feature, offered by both PLECS and PSIM, can be very useful for simulating the thermal performance of switching converters.

13.4 PLECS

PLECS is a product of Plexim GmbH and is available as two products, PLECS Blockset and PLECS Standalone. PLECS Blockset is an electrical toolbox for Simulink that is specifically designed for the simulation and analysis of power electronic systems. As in SPS, a PLECS Blockset model exists within the Simulink ecosystem, allowing Simulink components and additional Simulink toolboxes to be used in conjunction with PLECS.

PLECS Blockset models are created by placing components inside a PLECS circuit, a special subsystem that is placed on the Simulink worksheet. A buck converter that is modelled in PLECS Blockset is shown in Fig. 13.7. The control system can be modelled in either the PLECS circuit or on the Simulink worksheet, since PLECS includes control components in addition to electrical components. To model the control system in Simulink, connection ports are placed on the PLECS circuit, forming a direct connection to the Simulink worksheet.

A PLECS Blockset circuit is represented internally as a native Simulink S-function and is solved directly by the Simulink solver rather than relying on a co-simulation approach. On the other hand, PLECS Standalone is an independent program with its own inbuilt solver. PLECS Standalone shares the same component library and feature set as PLECS Blockset. Both types of PLECS circuit models are compatible with each other, and both PLECS programs are cross-platform versions. For both programs, additional functionality such as thermal modelling capabilities come bundled with the program rather than being offered as an add-on modules.

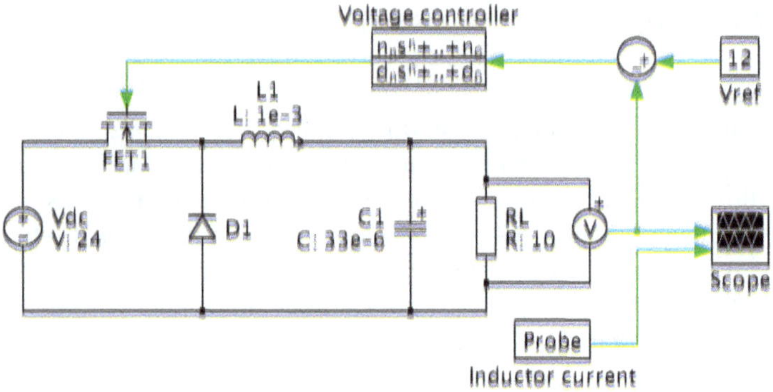

Fig. 13.7 PLECS Blockset model of a buck converter with the control system implemented using PLECS components

13.4.1 Features

In this section, an overview of the key features of PLECS is given. A complete description of the features can be found in the Plexim User Manual [8].

13.4.1.1 Library

The PLECS library includes individual electrical elements and switches as well as controller elements and complete converter and machine blocks. ANSI or DIN format can be selected for the appearance of certain components. Larger hierarchical models such as machine and converter models use an open subcircuit implementation, allowing easy inspection and customisation of the model. PLECS components support vectorisation, which is described in detail in Sect. 13.3.1.1.

Custom components can be created using the subsystem or configurable subsystem blocks. A subsystem can also be used for creating multi-sheet or hierarchical simulation models, and a configurable subsystem allows different physical implementations of custom blocks such as an input filter or controller to be created.

A Probe block provides a convenient method of making measurements without physically inserting a meter. A component is dragged inside the Probe block, and the appropriate signals are selected.

A C-Script block is included in the library for implementing advanced mathematical functions or custom control code. The code is directly entered into the inbuilt editor and compiled on the fly. The C-Script block is compatible with both continuous and discrete solvers.

13.4.1.2 Thermal Simulation

The PLECS library includes a thermal editor for describing semiconductor conduction and switching losses and dedicated thermal components for modelling thermal elements. The key component in the thermal library is the heatsink, which automatically absorbs any losses generated by a component that is placed within its bounds. Thus it conveniently eliminates the need for switching and conduction loss terminals on the semiconductors.

Thermal modelling in PLECS is based on the lookup table approach, which is fast and accurate [7]. With the lookup table approach, a switching loss surface for a component is defined in Joules as a function of voltage and current. Multiple loss-surfaces can be defined to account for the temperature-dependent nature of the losses. Directly after a switching event, the switching energy loss is obtained from the lookup table using the device voltage, current and temperature.

Conduction losses are defined by creating temperature-dependent conduction profiles of the forward voltage drop as a function of current. At each simulation step, the voltage is multiplied by the current to yield the conduction losses in Watts.

During a simulation run, the electrical and thermal circuits are solved simultaneously. The switching and conduction losses are obtained from the lookup table as functions of device voltage, current and junction temperature. These loss values are fed to the thermal circuit, which is solved to yield the junction temperatures. The junction temperatures are in turn fed back to the lookup tables.

13.4.1.3 Scope

Although the Simulink scope can be used for viewing simulation results, the PLECS scope has been specifically designed for power electronics engineers and includes a number of features that are useful for displaying and analysing electrical simulation results. Cursors are available for performing direct measurements, and advanced measurements such as total harmonic distortion (THD) and FFT of signals can be calculated. Zooming can be performed during a simulation, and results can be exported in graphical or 'csv' file format directly from the scope. Since a PLECS Blockset circuit natively resides in the Matlab/Simulink environment, the Matlab figure tool can also be used.

13.4.1.4 Analysis Tools

In addition to its simulation engine, PLECS includes analysis tools such as the Steady-State, AC Sweep and Impulse Response tools. The Steady-State Analysis

Tool is designed for quickly calculating the final value of a system rather than arriving at this value through simulation. Based on the Newton–Raphson iteration method, this tool is useful for reducing the simulation time in combined electrical–thermal simulations, where the slow thermal time constant would result in a long transient simulation.

The AC Sweep Tool is useful for obtaining converter transfer functions or frequency-dependent impedances within a system. During a simulation run, the model is perturbed with a sinusoidal signal over a range of frequencies. A Bode plot showing the composite frequency response is displayed at the end of the analysis run.

The Impulse Response Analysis Tool can also be used in control design workflows. It obtains the frequency response of an open-loop system using a special perturbation and analysis method [5]. A perturbation is applied to the system during a single switching period, from which the entire frequency response of the system can be obtained. The impulse response tool is an order of magnitude faster than the AC Sweep tool since the frequency response can be obtained from a single brief perturbation.

13.4.1.5 Other Features

Aside from the features that have been directly implemented in PLECS Blockset, virtually all of the features that are offered by Matlab/Simulink are compatible with PLECS due to its direct integration with Simulink. For example, a PLECS Blockset model can be configured and simulated from the Matlab command line, allowing advanced initialisation scripts and parameter sweeps to be executed.

An add-on feature for PLECS Blockset is the Real-Time Coder, which converts a fixed-step PLECS model into C-code that can be used for real-time simulations. Integration with Simulink's Real-Time Workshop, which provides the same functionality for Simulink models, is also possible.

PLECS Blockset is also compatible with Simulink's Accelerator and Rapid Accelerator simulation modes. In these simulation modes, a model is converted into C-code, then compiled and run as a native program. When either of these simulation modes are invoked, the PLECS model is also converted into C-code, which is seamlessly integrated with the C-code generated by Simulink.

13.4.2 Switch Models

Since PLECS is primarily designed for modelling at the circuit and system levels, most switches are modelled as ideal resistors. When the switch is ON, it has zero resistance, and when it is OFF, its resistance is infinite. An optional forward voltage drop and ON resistance can be added to each switch. Switching occurs instantly and numerical snubbers are not needed to ensure numerical convergence during a switching transition.

In addition to the ideal switch models, PLECS includes three non-ideal switch models: an IGBT and metal–oxide–semiconductor field-effect transistor MOSFET with limited $\frac{d}{dt}i$, and a power diode that implements the reverse recovery effect using a macro-model. Although these models are useful for studying component-level phenomena, they create additional computational overhead since the models are more detailed and additional time steps are required to simulate the transient response during switching.

13.4.3 Solver Operation

A PLECS circuit is modelled internally as a state-space system that can be solved using the continuous state-space or discrete state-space method. With the continuous state-space method, PLECS Blockset handles the switching events itself and uses Simulink to solve the system of differential equations. The circuit is treated as a linear system until a switching event occurs. The PLECS switch manager detects the exact time the switching event occurs by communicating with Simulink's zero-crossing detector. It then recalculates the state-space matrices for the new circuit topology. Switching occurs instantly, and the newly-calculated state-space matrices are cached for speed.

A PLECS circuit can be discretised and solved using the discrete state-space method as long as it contains no continuous blocks such as continuous integrators. PLECS discretises and solves the circuit using the bilinear transformation, which is equivalent to trapezoidal integration. In the discretised system, forced-commutation devices are switched at the sample instants; however, naturally commutated devices such as diodes may switch OFF between sample times. For accuracy, PLECS employs an efficient linear interpolation scheme to determine the instant at which a naturally commutated device turns off.

The discrete state-space method is compatible with both variable-step and fixed-step Simulink solvers. Both discrete and continuous control systems can be used with a discretised circuit model.

13.4.4 PLECS Standalone

PLECS Standalone is an additional product that is offered by Plexim. Providing the same functionality as PLECS Blockset, PLECS Standalone operates independently from Matlab/Simulink and therefore includes its own solvers. In contrast to PLECS Blockset, only control components from the PLECS library are used to model the control system. With PLECS Blockset, a controller can be implemented using Simulink or PLECS components.

PLECS Standalone includes both continuous and discrete solvers. Two types of continuous solvers are available for solving numerically stiff and non-stiff systems. For non-stiff systems, the Dopri solver, a fifth-order solver based on the Dormand–Prince method, is provided. The Dopri solver includes a stiffness detector to warn the user when numerical stiffness is detected. For stiff systems, a fifth-order Radau solver is used. This solver has particularly good stability characteristics [1]. For fixed-step, discrete simulations, a solver based on the trapezoidal integration method is used.

13.4.5 Summary

PLECS, available as a toolbox for Simulink or as a standalone product, is specifically designed for the modelling of switched electrical systems. In addition to its electrical modelling capabilities, the base package offers useful features such as thermal loss modelling and AC analysis. PLECS Blockset is hosted inside the Simulink environment and achieves native integration with Simulink through use of the S-function interface. The PLECS circuit is solved by Simulink, but the switching events are handled by PLECS itself.

PLECS Standalone is a leaner simulation tool that functions independently from Matlab and Simulink and therefore includes its own variable-step and fixed-step solvers. PLECS Standalone shares the same component library and core feature set as PLECS Blockset and allows models to be exported to PLECS Blockset. PLECS Standalone is a suitable alternative to PLECS Blockset when toolboxes or other functionality from the Simulink ecosystem are not needed.

13.5 PSIM

PSIM is a product of Powersim Inc. The base package comprises a schematic editor, a fixed-step simulation engine and a waveform viewer. The program

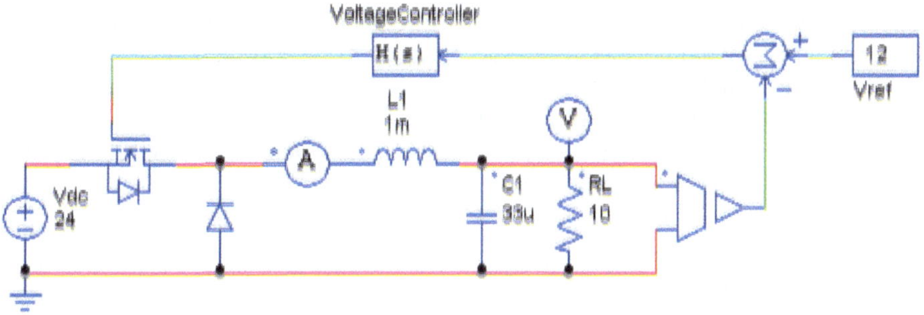

Fig. 13.8 Buck converter modelled in PSIM

is suitable for modelling switching power electronic converters and their controllers and loads at the circuit and control system levels, having a range of electrical and control components.

A buck converter modelled in PSIM is shown in Fig. 13.8 with the voltage controller implemented as a continuous transfer function. The simulation workflow is different to that of SPS and PLECS. Rather than using a scope to measure the results, the signals to be measured are probed. At the end of a simulation run, the probed signals are displayed and analysed using the Simview tool.

The PSIM base package is available in two versions, a standard version and professional version. The standard version has limitations on the model size and library features. The base packages can be extended with add-on modules that provide additional functionality or permit interfacing with third-party software tools. PSIM is available for the Windows platform only.

13.5.1 Features

An overview of the main features of PSIM is given here, and a complete description of the features can be found in the PSIM User Manual [9].

13.5.1.1 Library

The base PSIM package includes fundamental components such electrical and control elements, switches and meters, and additional components such as electrical machines and digital controller components can be accessed by installing add-on modules. Compared to SPS and PLECS, electrical machines have a closed implementation, which prevents inspection and customisation of the model. In addition, components are not vectorisable. However, the necessary building blocks for creating circuit and control level simulations are present, and magnetic elements for creating equivalent magnetic circuits are available. Other useful functionality such as thermal loss modelling can also be accessed by using add-on modules.

The professional version of PSIM includes a C-code block to allow the user to enter custom C-code for modelling custom controllers or components. The professional version also includes a subcircuit component, which can be used for creating custom components or hierarchical simulation models.

13.5.1.2 Thermal Simulation

With the thermal add-on module, device losses can be modelled based on the lookup table approach. Temperature-dependent descriptions of the device conduction and switching losses are defined for semiconductors using the device database editor. The database editor includes a convenient graph wizard for importing a conduction profile from a datasheet graph.

The conduction and switching losses are modelled internally as current sources that are connected to loss terminals on the device. The thermal circuit is created using electrical components, and thus thermal measurements are obtained from their equivalent electrical values. For example, the device junction temperature is equivalent to the voltage at the switching loss terminal.

13.5.1.3 Scope

PSIM offers two tools for viewing results. One is Simview, a viewer program or scope for post displaying and analysing the saved simulation results. With PSIM, any monitored signal is saved to file, allowing it to be accessed at the conclusion of the simulation using Simview. Simview allows the user to select which signals are plotted and to perform measurement functions on the data including THD or an FFT. For live viewing of simulation results, a scope component with one, two or four channels can be used. The scope component is not as comprehensive as Simview since only a single signal can displayed on each channel and the only persistent data is that which is displayed in the current view.

13.5.1.4 Analysis Tools

PSIM includes an AC sweep tool and parameter sweep tool in addition to its simulation engine. Based on a sinusoidal perturbation technique, the AC sweep tool is used for calculating the control-to-output transfer function of a converter or the closed loop gain of a control system. The parameter sweep tool increments a specified parameter over multiple simulation runs and plots the results against time and the last value against the swept parameter.

13.5.1.5 Other Features

The base product includes command line access, which can be used for scripting applications. However, because PSIM is an extension-based simulation tool, most additional functionality, interfaces to third-party products and additional library components are accessed by purchasing additional add-on modules. Library add-on modules include a motor drive component module, a renewable energy application example module and a digital control component module. Modules that expand the functionality of the program include the thermal module described in Sect. 13.5.1.2 and a control design module for visually tuning a control loop.

Included among the external interface modules is the SimCoupler module, which provides a co-simulation interface between PSIM and a Simulink model configured to provide a fixed-step output. The MagCoupler module provides a dynamic link between PSIM and JMAG, a finite element analysis program used for electromagnetic field analysis. The SimCoder and TI F28335 Target modules allow the user to generate C-code from a control schematic diagram and to port the code to a Texas Instruments TI F28335 floating-point digital signal processor (DSP).

13.5.2 Switch Models Implementation

PSIM offers two types of switches. The first type is the ideal switch, intended for use in switched converter models. These switches have an ON state and an OFF state that are modelled using resistors. Snubbers are not required to ensure numerical convergence. When the switch is ON, its resistance is $10\,\mu\Omega$, and when the switch is OFF, its resistance is set to $10\,\mathrm{M}\Omega$ [9]. The ideal switch model can be extended by adding a forward voltage drop and an ON resistance.

Linear models for a bipolar junction transistor (BJT) and MOSFET switch are also included in PSIM. These switch models include a behavioural representation of their respective linear operating regions. Rather than being designed for fast switching, they are intended for use in models such as linear amplifiers or voltage regulators.

13.5.3 Solver Operation

PSIM uses a fixed-step solver that is based on the trapezoidal integration method. The trapezoidal integration method, described in Sect. 13.2.2, uses a straight-line approximation between two data points. Although this is not as accurate for a given step size as the polynomial fit provided by variable-step solver, good accuracy can be assured with appropriate selection of the step size. As long as the time step is at least an order of magnitude smaller than the switching period or the fastest transient period in the system, good accuracy can be achieved as described in Sect. 13.2.2.

During a simulation, the solver implements an interpolation technique to update the system if switching devices commutate between sampling instants. The power circuit and the control circuit are also solved separately. There is one time step delay between the power and the control circuit to avoid an algebraic loop [9].

13.5.4 Summary

In terms of core functionality, there are more similarities between PSIM and PLECS than between PSIM and SPS. As with PLECS, PSIM provides modelling capabilities for switched electrical converters and control systems, and supports the modelling of semiconductor switching and conduction losses. Aside from these similarities, the main differences lie in the user interface and the solver implementation. PSIM uses a fixed-step simulation engine, whereas PLECS and SPS support both variable- and fixed-step simulations.

One notable omission from PSIM is the lack of component vectorisation for sources and control blocks, which allows multiphase systems to be represented in a simplified fashion. However, PSIM is supported by a number of add-on modules that expand the core functionality of the program and provide interfaces with other simulation tools.

13.6 A Numerical Test

To compare the speed of the three simulation tools, the neutral point clamped (NPC) converter shown in Fig. 13.9 was simulated. The NPC converter, connected to the grid via an inductor–capacitor–inductor (LCL) filter, employs a cascaded control strategy. The inner current control loop employs a proportional control strategy, and the outer voltage loop is based on a proportional–integral (PI) controller. The switch modulation strategy is based on multilevel PWM and switches at 10 kHz. The transient simulations, timed for a simulation run of 0.2 s, were carried out on an Intel dual-core personal computer with a memory capacity of 2 GB and a clock frequency of 2 GHz.

13.6.1 SimPowerSystems

Continuous and discrete models of the NPC converter system were created using Matlab/Simulink 2010b and SimPowerSystems 5.3 (2010). With the continuous

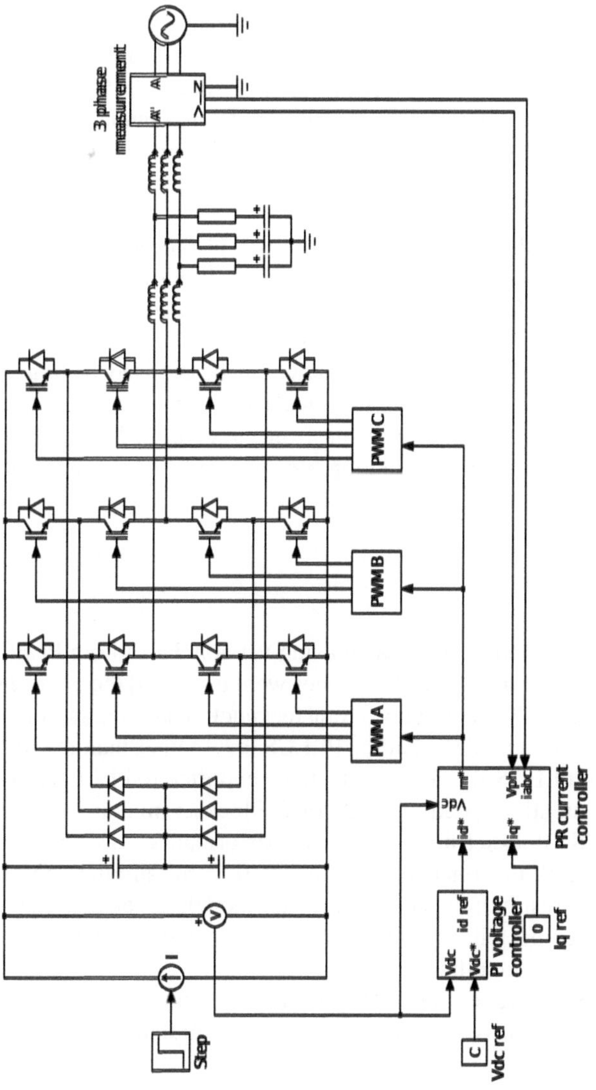

Fig. 13.9 NPC converter system

model, the ideal switching method was selected, and the switch snubbers were disabled. The control system was based on a continuous transfer function and integrator.

Several stiff and non-stiff solvers were tested, and it was found that the ode23tb stiff solver produced the best performance, closely followed by the ode15s stiff solver. With a relative tolerance of 10^{-3}, the simulation time was 81.3 s.

In a bid to improve the simulation speed, a discrete model using ideal switches was created. The model was discretised with a time step of 1 μs as a compromise between speed and accuracy. The circuit could be discretised automatically using the PowerGUI; however, the control system had to be manually recalculated and converted into a discretised form by replacing the continuous controllers with their equivalent discrete counterparts. The discretised model simulated over 50% faster than the continuous model, taking 35.8 s.

The use of the Simulink Rapid Accelerator was also tested. The simulation time for the continuous model was reduced slightly to 79.9 s and that for the discrete model increased to 46.0 s. These results can be attributed to the compile process, which took approximately 20 s in both cases.

To assess the potential speed gains that could be realised in larger simulation models where the simulation time is significantly longer than the compile process, the discretised simulation model was run from precompiled code. Neglecting the compile and build process, the simulation time was reduced from 35.8 s to 18.3 s.

13.6.2 PLECS

A continuous system model was created using version 3.1 of both PLECS Blockset and PLECS Standalone. Ideal IGBT switches were used, and the control system was modelled using the PLECS continuous transfer function and integrator components. The control system was modelled at the PLECS level rather than the Simulink level to allow the entire PLECS Blockset model to be exported to PLECS Standalone and to improve the simulation speed of the PLECS Blockset model.

For the PLECS Blockset model, optimal performance was obtained using the ode23tb stiff solver. With a relative tolerance of 10^{-3}, the simulation time was 9.4 s. Compared with the result of the continuous SPS model, which also implemented an ideal switching method, this would indicate the ideal switching technique implemented in PLECS is more efficient.

With PLECS Standalone, the fastest results were obtained using the non-stiff, Dopri solver. With a relative tolerance of 10^{-3}, the simulation time was 3.9 s.

13.6.3 PSIM

The NPC system was modelled using version 9 of PSIM. Idealised switching elements were used, and the control system was based on a continuous transfer function and integrator. Although continuous control elements can be used, these elements are discretised internally to ensure compatibility with the fixed-step trapezoidal solver. However, being able to enter a transfer function in its continuous form is more convenient when the control design is given as a continuous transfer function. Otherwise the controller must first be discretised.

The sample time was set to 1 μs as a compromise between speed and accuracy. The simulation time, measured to the nearest lowest second, was 17 s.

13.7 Summary

A commonly performed task in the design process for a power electronic converter is the functional modelling of the converter and its control system. The three programs presented in this chapter, SPS, PLECS and PSIM, are capable of this task, being circuit and control system-level simulators. At these simulation levels, the focus is on the functional behaviour of the converter and its interaction with the control system or other system modules. Therefore, switch models tend to be idealised for speed and simplicity. Some of the programs offer detailed macro-models of switches such as an IGBT with limited $\frac{\mathrm{d}}{\mathrm{d}t}i$; however, these programs are not designed for pure component-level simulation. This is the domain of other programs such as SPICE.

SPS is an electrical simulation toolbox for Simulink that allows circuits to be modelled within the control environment provided by Simulink. SPS can be used for functional modelling of a switched electrical circuit and control system and includes features that are targeted towards the simulation of power systems. Functionality can be extended with a large number of toolboxes from the Simulink ecosystem including the Real-Time Workshop.

PLECS Blockset is a third-party circuit simulation toolbox for Simulink that is integrated natively with Simulink and is designed specifically for simulating power electronic systems. In addition to its electrical simulation capabilities, it includes additional functionality such as thermal modelling and control design, which are inbuilt in the base package. Additionally, the Real-Time Coder tool is offered as an add-on tool. PLECS is also available as a standalone program that includes its own variable and fixed-step solvers.

PSIM is a fixed-step circuit simulation tool that also targets the simulation of power electronic converters and their controllers. Standard and professional versions of the program are available, and in both versions, the basic package can be extended with add-on modules that offer features such as thermal modelling, control design and integration with third-party tools such as JMAG.

Although there is a large crossover in the feature sets of the three tools, there are also some notable differences. The obvious differences are the inner workings of the solver algorithms and the features on offer, such as thermal modelling, embedded code generation and integration with third-party software products. These are the key factors that should be considered when comparing the simulation tools. Other differences include simulation speed, supported computer platforms and pricing structure. Care should be taken when comparing products that have different price structures as some features come included in the core product and some features must be accessed by purchasing additional toolboxes or modules. Care should also be taken when comparing listed features as not all implementations are necessarily equal in terms of functionality and usability.

Other more subjective differences are the usability of the scope and other features, the look and feel of the program and the ease of generating documentation from the circuit schematic diagram and simulation results. These are factors that are best evaluated by testing the respective software tools.

References

1. Cellier, F., Kofman, E.: Continuous System Simulation. Springer, Berlin (1996)
2. De Kelper, B., Dessaint, L.A., Al-Haddad, K., Nakra, H.: A comprehensive approach to fixed-step simulation of switched circuits. IEEE Trans. Power Electron. **17**(2), 216–224 (2002)
3. Dyer, S., Dyer, J.: By the numbers—numerical integration. IEEE Instrum. Meas. Mag. **11**(2), 47–49 (2008)
4. Kelper, B.D., Blanchette, H., Dessaint, L.: Switching time model updating for the real-time simulation of power-electronic circuits and motor drives. IEEE Trans. Energy Convers. **20**(1), 181–186 (2005)
5. Maksimovic, D.: Computer-aided small-signal analysis based on impulse response of DC/DC switching power converters. IEEE Trans. Power Electron. **15**(6), 1183–1191 (2000)
6. Mathworks, www.mathworks.com: SimPowerSystems User's Guide
7. Munk-Nielsen, S., Tutelea, L.N., Jaeger, U.: Simulation with ideal switch models combined with measured loss data provides a good estimate of power loss. In: Proc. of the IEEE Industrial Applications Society Annual Conference, Rome, Italy, pp. 2915–2922 (2000)
8. Plexim GmbH, www.plexim.com: Plexim User Manual
9. Powersim, www.powersimtech.com: Psim User Manual
10. Strunz, K.: Flexible numerical integration for efficient representation of switching in real time electromagnetic transients simulation. IEEE Trans. Power Deliv. **19**(3), 1276–1283 (2004)

Chapter 14
Time-Stepping via Complementarity

Vincent Acary

14.1 Introduction

In this chapter, the numerical time integration methods for switched electronic circuits are delineated. The chapter starts with a brief introduction to the simulation of switched electrical circuits. In this introduction, various approaches, among them the hybrid approach, the event-driven approach, the regularisation approach and the time-stepping method via complementarity are recalled. For each of these methods, their advantages and weaknesses are outlined, and the main conclusion is that the time-stepping methods based on complementarity are the most robust and efficient approach to simulate large switched electrical circuits together with a large number of events. A general formulation of dynamical complementarity systems (DCS) over cones is given as

$$\begin{cases} \dot{x}(t) = f(t, x(t), \lambda(t)), \\ y(t) = h(t, x(t), \lambda(t)), \\ K^* \ni y(t) \perp \lambda(t) \in K, \end{cases} \tag{14.1}$$

where K is a non-empty closed convex cone,[1] and K^* its dual cone.[2] This formulation serves as a basis for the design of the time-stepping methods developed in the sequel. The mathematical nature of the expected solution will also be discussed knowing that the nature of the solution has a strong influence on the choice and the consistency of the numerical method. The continuously differentiable (C^1) solution, the absolutely continuous solution and the solution of bounded variation (BV) will

[1] A set K is called a *cone* if for any $x \in K$ and any scalar $a \geq 0$, $ax \in K$.

[2] The *dual cone* K^* of the cone K is the set $K^* = \{y \mid y \cdot x \geq 0 \; \forall x \in K\}$.

V. Acary (✉)
INRIA Rhône-Alpes, Centre de recherche Grenoble, 655 avenue de l'Europe, Inovallée de Montbonnot, 38334 St Ismier Cedex, France
e-mail: Vincent.Acary@inrialpes.fr

F. Vasca, L. Iannelli (eds.), *Dynamics and Control of Switched Electronic Systems*, Advances in Industrial Control,
DOI 10.1007/978-1-4471-2885-4_14, © Springer-Verlag London Limited 2012

be described together with some basic mathematical results concerning existence and uniqueness. Then the formulation of the time-stepping methods via complementarity will be described. For each class of solutions, a suitable method is provided. The mathematical properties (convergence, stability, order) and the practical properties (numerical damping, computational cost) are discussed. Theses properties will be illustrated on simple electrical circuits. For instance, various configurations of the four-diode bridge circuit will be used to exhibit the interest of using such a time-stepping method. We also touch on wrong choices in the design of a time-stepping method which yield numerical instabilities and non-convergence issues. Some extensions of the basic time-stepping method via complementarity are surveyed. Mainly, the inclusion into the normal cone to a convex set will be introduced leading to the direct extensions of the previous time-stepping for relay feedback systems.

Some implementation details will be then explained. Especially, the complementarity solvers that are used at each time-step will be described recalling the main families of available solvers. Some insights on the software implementation will also be given. Finally, numerical applications and examples on more realistic circuits will be considered. We will mainly focus on the architecture of a direct current–direct current (DC–DC) power converter, the buck converter.

14.2 Strengths and Weaknesses of Different Simulation Techniques

The major difficulties that are inherent to the simulation of switched electronic systems are the stiffness and non-smoothness of the dynamical system that models the time evolution of the circuit. To handle such difficulties, four main approaches are listed in the sequel. Their strengths and their weaknesses are briefly described.

The first basic approach considers some regularised models of ideal components to end up with a standard system of smooth differential algebraic equations (DAE). This is the *standard analog modelling and simulation* approach commonly used in the family of analog simulators, e.g. SPICE® (Simulation Program with Integrated Circuit Emphasis)-like simulators. It enables the simulation of switched electronic circuits in the experienced framework of the analog simulation. The regularised model can be the result of a pure mathematical regularisation. Take, as an example, the regularisation of an ideal relay multi-function by using a sigmoid function. A smooth model can also be justified by more accurate physical models. Let us consider for instance the diode component depicted in Fig. 14.1a. A smooth model of the diode based on physical considerations can be invoked by using Shockley's law, an exponential model of the diode illustrated in Fig. 14.1b. Obviously, a better physical model is an advantage from the point of view of the modelling, but there is a price to pay in the numerical practise. The main weaknesses of such a smooth modelling of a switching electrical component are the resulting stiffness of the obtained DAE system and the introduction of small time-scales in the dynamics. The use

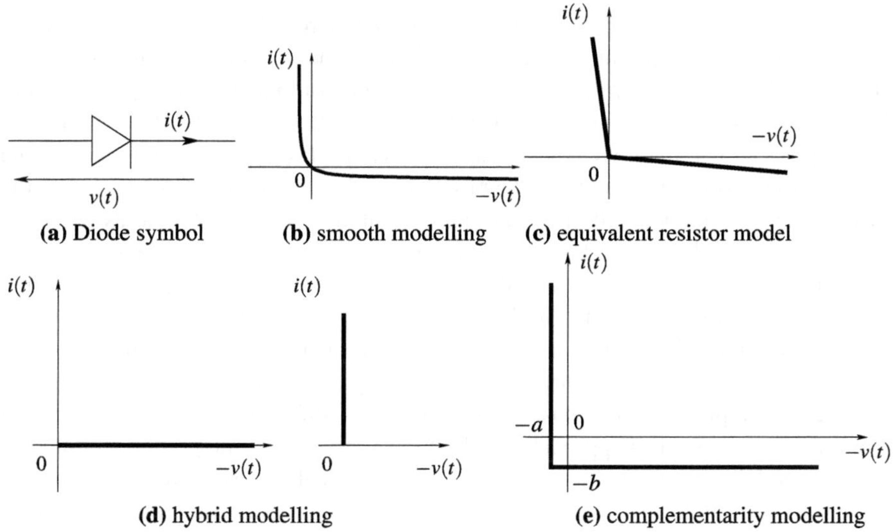

Fig. 14.1 Various modelling of a diode

of implicit time integration methods with a strict control of the time-step length is mandatory to ensure the consistency of the scheme and the convergence of Newton's method. An alternative model is the use an equivalent resistor model as depicted in Fig. 14.1c, which can be naively introduced in the SPICE framework. However, it has been shown in [39] that a standard Newton solver may fail on such a piecewise linear model. Furthermore, the intrinsic non-smoothness of the model destroys the order of accuracy of the integration scheme and disturbs the adaptive time-step strategy. More generally, the simulation of the switches with analog simulators becomes very time consuming and provides the user with very poor results which may contain chattering or even fail as reported in the literature [13, 20, 41, 53].

A substantial part of the stiffness of the regularised or smooth physical models can be removed by using ideal or idealised models. Such models often possess a discrete or digital character in the sense that they are stated with the help of conditional statements. They may be defined by single-valued piecewise-smooth functions or multi-valued mappings that may be naturally described by a set of distinct modes. The introduction of a discrete model with a set of continuous modes calls for the *hybrid modelling and simulation* approach. In Fig. 14.1d, a hybrid model of the diode is described, in which two modes are outlined. For instance, the Modelica description [23] of a diode given by

$$\begin{aligned} \mathsf{off} &= \mathsf{s} < 0 \\ v &= \text{if} \quad \mathsf{off} \quad \text{then} \quad \mathsf{s} \quad \text{else} \quad 0 \\ i &= \text{if} \quad \mathsf{off} \quad \text{then} \quad 0 \quad \text{else} \quad \mathsf{s} \end{aligned} \tag{14.2}$$

generates two distinct modes associated with the values of the boolean variable off. The hybrid simulation is based on a standard time integration of the smooth system in the current mode up to when an event is triggered. Then the following mode is selected, and the time integration is continued. The advantage of the hybrid approach is the possibility to take into account quite complicated models; even pure logical behaviours can be treated since digital modelling is implicitly contained in this approach. The main weaknesses are that we need: (a) to accurately trigger the events, (b) to know in advance all the possible modes and (c) to efficiently select the next mode. The knowledge of all the possible modes is not as simple as it seems. The presence of sliding modes for instance can be difficult to detect in a hybrid approach. The selection procedure of the next mode can also be very time-consuming. Let us consider a circuit with n diodes. We can have potentially 2^n modes to test, and this can be prohibitive. Fortunately, some heuristics can be practically implemented based for instance on mixed-integer programming techniques.

The last approaches are based on the *non-smooth modelling and simulation* framework. The two last approaches only differ in their time-integration techniques. For the modelling part, the goal is to design some ideal electrical models by means of formulations extensively used in the mathematical programming theory. To list a few of them, the models that are based on variational inequalities (VI) [24], complementarity problems (CP) [22], inclusions into a normal cone to a convex set [46] and non-smooth and convex optimisation problems [33] are very good candidates. The main difference with the hybrid approach is the intrinsic functional description (possibly multi-valued) of the component rather than a discrete/modal description. In Fig. 14.1e, a complementarity modelling of the diode with possible residual current b and voltage a is depicted. It can be defined by the following complementarity condition, or inclusion into a normal cone, as

$$0 \leq i + b \perp a - v \geq 0 \quad \Longleftrightarrow \quad -(i+b) \in \mathrm{N}_{[-a,+\infty)}(-v). \tag{14.3}$$

The notation $x \perp y$ means that $x^\top y = 0$. Inequalities involving vectors are understood to hold component-wise. The normal cone to K at $x \in K$ is $\mathrm{N}_K(x) = \{z \in \mathbb{R}^n \mid \langle z, \zeta - x\rangle \leq 0 \text{ for all } \zeta \in K\}$. From the definition of the normal cone it follows that an equivalent definition of the model of the diode as a VI is

$$(i+b)(v+u) \geq 0 \quad \text{for all } u \in [-a, +\infty). \tag{14.4}$$

The first time integration technique is half way between the hybrid approach and the non-smooth approach; we will call it *event-tracking time-stepping schemes* or shortly *event-driven schemes*. As in the hybrid approach, the time integration is performed with standard integration methods for DAE [29] for smooth periods up to an event is triggered. To select the following mode, contrary to the hybrid approach, one of the problems of mathematical programming theory that has been cited above (VI, CP, etc.) is solved. In most favourable cases, these problems can be solved very efficiently in polynomial time. For more details on event-driven schemes, we refer to [1, Chaps. 7 and 8]. When the number of events is too large, a second time-integration technique is more suitable, the so-called *event-capturing time-stepping schemes* or simply *time-stepping schemes*. In this family of methods, the events are not accurately located, and the time integration is performed without explicitly taking into account the changes of modes. The price to pay is their intrinsic low order, but this is balanced by very good properties of robustness and consistency. In particular, when the number of events is huge, even worse infinite (finite accumulation of events or Zeno-behaviour), or when a sliding-mode motion is encountered, the time-stepping schemes remain really efficient. In the remaining part of the chapter, we will detail the design and the properties of the time-stepping schemes mainly based on complementarity formulations of switching electrical components.

14.3 Solution Concepts and Well-Posedness for Complementarity Systems

The aim of this section is to give a flavour of what types of solutions are expected when some non-smooth components are in the loop. The goal is not to enter into the deepest details of the mathematical nature of solutions, but to give a sufficient insight when we want to deal with the numerical time integration of switched electrical circuits with non-smooth components.

Complementarity modelling of electrical components was introduced in the pioneering works of the Eindhoven school led by W. M. G. van Bokhoven [37, 48–52]. Their aim was to compute the steady-state of piecewise linear circuit by means of the complementarity models. See the book [36] for a good account. The notion of solution was extremely clear since it reduces to the question of the existence and possible uniqueness of solutions as a vector, say x, in a finite-dimensional space, say $\mathbb{R}^n$. When we deal with dynamics, the question of the solution as a function of time $x(t) \in \mathbb{R}^n$ is more difficult since we deal with some functional spaces of functions of time.

The more general systems that we addressed in this chapter may be defined by

$$\begin{cases} \dot{x}(t) = f(t, x(t), \lambda(t)), \\ y(t) = h(t, x(t), \lambda(t)), \\ -y(t) \in \mathrm{N}_X(\lambda(t)), \end{cases} \tag{14.5}$$

where X is a non-empty closed set of $\mathbb{R}^n$. The mappings f and h are assumed to be sufficiently smooth. If the set X is a closed cone denoted by K, the following dynamical complementarity system over cones is obtained:

$$\begin{cases} \dot{x}(t) = f(t, x(t), \lambda(t)), \\ y(t) = h(t, x(t), \lambda(t)), \\ K^* \ni y(t) \perp \lambda(t) \in K. \end{cases} \tag{14.6}$$

If K is the non-negative orthant of $\mathbb{R}^m$, a cone which has a central role in the complementarity theory, a standard dynamical complementarity system is derived:

$$\begin{cases} \dot{x}(t) = f(t, x(t), \lambda(t)), \\ y(t) = h(t, x(t), \lambda(t)), \\ 0 \leq y(t) \perp \lambda(t) \geq 0. \end{cases} \tag{14.7}$$

If the mappings f and h are affine, we get a Linear Complementarity System (LCS).

Well-posedness properties of LCS have been extensively studied in [14, 15, 17, 32]. Some instances of system (14.5) where X is not cone are also very interesting in practise. Indeed, note that

$$-y(t) \in \mathbb{N}_{[-1,1]}\big(\lambda(t)\big) \quad \Longleftrightarrow \quad -\lambda(t) \in \operatorname{sgn}\big(y(t)\big), \tag{14.8}$$

where the function $\operatorname{sgn}(\cdot)$ is the multi-valued signum function defined by

$$\operatorname{sgn}\big(y(t)\big) = \begin{cases} 1 & \text{if } y > 0, \\ -1 & \text{if } y < 0, \\ [-1, 1] & \text{if } y = 0. \end{cases} \tag{14.9}$$

For a vector $y \in \mathbb{R}^m$, $\operatorname{sgn}(y)$ holds component-wise. Let us consider for instance that $X = [-1, 1]^m$ in (14.5). We end up with a dynamical relay system

$$\begin{cases} \dot{x}(t) = f(t, x(t), \lambda(t)), \\ y(t) = h(t, x(t), \lambda(t)), \\ -\lambda(t) \in \operatorname{sgn}(y(t)). \end{cases} \tag{14.10}$$

In the affine case, the well-posedness of linear relay systems has been studied in [2, 45].

In order to say more on the mathematical properties of (14.5), we note that the inclusion in the third line of (14.5) is equivalent to the following VI:

$$y(t)\big(\tau - \lambda(t)\big) \geq 0 \quad \text{for all } \tau \in X, \tag{14.11}$$

that is

$$h\big(t, x(t), \lambda(t)\big)\big(\tau - \lambda(t)\big) \geq 0 \quad \text{for all } \tau \in X. \tag{14.12}$$

Thanks to (14.12), system (14.5) can be recast into the Differential Variational Inequalities (DVI) framework introduced by [44]. As it is noted by the authors, there can be substantial variations of the characteristics of DVIs regarding the existence and regularity of solutions depending mainly on the solution of the VI (14.12). Let us denote by $\lambda(t) \in \mathrm{SOL}(X, h(t, x(t), \cdot))$ an element of $\mathbb{R}^m$ solution of (14.12). Depending on the mathematical nature of the mapping $(x, t) \mapsto \mathrm{SOL}(X, h(t, x, \cdot))$, various types of solutions to (14.5) are obtained. A brief description is given in the following sections.

14.3.1 Continuously Differentiable Solutions

Let us assume that the mapping $(x, t) \mapsto \mathrm{SOL}(X, h(t, x, \cdot))$ is a single-valued Lipschitz function. The substitution of $\lambda(t)$ into (14.5) yields an ordinary differential equation (ODE) with a Lipschitz right-hand side. The standard theory of ODEs [21] can be then applied, and the solution $x(t)$ is sought as a continuously differentiable function of time, that is a function of class C^1. In the general VI case (14.12), the conditions to obtain a Lipschitz mapping are quite technical and mainly rely on the Lipschitz property and the strong monotonicity of h. In the simplest case of the non-negative orthant $K = \mathbb{R}^m$ and an affine mapping $h(t, x, \lambda) = D\lambda + Cx + q$, the solution $\lambda(t)$ of the linear complementarity system

$$0 \leq \lambda \perp D\lambda + Cx + q \geq 0 \tag{14.13}$$

is unique for all $Cx + q$ if and only if D is a P-Matrix [22], and moreover λ is a Lipschitz function of x.

When the solutions are expected to be of class C^1, system (14.5) is said to be of index *one* [44] or of relative degree *zero* [4]. The notion of index in DVI is very similar to the notion of index in DAE theory. The naive definition of the (differential) index is based on the number of times that the algebraic equations need to be differentiated in order to obtain a solvable ODE in x [29]. The relative degree is a familiar notion in Control Theory [35]. When it is uniform, it is also based on the number of differentiation of the output $y(t)$ to obtain explicitly the input $\lambda(t)$ in function of $x(t)$. For simple linear time invariant systems, the index is given by the relative degree plus one.

14.3.2 Absolutely Continuous Functions

As in the DAE theory, higher index or relative degree systems are obtained if the mapping h is not a one-to-one mapping of λ. For instance, if the Jacobian matrix $\nabla_\lambda^\top h(t, x(t), \lambda(t))$ is singular or worse if the λ does not explicitly appear in the definition of h, indices higher than one are obtained. It is difficult to say more on the nature of solutions without giving more structure to the functions h and f. In [44], some results are given with a function f affine in λ and a function h separable in λ. Without entering into deeper details, the solution $x(t)$ is sought in the sense of Carathéodory: the trajectory $x(t)$ is an absolutely continuous function given by

$$x(t) = x(0) + \int_0^t \dot{x}(s)\,\mathrm{d}s, \tag{14.14}$$

where $\dot{x}(t)$ is the solution of (14.5) ds-almost everywhere. Usually, $\dot{x}(t)$ is sought as a function of Bounded Variation (BV).

If we consider the affine case (14.13) with $D = 0$, we get

$$\begin{cases} \dot{x}(t) = f(t, x(t)) + B\lambda(t), & x(0) = x_0, \\ y(t) = Cx(t) + q, \\ 0 \leq y(t) \perp \lambda(t) \geq 0. \end{cases} \tag{14.15}$$

If CB is a positive definite matrix (relative degree *one*) and $Cx_0 + q \geq 0$ (consistent initial condition), the unique solution of (14.15) is an absolutely continuous function. The time-derivative of the state $\dot{x}(t)$ and $\lambda(t)$ are expected to be, in this case, discontinuous functions of time. Indeed, if the output $y(t)$ reaches the boundary of the feasible domain at time t_*, i.e. $y(t_*) = 0$, the time-derivative $\dot{y}(t)$ needs to jump if $\dot{y}(t_*^-) < 0$.

More generally, if (a) the relative degree of the system is *one*, or the index *two*, and if (b) the initial conditions are consistent, one looks for an absolutely continuous function $x(t)$ and a BV function $\lambda(t)$. If one of the conditions (a) or (b) is not satisfied, a stronger non-smoothness in $x(t)$ and/or $\lambda(t)$ has to be expected. Such a situation is detailed in the next section.

14.3.3 Functions of Bounded Variation

When discontinuities (jumps) are encountered in the solution $x(t)$, we often consider the solutions as functions of Bounded Variation (BV) [43]. A first source of jumps in such systems is the inconsistency of the initial conditions. Let us consider the previous example (14.15) with $Cx_0 + q < 0$. At the initial time, the solution has to jump to a consistent value with respect to the inequality. More generally, if λ does not appear explicitly in the function h in (14.5), the initial condition must meet a consistency condition such that

$$-y(t_0) = -h(t_0, x_0) \in \mathrm{N}_X\big(\lambda(t_0)\big) \tag{14.16}$$

must have at least one solution for $\lambda(t_0)$. Obviously, a first necessary condition is

$$-h(t_0, x_0) \in \mathrm{range}\big(\mathrm{N}_X(\cdot)\big), \tag{14.17}$$

where range($\mathrm{N}_X(\cdot)$) is the range of $\mathrm{N}_X(\cdot)$. If it is not the case, the state needs to jump to satisfy the inclusion (14.16) at t_0^+. Note that the nature of the set X is also very important in the consistency of the initial conditions. For instance, if $X = [-1, 1]$, then range($\mathrm{N}_{[-1,1]}(\cdot)$) $= \mathbb{R}$, and condition (14.16) is trivially solved for any value of x_0. If $X = \mathbb{R}_+$, range($\mathrm{N}_{\mathbb{R}_+}(\cdot)$) $= \mathbb{R}_-$, and condition (14.17) is not trivial. A second source of state discontinuities can also be caused by non-smooth external perturbations. If the output is given by $y(t) = Cx(t) + q(t)$ with $q(\cdot)$ a BV function, we may expect to have jumps at any time when the output is moved outside of range($\mathbb{N}_{[-1,1]}(\cdot)$). The reader will find a good account on state discontinuities, jump rules and conditions to obtain absolutely continuous solutions in the context of linear complementarity systems in [17, 18, 32], in the context of linear switched systems (LSS) in [25, 26, 34] and in [15] for differential inclusions.

If the trajectory $x(t)$ has jumps in time, the time-derivative $\dot{x}(t)$ cannot be defined as the standard derivative with respect to time. The state $x(t)$ is usually assumed to be a right-continuous function of bounded variation (RCBV). For relative degree *one* system, the variable λ has to be replaced by a measure that contains Dirac distributions. In the same vein, the time-derivative of the state $x(t)$ cannot be considered in the usual sense but as a differential measure $\mathrm{d}x$ associated with an RCBV function [15, 40, 42]. The dynamics in problem (14.15) is written in terms of a measure differential equation as

$$\mathrm{d}x = f(t, x(t))\,\mathrm{d}t + B\,\mathrm{d}i, \tag{14.18}$$

where $\mathrm{d}x$ is the differential measure associated with the RCBV function $\dot{x}(t)$, and $\mathrm{d}i$ is also a measure. The absolutely continuous function $\lambda(t)$ is the Radon–Nikodym derivative of $\mathrm{d}i$ with respect to the Lebesgue measure, i.e.

$$\frac{\mathrm{d}i}{\mathrm{d}t} = \lambda(t). \tag{14.19}$$

If the singular part of the differential measure is neglected, a decomposition of the measure can be written as

$$\mathrm{d}i = \lambda(t)\,\mathrm{d}t + \sum_l \sigma_l \delta_{t_l}, \tag{14.20}$$

where δ_{t_l} is the Dirac measure at times of discontinuities t_l, and σ_l is the magnitude. Thanks to (14.20), the differential measure equation (14.18) is decomposed in a smooth dynamics,

$$\dot{x}(t) = f\bigl(t, x(t)\bigr) + B\lambda(t) \quad \mathrm{d}t\text{-almost everywhere}, \tag{14.21}$$

and in a jump dynamics at t_l,

$$x\bigl(t_l^+\bigr) - x\bigl(t_l^-\bigr) = B\sigma_l. \tag{14.22}$$

14.3.4 Summary

The higher the relative degree (or index), the higher the level of non-smoothness. Let us consider the following dynamical system:

$$\begin{cases} x(0) = x_0, \\ x^{(3)}(t) = -1 + \lambda(t), \\ 0 \le x(t) \perp \lambda(t) \ge 0. \end{cases} \tag{14.23}$$

The relative degree of (14.23) is equal to 3. If the initial conditions $x_0 \geq 0$ is consistent at the initial time, the solution is sought as an absolutely continuous function, but λ is no longer a BV function. When the state hits the boundary $x = 0$, λ must contain a derivative of the Dirac distribution $\delta^{(1)}$, and the meaning of the inequality $\lambda(t) \geq 0$ has to be rethought. Without going into deeper details, we refer to [4] for the analysis of higher relative degree systems.

In this chapter, only switched electrical systems with relative degree less than or equal to *one* will be considered. Three kinds of solutions can be therefore expected:

- $x(t)$ is a solution of class C^1, and $\lambda(t)$ is of class C^0 (see conditions in Sect. 14.3.1),
- $x(t)$ is an absolutely continuous function of time, and $\lambda(t)$ is a BV function (see conditions in Sect. 14.3.2),
- $x(t)$ is a BV function of time, and $\lambda(t)$ is the density of a differential measure di with respect to the Lebesgue measure (see conditions in Sect. 14.3.3).

It is noteworthy that it can been shown that passive systems have a relative degree less than or equal to *one*. As for smooth passive electrical systems for which the DAE index is no more than *two*, we can conclude on the nature of solutions. Otherwise, each circuit configuration must be cautiously considered.

14.4 Time-Stepping Schemes via Complementarity

In this section, an event-capturing time-stepping scheme, which has been proved to be efficient and robust in many practical situations (see for instance [6, 8]), is proposed for each kind of expected solutions (C^1, absolutely continuous or BV). For convergence proofs, we refer to [2, 11, 16, 30, 31, 40, 42, 44]. In order to avoid some useless complexity about the treatment of standard nonlinearities, the affine instance of (14.5),

$$\begin{cases} \dot{x}(t) = Ax(t) + B\lambda(t) + u(t), \qquad x(0) = x_0, \\ y(t) = Cx(t) + D\lambda(t) + a(t), \\ K^* \ni y(t) \perp \lambda(t) \in K, \end{cases} \tag{14.24}$$

is first considered in this section. The design of the following time-stepping schemes is based on the following principle: a consistent evaluation of the unknown variables and their derivatives according to their smoothness. For instance, time-stepping schemes must not approximate high-order time-derivatives of functions which are not sufficiently smooth or must not try to point-wisely evaluate distributions or measures.

The following notation is used throughout this part. We denote by $0 = t_0 < t_1 < \cdots < t_k < \cdots < t_N = T$ a finite partition (or a subdivision) of the time interval $[0, T]$ $(T > 0)$. The integer N stands for the number of time intervals in the subdivision. The length of a time step is denoted by $h_k = t_{k+1} - t_k$. For simplicity, in the sequel we consider only a constant time length $h = h_k$ $(0 \le k \le N-1)$. Then $N = T/h$. The approximation of $f(t_k)$, the value of a real function $f(\cdot)$ at the time t_k, is denoted by f_k. For $\theta \in [0, 1]$, the notation $f_{k+\theta}$ stands for $\theta f_{k+1} + (1-\theta) f_k$, and respectively for $\gamma \in [0, 1]$.

14.4.1 *Time-Stepping for Continuously Differentiable Solutions*

The following time-stepping scheme is used for (14.24) when a solution $x(t)$ of class C^1 with $\lambda(t)$ continuous is expected:

$$\begin{cases} x_{k+1} - x_k = h(Ax_{k+\theta} + u_{k+\theta} + B\lambda_{k+\gamma}), \\ y_{k+\gamma} = Cx_{k+\gamma} + D\lambda_{k+\gamma} + a_{k+\gamma}, \\ K^* \ni y_{k+\gamma} \perp \lambda_{k+\gamma} \in K, \end{cases} \tag{14.25}$$

where $\theta \in [0, 1]$ and $\gamma \in [0, 1]$. The initial value of $\lambda_0 = \lambda(t_0)$ is given by the solution of

$$K^* \ni Cx_0 + D\lambda_0 + a_0 \perp \lambda_0 \in K. \tag{14.26}$$

The discretised system (14.25) amounts to solving at each time-step the following one-step problem:

$$\begin{cases} y_{k+\gamma} = M\lambda_{k+\gamma} + q, \\ K^* \ni y_{k+\gamma} \perp \lambda_{k+\gamma} \in K, \end{cases} \tag{14.27}$$

with

$$\begin{aligned} M &= D + h\gamma C(I - h\theta A)^{-1} B, \\ q &= a_{k+\gamma} + \gamma C(I - h\theta A)^{-1}\big[\big(I + h(1-\theta)A\big)x_k + hu_{k+\theta}\big] + C(1-\gamma)x_k. \end{aligned} \tag{14.28}$$

Example 14.4.1 (four-diode bridge rectifier supplied by an LC oscillator) Let us illustrate the influence of the parameters θ and γ on the behaviour of the numerical scheme (14.25). Let us consider first the configuration of the four-diode bridge in Fig. 14.2. The following values are taken for the physical parameters: $R = 1\,\mathrm{k}\Omega$,

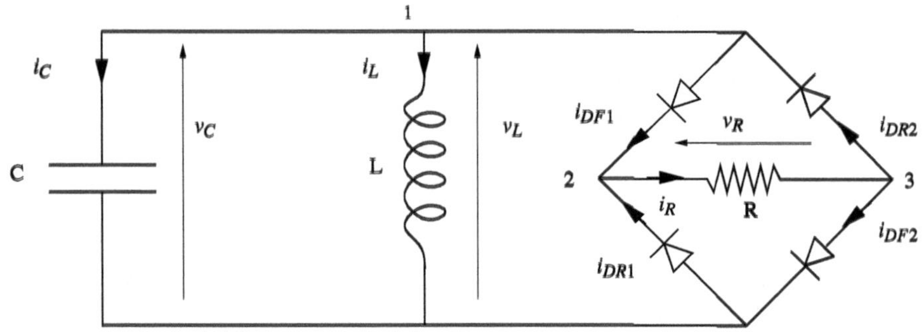

Fig. 14.2 The four-diode bridge rectifier. Inductor–capacitor (LC) oscillator with a load resistor

$L = 10^{-2}$ H, $C = 1$ μF. The resistor inside the bridge is supplied by an LC oscillator. The dynamical equations (14.24) are stated choosing:

$$x = \begin{pmatrix} v_L \\ i_L \end{pmatrix}, \qquad y = \begin{pmatrix} i_{DR1} \\ i_{DF2} \\ v_2 - v_1 \\ v_1 - v_3 \end{pmatrix}, \qquad \lambda = \begin{pmatrix} v_2 \\ -v_3 \\ i_{DF1} \\ i_{DR2} \end{pmatrix}, \tag{14.29}$$

and with

$$A = \begin{pmatrix} 0 & -1/C \\ 1/L & 0 \end{pmatrix}, \qquad B = \begin{pmatrix} 0 & 0 & -1/C & 1/C \\ 0 & 0 & 0 & 0 \end{pmatrix}, \qquad u = 0,$$
$$C = \begin{pmatrix} 0 & 0 \\ 0 & 0 \\ -1 & 0 \\ 1 & 0 \end{pmatrix}, \qquad D = \begin{pmatrix} 1/R & 1/R & -1 & 0 \\ 1/R & 1/R & 0 & -1 \\ 1 & 0 & 0 & 0 \\ 0 & 1 & 0 & 0 \end{pmatrix}, \tag{14.30}$$
$$a = 0, \qquad K = K^* = \mathbb{R}^4_+.$$

Note that the matrix D has full rank and is semi-definite positive. The solution $x(t)$ is of class C^1 since $x \mapsto \text{BSOL}(\mathbb{R}^4_+, D\lambda + Cx + a)$ is a Lipschitz function of x. The results given by the scheme (14.25) are depicted in Fig. 14.3 with time-step $h = 10^{-6}$ and initial condition $x_0 = (1 0 0)^\top$. As we can expect, the scheme correctly approximates the trajectory $x(t)$ and the variables $\lambda(t)$ and $y(t)$. We note that the use of a midpoint rule decreases the numerical damping and improves the quality of the solution for a fixed time-step.

As it is illustrated in Fig. 14.3, the scheme does not generate spurious oscillations although switches occur during the time steps. The fact that the matrix D is semi-definite positive yields a very easy solving of the LCP at each time step. The computational effort is of the same order as in solving a comparable linear system.

Index

F. Vasca, L. Iannelli (eds.), *Dynamics and Control of Switched Electronic Systems*, Advances in Industrial Control,
DOI 10.1007/978-1-4471-2885-4,

L

M

N

O

P

Q

MIX
Papier aus verantwortungsvollen Quellen
Paper from responsible sources
FSC® C105338
FSC
www.fsc.org

If you have any concerns about our products,
you can contact us on
ProductSafety@springernature.com

In case Publisher is established outside the EU,
the EU authorized representative is:
Springer Nature Customer Service Center GmbH
Europaplatz 3, 69115 Heidelberg, Germany

Printed by Libri Plureos GmbH
in Hamburg, Germany